Microwave Polarizers, Power Dividers, Phase Shifters, Circulators, and Switches

Microwave Polarizers, Power Dividers, Phase Shifters, Circulators, and Switches

Joseph Helszajn
Heriot Watt University, Edinburgh, UK

WILEY

IEEE PRESS

This edition first published 2019

Registered Office
John Wiley & Sons, Inc., 111 River Street, Hoboken, NJ 07030, USA

Editorial Office
111 River Street, Hoboken, NJ 07030, USA

For details of our global editorial offices, customer services, and more information about Wiley products visit us at www.wiley.com.

Wiley also publishes its books in a variety of electronic formats and by print-on-demand. Some content that appears in standard print versions of this book may not be available in other formats.

Library of Congress Cataloging-in-Publication Data

Names: Helszajn, J. (Joseph) author.
Title: Microwave polarizers, power dividers, phase shifters, circulators, and switches / authored by Joseph Helszajn.
Description: First edition. | Hoboken, NJ : John Wiley & Sons, Inc., [2019] | Includes bibliographical references and index. |
Identifiers: LCCN 2018027839 (print) | LCCN 2018038875 (ebook) | ISBN 9781119490081 (Adobe PDF) | ISBN 9781119490074 (ePub) | ISBN 9781119490050 (hardcover)
Subjects: LCSH: Microwave devices.
Classification: LCC TK7876 (ebook) | LCC TK7876 .M267 2019 (print) | DDC 621.381/33–dc23
LC record available at https://lccn.loc.gov/2018027839

Cover design: Wiley
Cover image: © iStock.com/Ivanastar

Set in 10/12pt Warnock by SPi Global, Pondicherry, India

Printed in the United States of America

V10005879_110618

Contents

Preface

Ferrite phase-shifter and control devices are widely used in conjunction with passive microwave circuits in beam shaping and steering of array antennas and in multichannel switching. The intention of this text is to provide the reader with some preliminary insight into the operation of some basic ferrite control devices and to note some system uses. In the beam steering application, variable phase-shifters are employed to tilt the beam of a simple one-dimensional array or more sophisticated two- and three-dimensional ones. Beam shaping is achieved by using variable power dividers and switches. At modest microwave wavelengths, this is often done with the aid of semiconductor devices, but at very high power levels and at millimeter wavelengths, ferrite devices are used almost exclusively. A drawback of the ferrite control device is its longer switching time; its microwave power rating is, however, usually superior. Although many ferrite devices are nonreciprocal, this is often not essential or indeed desirable in the control area. Mechanically actuated passive switches and variable phase shifters using rotatable half-wave plates are other possibilities. Multichannel switching may consist of making provisions for switching on a standby transmitter in case of a failure mode in some simple radar or satellite equipment or it may involve the control of a high-power signal using Butler matrices; it may also be utilized in the construction of multiport power combiners. The three-port junction circulator is, of course, also ideally suited for switching a signal at one port to any of $n - 1$ others. Frequency reuse where spatially isolated beams operate in the same frequency band is another area where power dividers and variable phase shifters are required. Switching of the hand of polarization of a wave or rotating its polarization are other applications. Microwave ferrite phase-shifters and other devices essentially rely for their operation on the different birefringences exhibited by a magnetized magnetic insulator under the influence of different direct and alternating magnetic fields. Nonlinear effects or spinwave instabilities at large signal levels are a separate consideration.

Acknowledgments

This text is dedicated to my colleagues and friends without whom this book would not have been possible: Mark McKay, Marco Caplin, Henry Downs, David J. Lynch, John Sharp, William D'Orazio, Alicia Casanueva, and Angel Mediavilla Sánchez.

List of Contributors

Mr. Marco Caplin
RF Designer
Apollo Microwaves Ltd
Dorval, Quebec
Canada

Professor Alicia Casanueva
Communication Engineering
Department
University of Cantabria
Santander
Spain

Mr. William D'Orazio
RF Designer
Apollo Microwaves Ltd
Dorval, Quebec
Canada

Mr. Henry Downs
Chief Science Officer – EVP
Engineering
Mega Industries, LLC
Gorham, ME
USA

Dr. David J. Lynch
Director
Filtronic Wireless Ltd
Salisbury, MD
USA

Dr. Mark McKay
Principal Engineer
Honeywell
Edinburgh
UK

Professor Angel Mediavilla Sánchez
Communication Engineering
Department
University of Cantabria
Santander
Spain

Professor John Sharp
Professor Emeritus
Napier University
Edinburgh
UK

1

Microwave Switching Using Junction Circulators

Joseph Helszajn

Heriot Watt University, Edinburgh, UK

1.1 Microwave Switching Using Circulators

Since the direction of circulation of a circulator is determined by that of the direct magnetic field, it may be employed to switch an input signal at one port to either one or the other two. Switching is achieved by replacing the permanent magnet by an electromagnet or by latching the microwave ferrite resonator directly by embedding a current carrying wire loop within the resonator.

The schematic diagram of a switched junction is shown in Figure 1.1a. It is particularly useful in the construction of Butler-type matrices in phase array systems. A single-pole three throw version is depicted in Figure 1.1b.

Two common arrangements in which ferrite circulators may be employed to obtain microwave switching are separately illustrated in Figure 1.1c and 1.1d. The first uses a circulator in conjunction with a pin diode switch to vary the short-circuit plane terminating port 2. A transmission analog phase shifter is therefore obtained between ports 1 and 3 with this mode of operation. The second version is also a transmission configuration but now a switchable circulator is used to control the path between ports 1 and 3 of the circulator. The switching speed of the pin device is normally the faster one.

1.2 The Operation of the Switched Junction Circulator

The adjustment of a fixed field circulator or a switched circulator is a two-step procedure. The first fixes its midband frequency and the second its gyrotropy. A phenomenological description of these two operations is illustrated in

Microwave Polarizers, Power Dividers, Phase Shifters, Circulators, and Switches,
First Edition. Joseph Helszajn.

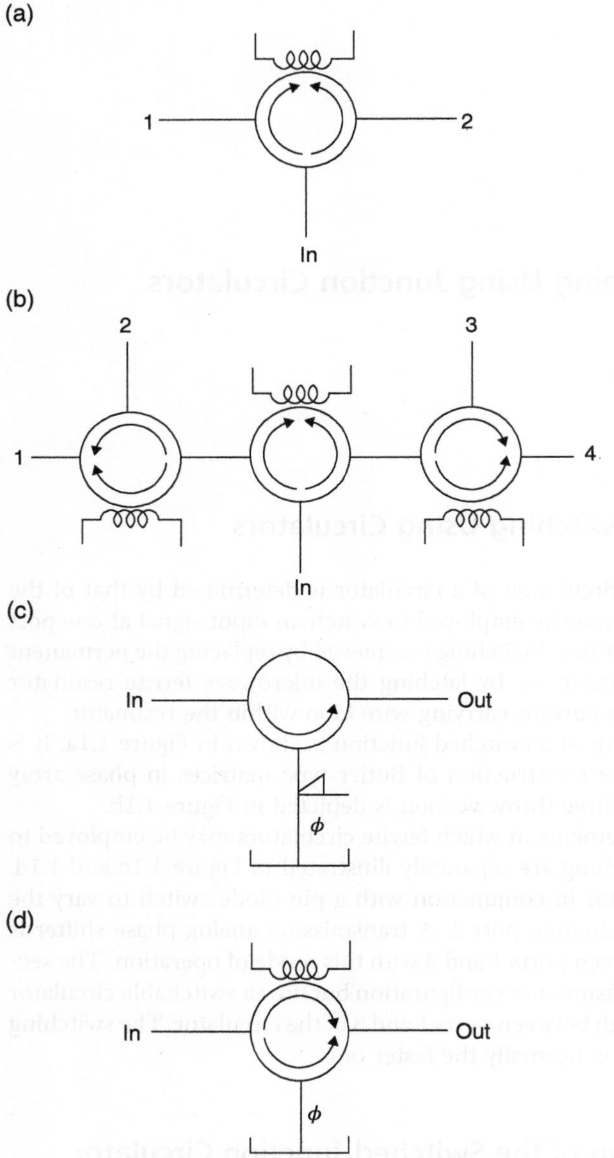

Figure 1.1 Microwave phase shifter using (a) schematic of circulator switch, (b) SP4T Butler switch using circulators, (c) pin dioded switch and fixed circulator, and (d) switched circulator.

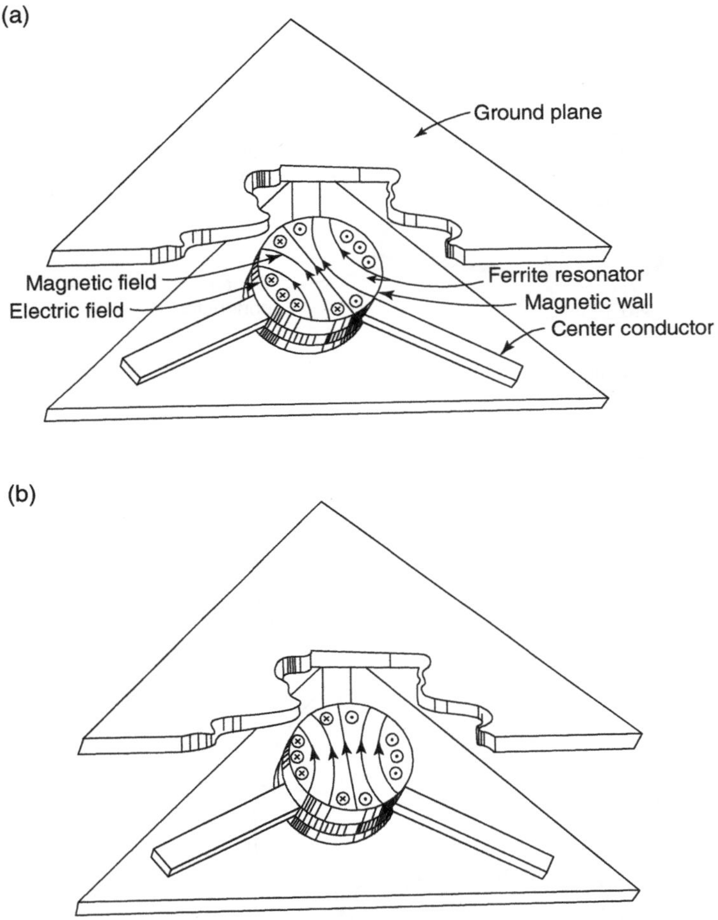

Figure 1.2 Standing wave patterns in (a) demagnetized stripline junction and (b) magnetized stripline junction.

Figure 1.2a and b in the case of a stripline geometry. The direction of circulation is here fixed by the sense of the direct magnetic field intensity along the axis of the resonator. This may be done by either internally latching the hysteresis loop of the magnetic insulator between its two remanent states or by having recourse to an external magnetic circuit. The electric field pattern may be rotated either clockwise or anticlockwise by splitting the degeneracy of the counterrotating field patterns of the resonator. A latched stripline geometry is indicated in Figure 1.3.

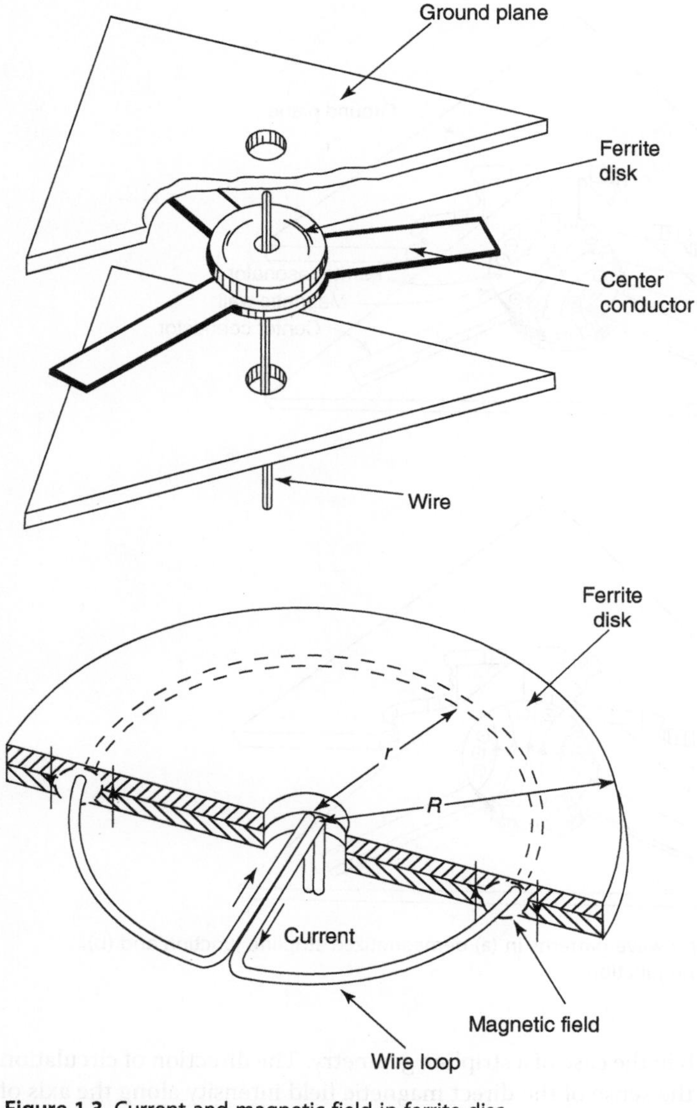

Figure 1.3 Current and magnetic field in ferrite disc.

1.3 The Turnstile Circulator

The waveguide junction switch is usually but not exclusively based on a Faraday rotation effect along a quarter-wave long cavity resonator open-circuited at one flat face and short-circuited at the other. Its first circulation condition is a 90°

cavity with no rotating of the electric field pattern, which is again a figure of eight pattern. Its second circulation condition is obtained by replacing the dielectric resonator by a gyromagnetic insulator. The effect is to rotate the polarization of the electric field by a 15° angle in the positive direction of propagation and a further 15° in the opposite direction. The total rotation places an electric null at a typical output port.

Figure 1.4a and b are sketches of the electric and magnetic HE_{11} standing wave patterns about midway along the cavity. The electric field is zero at the electric wall of the cavity, whereas the magnetic field is zero at its magnetic flat wall.

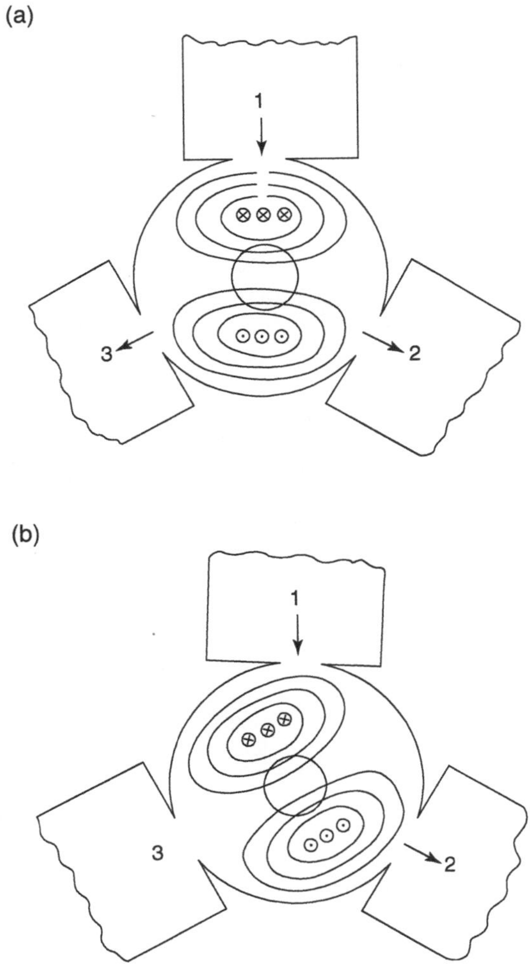

Figure 1.4 (a) Ferrite unmagnetized; first circulation condition. (b) Ferrite magnetized; second circulation condition.

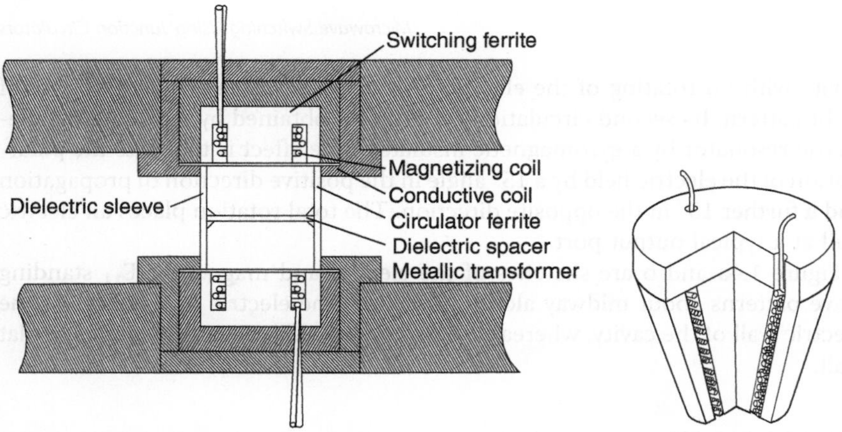

Figure 1.5 Schematic diagram of externally latched circulator using a post-resonator waveguide junction.

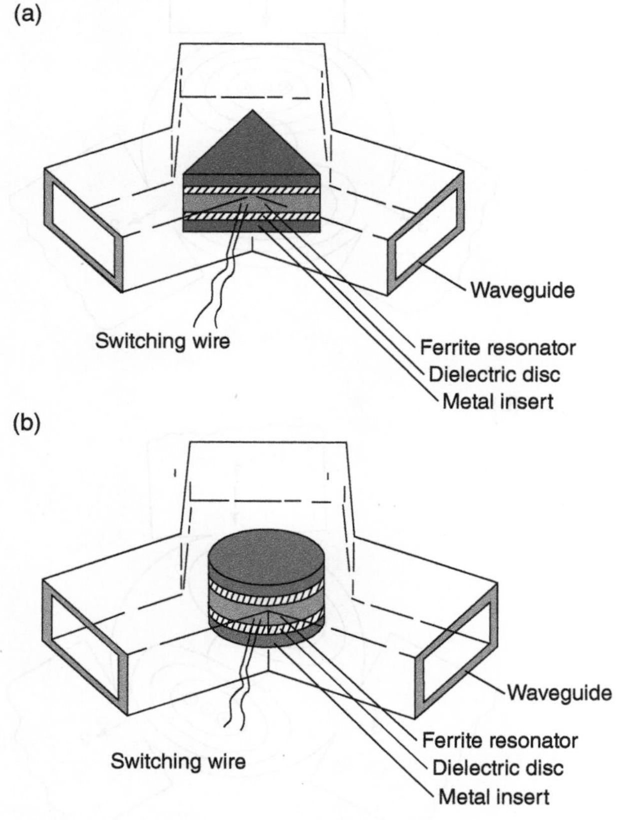

Figure 1.6 Schematic diagram of waveguide junction circulator using a partial height: (a) triangular and (b) circular resonator with a wire loop.

1.4 Externally and Internally Latched Junction Circulators

Circulators may be either actuated by an electromagnet or they may be operated by internally or externally latching the ferrite resonator. Figure 1.5 illustrates one externally latched arrangement. Figure 1.6a and b depict internally latched waveguide devices using half-wave or quarter-wave long resonators.

Figure 1.7 indicates the two possible wire configurations met in the construction of a waveguide switch using a prism resonator.

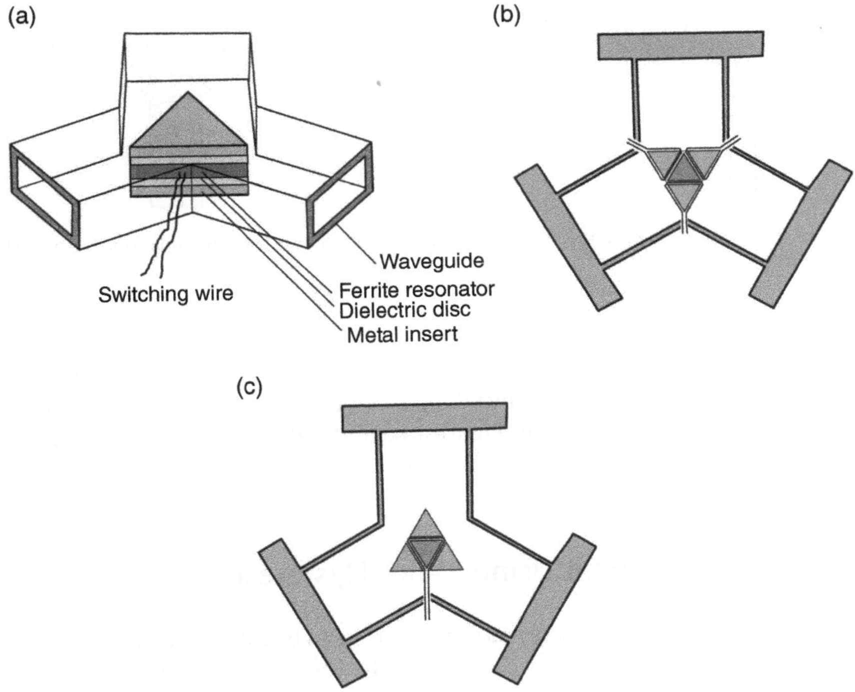

Figure 1.7 Schematic diagrams of waveguide circulators showing different switching wire configurations.

1.5 Standing Wave Solution of Resonators with Threefold Symmetry

Two resonators met in the design of switched circulators with threefold symmetry are the equilateral triangle structure and the quasi WYE geometry.

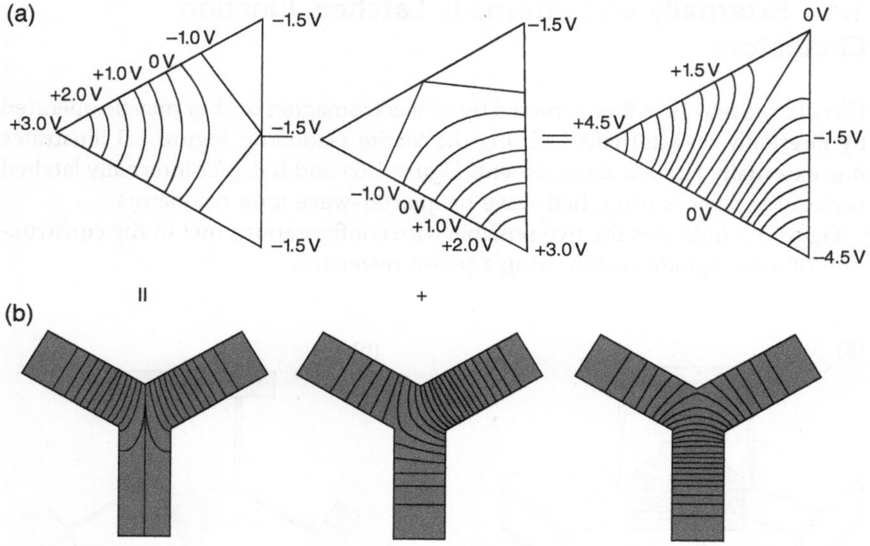

Figure 1.8 Standing wave solution of three-port circulators using (a) triangular resonator and (b) WYE resonator.

The standing wave solution of the second circulation solutions is here not obvious but each may be constructed by taking suitable linear combinations of those of the first circulation condition. Figure 1.8a and b illustrate the equipotential lines of the standing wave patterns in each situation.

1.6 Magnetic Circuit Using Major Hysteresis Loop

The direct magnetic field in a junction circulator can be established using either an external electromagnet or it can be switched by current pulses through a magnetizing wire between the two remanent states of the major or indeed of a minor hysteresis loop of a closed magnetic circuit. The former arrangement requires a holding current to hold the device in a given state.

In the latter one, however, no such current is necessary; the device remains latched in a given state until another switching operation is required. The advantages and disadvantages of each type of circuit are understood.

Operation on the major hysteresis loop may be understood by scrutinizing the hysteresis loop in Figure 1.9, providing it is recognized that the size and shape of this loop may vary with the speed of the switching process. In this situation, the magnetization of the toroid is driven between two remanent states ($\pm 4\pi M_r$)

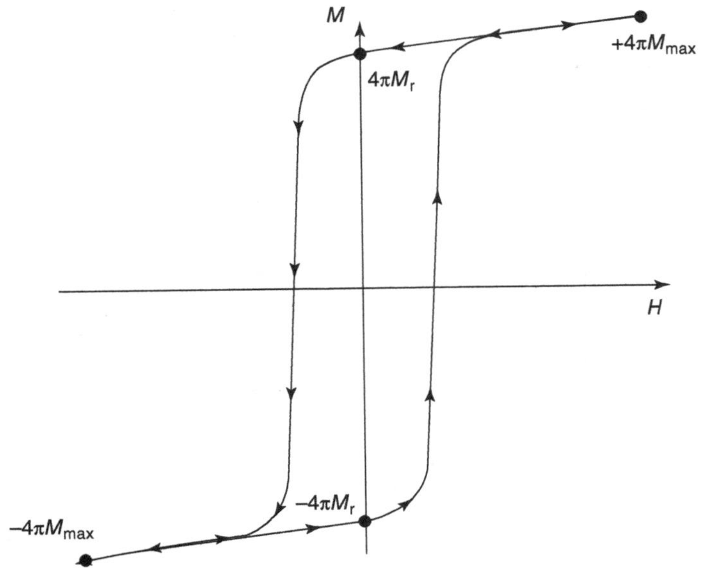

Figure 1.9 Typical hysteresis loop of a latching phase shifter operating with a major hysteresis loop switching.

equidistant from the origin by the application of a current pulse sufficiently large to produce a field perhaps three or five times that of the coercive force.

After this point is reached, the current pulse is removed and the magnetization will move to the remanent value ($\pm 4\pi M_r$) and remain there until another switching operation is desired. This sort of electronic driver circuit is relatively simple since it is only required that the toroids be driven back and forth between the major remanent states of the hysteresis loop.

1.7 Display of Hysteresis Loop

The magnetic properties and parameters of a magnetic core or toroid under different operating conditions, such as temperature, say, are best discussed in terms of the details of its hysteresis loop.

Some experimental quantities that are of particular interest include the saturation magnetization (M_0), the remanent magnetization (M_r), and the coercive force (H_c). The experimental display of such loops is therefore of some interest. One circuit that may be used for this purpose is outlined in Figure 1.10. This

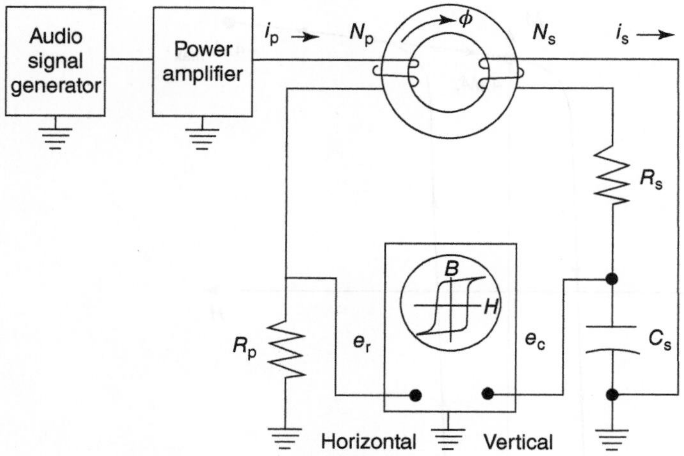

Figure 1.10 Schematic diagram of hysteresis display.

arrangement develops voltage V_p and V_i that are proportional to B and H, respectively.

The magnetic field (H) in the core is monitored by measuring the voltage (V_p) across a resistor in series with the primary winding, see Figure 1.10.

$$H = \frac{N_p}{I_p}\left(\frac{V_p}{R_p}\right), \mathrm{A\,m^{-1}} \tag{1.1}$$

where I_p is the effective of the primary winding, N_p is the number of turns of the primary winding (10–30), and R_p is the resistor in series with the primary coil (10 Ω). The magnetization (B) is likewise evaluated by forming the voltage (V_i) across the capacitance of the RC integrator in the secondary circuit.

$$B \approx \frac{-V_i R_i C_i}{N_s A} \tag{1.2}$$

where R_i is the series resistance of the integrator (100 kΩ), C_i is the capacitance of the integrator (0.10 μF), N_s is the number of turns of the secondary winding (10–30), and A is the cross-sectional area of the core.

The data shown in Figure 1.11 on the effects of small air gaps on the squareness of the hysteresis loop have been obtained using the arrangement outlined here.

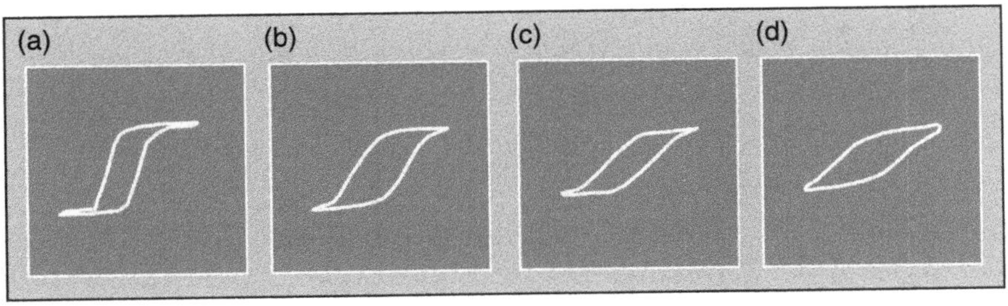

Figure 1.11 Hysteresis loops showing the effect of gaps in a magnetic circuit. (a) No gap, (b) gap of 2½ thou, (c) gap of 5 thou, and (d) gap of 10 thou.

1.8 Switching Coefficient of Magnetization

The change of magnetization in a ferrite core consists usually in reversal of the magnetization, e.g. from negative remanence to the positive remanence corresponding to the magnetic field applied. The ultimate state of the magnetization that is set up (after the passage of the current pulse) is always symmetrical here, with respect to zero. It is observed that in most cases, the change in the magnetization produced by this field cannot follow the increase in the current.

The general situation is quite complicated but for an applied magnetic field slightly in excess of the coercive force, H_c, domain wall motion will, in general, be the predominant reversal mechanism. In this case, the flux change is accomplished by the motion of Bloch walls, which separate the domains of differently oriented magnetization.

For suitable oriented single crystals of ferrite, a very simple domain configuration may be achieved, which makes it possible to obtain information on the behavior of moving domain wall.

Studies on single crystals of ferrite have demonstrated that, under this condition, the wall velocity depends linearly on the applied magnetic field. This leads to a linear relation between the direct field and the reciprocal of the switching time. Such a relationship is also noted experimentally for polycrystalline ferrites although the actual domain configuration is not known.

The switching time is usually measured by using a core with two windings as shown in Figure 1.12. The output voltage pulse appearing at the termination of the secondary winding exhibits a characteristic shape with two separate maxima when a current pulse is passed through the magnetization winding.

For ferrite with hysteresis loops, the first maximum in the output voltage represents a small percentage of the total area under the curve and hence of the shape in magnetization.

The duration T of a voltage pulse is defined as the time (counted from the beginning of the current pulse) that elapses before the voltage has dropped

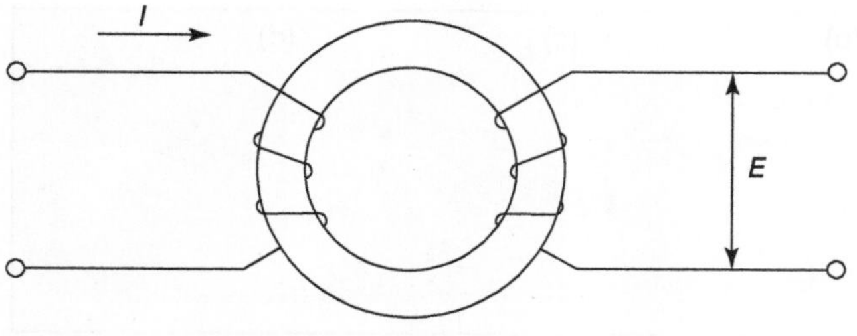

Figure 1.12 Ferrite core with two windings for measuring switching time.

to 10% of the maximum value; for the maximum value the second peak is considered. The dependence of the switching time on the magnetization producing flux reversal is most clearly represented by plotting $1/T$ as a function of the magnetizing strength H, in the manner indicated in Figure 1.13. Over the majority of the range shown, $1/T$ has a linear dependence on H, which may be adequately represented by

$$T(H - HH_0) = S \tag{1.3}$$

In this expression, S is known as the switching coefficient, and H_0, which is of the same order of magnitude as the coercive force H_c of the material, may be termed the threshold field for irreversible magnetization. It should be noted that, although the curve is continued to values of H less than H_0, the switching of the core under these conditions produces a smaller hysteresis loop: i.e. the material is not driven to magnetic saturation and such operation is not desirable.

The optimum squareness ratio R_s occurs for values of H fractionally different to H_0, but it is usual to adopt magnetizing fields of between $2H_0$ and $5H_0$, the slight deterioration in squareness being accepted in the interests of faster switching.

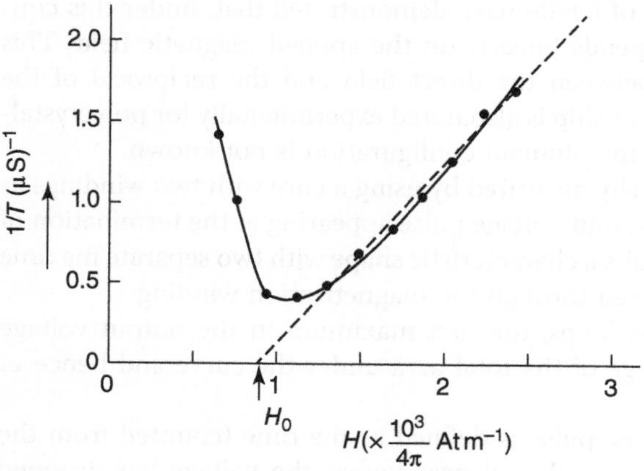

Figure 1.13 Reciprocal reversal time $1/T$ as a function of direct magnetic field H for ferroxcube. *Source:* Reprinted with permission Van der Heide et al. (1956).

It is found that the great majority of ferrites with the squareness and the coercive force usual for switching elements all have a reversal constant $S = (H - H_0)T$ of the same order of magnitude.

In the case of ferroxcube 6 A considered above, the resultant value is $S_0 = 80$ (μs)(A m^{-1}).

1.9 Magnetostatic Problem

One way to explore the internal direct magnetization of the magnetic insulator in the presence of one or more loops is to have recourse to a magnetostatic solver. Figure 1.14 shows the magnetizing effect of a single circular wire loop of radius *r*, carrying 10 A, on a cylindrical resonator with radius *R*. One feature of this result is that the magnetization on the axis of the loop is inversely proportional to its radius so that such switches are more readily realized at high frequencies than at lower ones.

One possible first-order model of such a resonator is one with a narrow demagnetized concentric region, a second with a magnetization in one sense, and a third with a magnetization in the other sense with still another value.

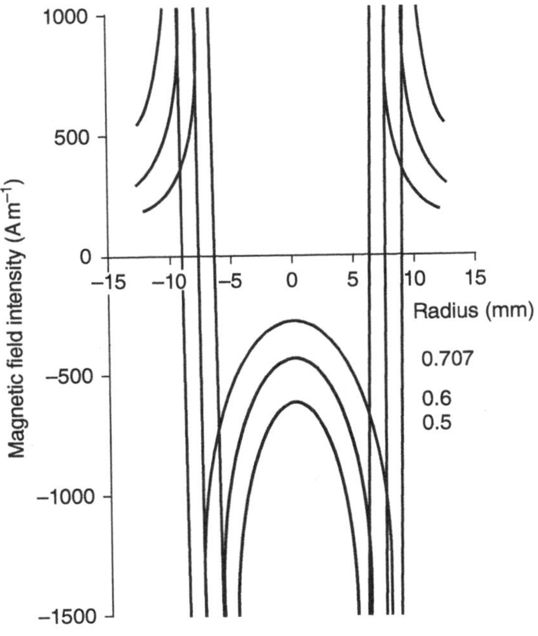

Figure 1.14 Up and down direct magnetic field strength in a cylindrical resonator using a single wire loop using a magnetostatic solver ($r/R = 0.5$, $r/R = 0.6$, $r/R = 0.707$) (Helszajn and Sharp 2012).

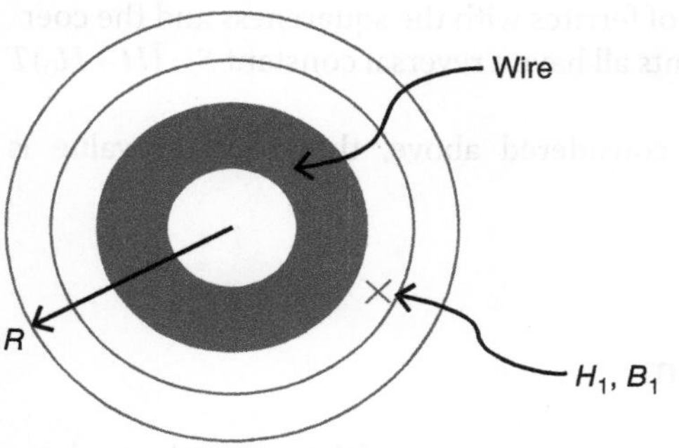

Figure 1.15 Plan view of a cylindrical resonator subdivided in concentric rings.

The more general problem divides the cross-section of the resonator into a number of concentric rings as shown in Figure 1.15.

Such a model can readily be set up, in the case of a cylindrical geometry in closed form or, in general, using a commercial FE package.

1.10 Multiwire Magnetostatic Problem

Figure 1.16 depicts the situation in the case of the pair of stacked circular loops. The spacing between the wires is half the thickness of the resonator. The inductance of the wire configuration using two wire loops is four times

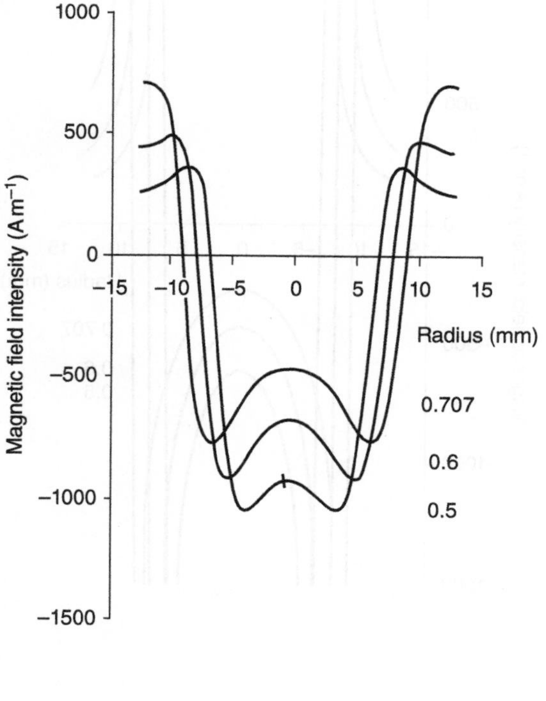

Figure 1.16 Up and down direct magnetic field strength in a cylindrical resonator using a pair of wire loops using a magnetostatic solver ($r/R = 0.5$, $r/R = 0.6$, $r/R = 0.707$) (Helszajn and Sharp 2012).

that of the single loop. The stored energy is likewise increased by a factor of four. This energy, divided by the switching time, is related to the instantaneous power required from the driver. The exact problem requires the discretization of the resonator both across and along the geometry and the assignment of the local gyrotropies in each region. This is obviously not a realistic approach without a three-dimensional solver in conjunction with a magnetostatic analyzer. It is of note, however, that the alternating magnetic field goes to zero at the open flat faces of the resonator. This means that the precise gyrotropy has no significant impact on the gyrotropy of the resonator.

1.11 Shape Factor of Cylindrical Resonator

The ratio of the two oppositely magnetized regions is defined by a shape factor q:

$$q = \frac{\text{Surface area of the inner region magnetized along the} + z\,\text{direction}}{\text{Surface area of one typical outer region magnetized along the} - z\,\text{direction}} \tag{1.4}$$

If the inner radius of the resonator is taken as r_i and the outer radius as r_0, then

$$q = \frac{r_\mathrm{i}^2}{r_0^2 - r_\mathrm{i}^2} \tag{1.5}$$

The condition $q = 1$ corresponds to that for which the cross-sectional areas of the two regions are equal.

The shape factor of a prism resonator with an equilateral core is unity. The shape factor of equilateral prism resonator employing a regular hexagonal wire is

$$q = \frac{A^2 - 3L^2}{L^2} \tag{1.6}$$

For a wye cavity with the wire loop located at the terminals between the circular region and a typical strip is

$$q = \frac{W(R - r)}{\pi r^2} \tag{1.7}$$

The work assumes a constant gyrotropy κ_pi in the inner region equal to 0.707 that of the saturated material, κ_0, which is taken here as 0.7. This means that the gyrotropy in the outer region lies between 0 and κ_0. When q equals unity the

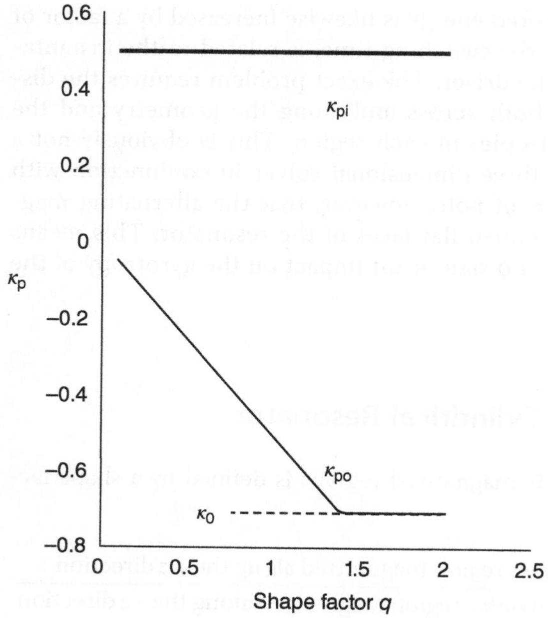

Figure 1.17 Uniform partial gyrotropy in the inner region of a composite gyromagnetic resonator against factor q (ratio of up and down surfaces) with $\kappa_{pi} = 0.5$ ($\kappa_0 = 0.7$).

partial gyrotropies in the outer region is equal to one third. The diagonal element of the tensor permeability in this work is taken as unity for simplicity sake. Figure 1.17 indicates the variation of the gyrotropy in the outer region with the shape factor q, assumed in this work.

Bibliography

Betts, J., Temme, D.H., and Weiss, J.A. (1966). A switching circulator s-band, stripline, 15 kilowatts, 10 microseconds, temperature stable. *IRE Trans. Microw. Theory Tech.* **MTT-14**: 665–669.

Bosma, H. (1962). On the principle of stripline circulation. *Proc. IEE* **109** (Part B, Suppl. 21): 137–146.

Clavin, A. (1963). Reciprocal and nonreciprocal switches utilizing ferrite junction circulators. *IEEE Trans. Microw. Theory Tech.* **MTT-11**: 217–218.

Freiberg, L. (1961). Pulse operated circulator switch. *IRE Trans. Microw. Theory Tech.* **MTT-9**: 266.

Goodman, P.C. (1965). A latching ferrite junction circulator for phased array-switching applications. IEEE G-MTT Symposium, Clearwater, FL (5–7 May 1965). Piscataway: IEEE.

Helszajn, J. (1960). Switching criteria for waveguide ferrite devices. *Radio Electron. Eng.* **30**: 289–296.

Helszajn, J. and Hines, M.L. (1968). A high speed TEM junction ferrite modulator using a wire loop. *Radio Electron. Eng.* **35**: 81–82.

Helszajn, J. and Sharp, J. (2012). Cut-off space of a gyromagnetic planar disk resonator with a triplet of stubs with up and down magnetization. *IET Microw. Antennas Propag.* **6**: 569–576.

King, L.V. (1933). Electromagnetic shielding at radio frequencies. *Philos. Mag.* **15** (Series 7): 201.

Laplume, J. (1945). Shielding effects of a cylindrical tube placed in a uniform magnetic field perpendicular to its axis. *Ann. Radioélectr.* **1** (1): 65–73.

Levy, L. and Silber, L.M. (1960). A fast switching X-band circulator utilizing ferrite toroids. *WESCON/60 Conference Record*, Part 1, pp. 11–20.

Lyons, W. (1933). Experiments on electromagnetic shielding at frequencies between one and thirty kilocycles. *Proc. IRE* **21** (4): 574–590.

Siekanowicz, W.W. and Schilling, W.A. (1968). A new type of latching switchable ferrite junction circulator. *IEEE Trans. Microw. Theory Tech.* **MTT-16**: 177–183.

Siekanowicz, W.W., Paglione, R.W., and Walsh, T.E. (1970). A latching ring-and-post ferrite waveguide circulator. *IEEE Trans. Microw. Theory Tech.* **MTT-18**: 212–216.

Stern, R.A. and Ince, W.J. (1967). Design of composite magnetic circuits for temperature compensation of microwave ferrite devices. *IEEE Trans. Microw. Theory Tech.* **MTT-15**: 295–300.

Taft, D.R. and Hodges, L.R. Jr. (1965). Square loop materials for digital phase shifter applications. *J. Appl. Phys.* **36**: 1263–1264.

Treuhaft, M.A. and Silber, L.M. (1958). Use of microwave ferrite toroids to eliminate external magnets and reduce switching power. *Proc. IRE* **46** (8): 1538.

Uebele, G.S. (1957). High speed ferrite microwave switch. *IRE Natl. Conv. Rec.* **5** (Part I): 227–234.

Van der Heide, H., Bruijning, H.G., and Wijn, H.P.J. (1956). Switching time of ferrites with rectangular hysteresis loop. *Philips Tech. Rev.* **18**: 339.

2

The Operation of Nonreciprocal Microwave Faraday Rotation Devices and Circulators

Joseph Helszajn

Heriot Watt University, Edinburgh, UK

2.1 Introduction

A classic phenomenon encountered in connection with propagation in a gyromagnetic medium is that of the rotation of the plane of polarization of an alternating radio frequency wave in the plane transverse to that of propagation. This effect is displayed by the medium, provided the direct magnetic field intensity is along the direction of propagation. Faraday rotation is nonreciprocal and is responsible for a number of important ferrite devices.

The Faraday rotation effect describes the rotation of the plane of polarization of a linearly polarized wave in an infinite gyromagnetic medium or circular waveguide. This effect is nonreciprocal. This means that, if a wave is rotated by $\theta°$ in the positive z-direction of propagation, it is rotated a further $\theta°$ in the negative z-direction. Faraday rotation may be understood by either forming a linear combination of the counterrotating magnetic circularly polarized normal modes of the medium or in terms of coupled wave theory. Both approaches are developed in detail. The origin of the splitting between the counterrotating modes may be understood by recalling that a gyromagnetic region displays different scalar permeabilities under the influence of such alternating fields. If the input waves into the two coupled waves correspond to one or the other of the two normal modes, no coupling takes place, and the output waves are emergent in the same normal mode. The chapter includes a definition of the gyrator network and the description of some common ferrite devices in round waveguides based on Faraday rotation effect, such as Faraday rotation isolators, circulators, nonreciprocal phase-shifters, and amplitude modulators and switches. Three and four-port circulators based on turnstile junctions are also described.

Microwave Polarizers, Power Dividers, Phase Shifters, Circulators, and Switches,
First Edition. Joseph Helszajn.

2.2 Faraday Rotation

The simplest problem involving the tensor permeability is that of propagation in an infinite medium. This simple introduction serves to demonstrate the well-known Faraday effect on which a number of different ferrite devices rely. The arrangement considered is that where the direction of propagation and the direct magnetic field are both in the z-direction and the microwave field is in the transverse plane. Propagation in the medium is described by replacing μ_r by $[\mu]$ in Maxwell's equations

$$\nabla \times \vec{E} = -j\omega\mu_0[\mu]\vec{H} \tag{2.1a}$$

$$\nabla \times \vec{H} = j\omega\varepsilon_0\varepsilon_r\vec{E} \tag{2.1b}$$

$$\nabla \cdot \vec{B} = 0 \tag{2.1c}$$

$$\nabla \cdot \vec{D} = 0 \tag{2.1d}$$

where $\vec{B} = [\mu]\vec{H}, \vec{D} = \varepsilon_0\varepsilon_r\vec{E}$, and

$$[\mu] = \begin{bmatrix} \mu & -j\kappa & 0 \\ j\kappa & \mu & 0 \\ 0 & 0 & 1 \end{bmatrix} \tag{2.2}$$

The formal derivation of plane wave propagation in this medium begins by forming the wave equation for $\vec{H}$. Taking the rotational of Eq. (2.1b), and use of Eq. (2.1a) gives

$$\nabla \times \nabla \times \vec{H} - \omega^2\varepsilon_0\varepsilon_r\mu_0[\mu]\vec{H} = 0 \tag{2.3}$$

with the alternating magnetic field in the transverse plane

$$\vec{h} = \begin{bmatrix} H_x \\ H_y \\ 0 \end{bmatrix} \tag{2.4}$$

It continues by assuming propagation of the form exp $(-j\beta z)$ with no variation of the fields in the transverse x–y plane

$$\nabla \times \nabla \times \vec{H} = -\frac{\partial^2}{\partial z^2}\vec{H} = \beta^2\vec{H} \tag{2.5}$$

Specializing the wave equation in this situation in the transverse plane gives

$$\beta^2\begin{bmatrix} H_x \\ H_y \end{bmatrix} = \omega^2\varepsilon_0\varepsilon_r\mu_0\begin{bmatrix} \mu & -j\kappa \\ j\kappa & \mu \end{bmatrix}\begin{bmatrix} H_x \\ H_y \end{bmatrix} \tag{2.6}$$

The roots or eigenvalues of this equation are determined by

$$\begin{bmatrix} \mu - \dfrac{\beta^2}{\omega^2 \varepsilon_0 \varepsilon_r \mu_0} & -j\kappa \\ j\kappa & \mu - \dfrac{\beta^2}{\omega^2 \varepsilon_0 \varepsilon_r \mu_0} \end{bmatrix} = 0 \tag{2.7}$$

or

$$\beta_{\pm}^2 = \omega^2 \varepsilon_0 \varepsilon_r \mu_0 (\mu \mp \kappa) \tag{2.8}$$

The magnetic fields corresponding to the eigenvalues $\beta_{\pm}^2$ may now be derived by introducing $\beta_{\pm}$ one at a time into Eq. (2.6). The result is

$$H_y^{\pm} = \pm j H_x^{\pm} \tag{2.9}$$

In the transverse plane the solution consists of two plane circularly polarized magnetic waves rotating in opposite directions with propagation constants $\gamma_{\pm}$ and scalar permeabilities $\mu \pm \kappa$. The superscripts or subscripts refer to the sense of the circular polarization; the plus sign indicating clockwise rotation when viewed in the direction of propagation and the minus sign anticlockwise rotation.

If the direction of the direct magnetic field is reversed, the propagation constants of the two normal modes are interchanged. This observation can be quite simply understood by investigating the dependence of μ and κ upon the direction of the direct magnetic field. This indicates that μ is an even function of the direct field and κ is an odd function of the direct field. The direction of propagation has no bearing on the sign of the propagation constants since β appears only as an even power in Eq. (2.8). Figure 2.1 summarizes the different situations.

The nature of the electric field may be separately deduced by making use of the first of Maxwell's two curl equations

$$\begin{bmatrix} \hat{x} & \hat{y} & \hat{z} \\ 0 & 0 & -j\beta \\ E_x & E_y & 0 \end{bmatrix} = -j\omega\mu_0 \begin{bmatrix} \mu & -j\kappa & 0 \\ j\kappa & \mu & 0 \\ 0 & 0 & 1 \end{bmatrix} \begin{bmatrix} H_x \\ H_y \\ 0 \end{bmatrix} \tag{2.10}$$

or

$$j\beta E_y = j\omega\mu_0 (\mu H_x - j\kappa H_y) \tag{2.11a}$$

$$j\beta E_x = j\omega\mu_0 \left(j\kappa H_x - \mu H_y\right) \tag{2.11b}$$

where for a loss-free medium γ is replaced by $j\beta$. Adopting the boundary condition with the upper sign in Eq. (2.9), readily gives

$$\beta_+ E_y = -\omega\mu_0 (\mu - \kappa) H_x \tag{2.12a}$$

$$\beta_+ E_x = -\omega\mu_0 (\mu - \kappa) H_y \tag{2.12b}$$

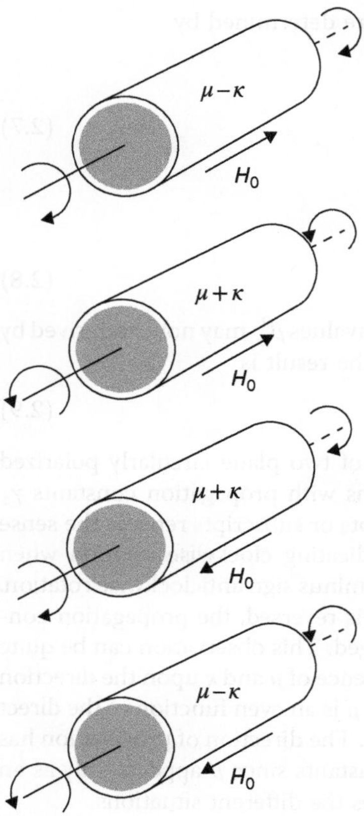

Figure 2.1 Schematic diagrams of normal modes of propagation in longitudinally magnetized ferrite medium.

This result indicates that the electric field is also circularly polarized in that

$$E_y = jE_x \tag{2.13}$$

and that the wave impedance in such a medium is a scalar quantity described by

$$Z^+ = \frac{E_y}{-H_x} = \frac{E_x}{H_y} = \frac{\omega\mu_0(\mu-\kappa)}{\beta_+} \tag{2.14}$$

Taking the lower sign in Eq. (2.9) indicates that

$$E_y = -jE_x \tag{2.15}$$

and

$$Z^- = \frac{E_y}{-H_x} = \frac{E_x}{H_y} = \frac{\omega\mu_0(\mu+\kappa)}{\beta_-} \tag{2.16}$$

Expressing Eqs. (2.14) and (2.16) in terms of the original variables also indicates that

$$Z^{\pm} = \sqrt{\frac{\mu_0(\mu \mp \kappa)}{\varepsilon_0 \varepsilon_r}} \tag{2.17}$$

To determine the behavior of a linearly polarized electric wave in this type of gyromagnetic medium, it is sufficient to take a linear combination of the two circularly polarized waves defined by Eqs. (2.13) and (2.15). This gives

$$E_y(z) = \frac{1}{2}\exp(-j\beta_- z) + \frac{1}{2}\exp(-j\beta_+ z) \tag{2.18a}$$

$$E_x(z) = \frac{j}{2}\exp(-j\beta_- z) - \frac{j}{2}\exp(-j\beta_+ z) \tag{2.18b}$$

The preceding equations satisfy the boundary conditions $E_y(0) = 1$ and $E_x(0) = 0$. These may be simplified by taking out a common factor

$$\exp\left(-j\frac{\beta_- + \beta_+}{2}z\right) \tag{2.19}$$

The required result is

$$E_y(z) = \cos\left(\frac{\beta_- - \beta_+}{2}z\right)\exp\left(-j\frac{\beta_- + \beta_+}{2}z\right) \tag{2.20a}$$

$$E_x(z) = \sin\left(\frac{\beta_- - \beta_+}{2}z\right)\exp\left(-j\frac{\beta_- + \beta_+}{2}z\right) \tag{2.20b}$$

It is readily appreciated that the vector addition of these two quantities produces a linearly polarized wave rotating with the quantity $(\beta_- - \beta_+)/(2z)$. The angle of polarization θ is described by

$$\theta = \frac{\beta_- - \beta_+}{2}z \tag{2.21}$$

The notation used here is consistent with the fact that $\beta_- > \beta_+$ below the so-called ferromagnetic resonance.

This result indicates that a wave propagating a certain distance in one direction is rotated through an angle θ with respect to the input polarization. When it is reflected and returns to its starting point it is again rotated by θ. This feature may be understood by recognizing that the signs of β_+ and β_- are reversed for propagation in the $-z$ direction and furthermore, that the phase constants of $\beta_\pm$ are separately interchanged.

The total rotation of the reflected wave is therefore 2θ with respect to the outgoing wave, i.e. it is not rotated back to its original orientation. Figure 2.2 illustrates the situation for a 90° section. Thus, Faraday rotation is nonreciprocal

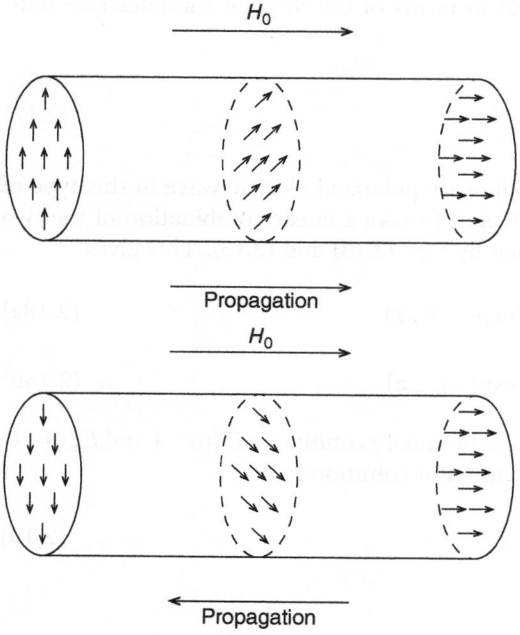

Figure 2.2 Schematic diagrams of Faraday rotation in positive and negative directions of propagation.

and leads to a number of nonreciprocal devices. Some practical cross-sections are illustrated in Figure 2.3.

It is also possible to take a linear combination of the normal modes so that $E_x(0) = 1$ and $E_y(0) = 0$. Applying these boundary conditions indicates that

$$E_y(z) = -\sin\left(\frac{\beta_- - \beta_+}{2}z\right)\exp\left(-j\frac{\beta_- + \beta_+}{2}z\right) \tag{2.22a}$$

$$E_x(z) = \cos\left(\frac{\beta_- - \beta_+}{2}z\right)\exp\left(-j\frac{\beta_- + \beta_+}{2}z\right) \tag{2.22b}$$

and combining the results given by Eqs. (2.20) and (2.22) gives

$$\begin{bmatrix} E_y(z) \\ E_x(z) \end{bmatrix} = \exp\left(-j\frac{\beta_- + \beta_+}{2}z\right)\begin{bmatrix} \cos\left(\frac{\beta_- - \beta_+}{2}z\right) & \sin\left(\frac{\beta_- - \beta_+}{2}z\right) \\ -\sin\left(\frac{\beta_- - \beta_+}{2}z\right) & \cos\left(\frac{\beta_- - \beta_+}{2}z\right) \end{bmatrix}\begin{bmatrix} E_y(0) \\ E_x(0) \end{bmatrix} \tag{2.23}$$

If the signals in the two polarizations coincide with one of the normal modes, no coupling between the two takes place and the outgoing waves will be emergent in the same mode. Because the prototype is nonreciprocal, the wave will be

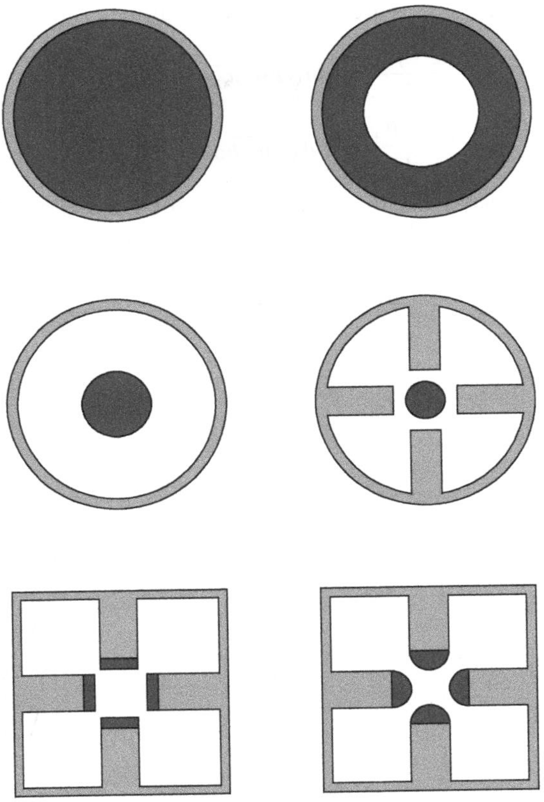

Figure 2.3 Some waveguides with required symmetry for use in Faraday rotation devices.

phase-shifted through either $\beta_+ z$ or $\beta_- z$ depending upon the direction of the dc biasing magnetic field.

2.3 Magnetic Variables of Faraday Rotation Devices

The difference and sum of the split-phase constants are given in terms of the original variables in Eq. (2.8) by

$$\frac{\beta_- - \beta_+}{2} = \frac{1}{2}\omega\sqrt{\varepsilon_0\varepsilon_r\mu_0\mu}\left[\left(1+\frac{\kappa}{\mu}\right)^{1/2} - \left(1-\frac{\kappa}{\mu}\right)^{1/2}\right] \tag{2.24}$$

and

$$\frac{\beta_- + \beta_+}{2} = \frac{1}{2}\omega\sqrt{\varepsilon_0\varepsilon_r\mu_0\mu}\left[\left(1+\frac{\kappa}{\mu}\right)^{1/2} + \left(1-\frac{\kappa}{\mu}\right)^{1/2}\right] \tag{2.25}$$

If κ/μ are appreciably less than unity, then

$$\frac{\beta_- - \beta_+}{2} = \frac{1}{2}\omega\sqrt{\varepsilon_0\varepsilon_r\mu_0\mu}\frac{\kappa}{\mu} \tag{2.26}$$

$$\frac{\beta_- + \beta_+}{2} = \frac{1}{2}\omega\sqrt{\varepsilon_0\varepsilon_r\mu_0\mu} \tag{2.27}$$

Faraday rotation is therefore proportional to the ratio of the off-diagonal and diagonal elements of the tensor permeability. The variation of κ/μ as a function of the direct magnetic field for a finite medium is shown in Figure 2.4.

For a saturated material,

$$\kappa = \frac{\omega_m}{\omega} \tag{2.28a}$$

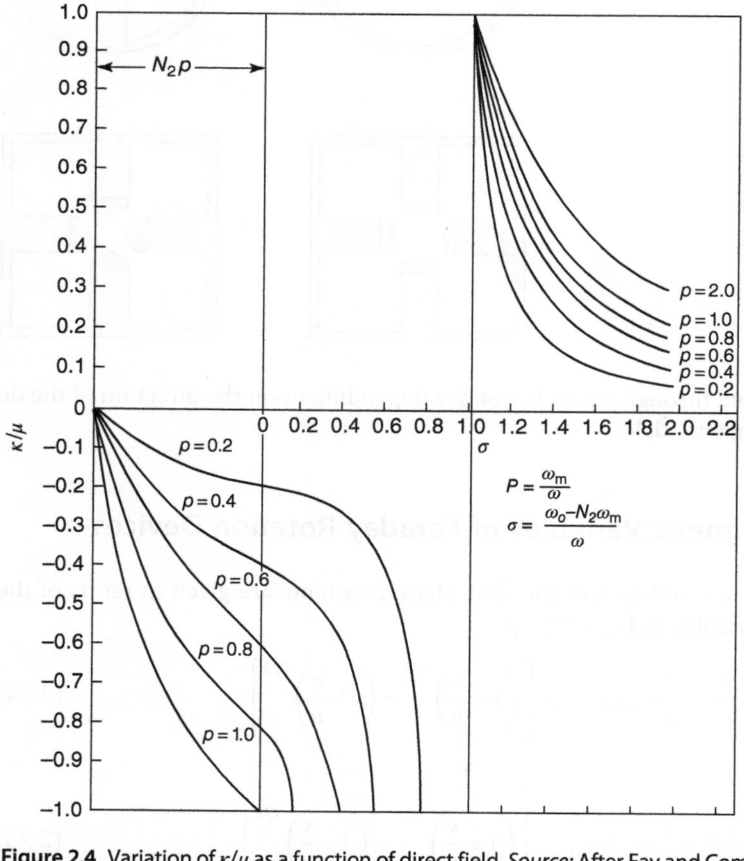

Figure 2.4 Variation of κ/μ as a function of direct field. *Source:* After Fay and Comstock (1965).

$$\mu \approx 1 \tag{2.28b}$$

and

$$\frac{\beta_- - \beta_+}{2} = \frac{1}{2}\omega\sqrt{\varepsilon_0\varepsilon_r\mu_0\mu}\omega_m \tag{2.29a}$$

$$\frac{\beta_- + \beta_+}{2} = \frac{1}{2}\omega\sqrt{\varepsilon_0\varepsilon_r\mu_0} \tag{2.29b}$$

Faraday rotation in an infinite saturated medium is therefore frequency independent; an observation that has been utilized in the construction of octave-band wide Faraday rotation devices.

2.4 The Gyrator Network

The simplest component that illustrates the nonreciprocal property of a Faraday rotator is the gyrator circuit. This element is a four-terminal two-port device that has zero relative phase shift in one direction of propagation and 180° relative phase shift in the other. It is characterized by the following scattering matrix:

$$S = \begin{bmatrix} 0 & -1 \\ 1 & 0 \end{bmatrix} \tag{2.30}$$

The fact that such as circuit is realizable as a lossless network is readily verified by having recourse to the unitary condition:

$$S^{\mathrm{T}}S^* - I = 0$$

One realization of this network consists of a rectangular waveguide with a 90° twist followed by a 90° Faraday rotator section. The output port is oriented in the same plane as the input one in the manner indicated in Figure 2.5.

A vertically polarized wave propagating from left to right has its polarization rotated 90° by the twisted rectangular waveguide and a further 90° by the Faraday rotation; it therefore emerges at the output port having being rotated 180° with respect to the input port. A vertically polarized wave propagating in the opposite direction is rotated by the 90° Faraday rotator in the same sense as before. In this case, however, the waveguide twist will cancel the Faraday rotation instead of adding to it; the wave therefore displays no rotation in this direction or propagation. The effective length of this gyrator is an odd integral number of half wavelength for transmission in one direction and an even number in the other direction of propagation.

One application of the gyrator network is in the construction of the four-port differential phase-shift circulator depicted in Figure 2.6.

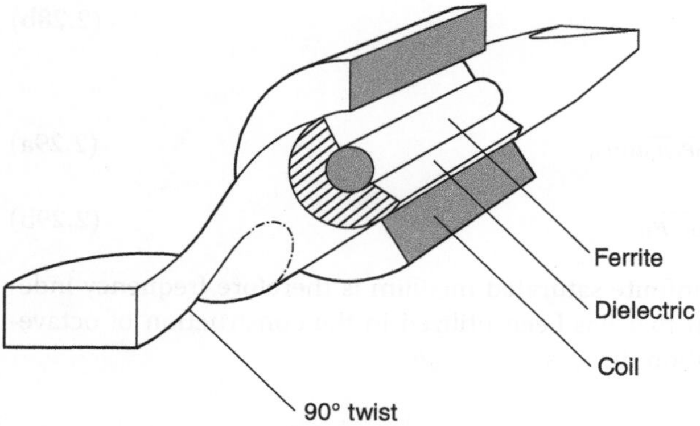

Figure 2.5 Schematic diagram of a gyrator network. *Source:* After Hogan (1952).

This circuit consists of two magic-tee-type power dividers with a gyrator network at the plane of symmetry of one of its two connecting waveguides. The operation of this device may be readily understood by noting that a signal at the *E*-plane port of this network is divided into equal out-of-phase signals at the two symmetric *H*-plane ones, whereas a signal at the *H*-plane port is divided into equal in-phase signal at the two symmetric *H*-plane ports. A signal at port 1 of the overall arrangement illustrated in Figure 2.6 therefore is divided into equal in-phase signals, which recombine at port 2. A signal at port 2 is also

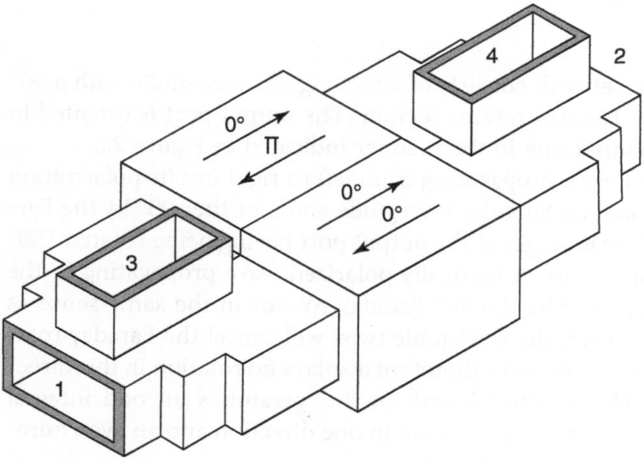

Figure 2.6 Four-port differential circulator employing a Faraday gyrator network. *Source:* After Hogan (1952).

divided into equal in-phase signals but due to the additional 180° differential phase-shift of the gyrator network it recombines at port 3. In a likewise manner, a signal at port 3 is transmitted to port 4 and one at port 4 is routed to port 1 and so on in a cyclic manner. Figure 2.7 depicts the schematic diagram of the four-port circulator.

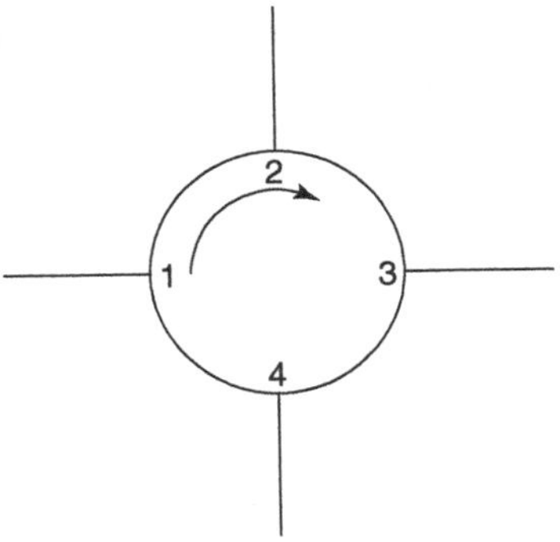

Figure 2.7 Equivalent circuit of four-port circulator.

2.5 Faraday Rotation Isolator

The operation of this type of isolator can be understood with the help of Figure 2.8. The rotator prototype is matched to the rectangular waveguide at the two ends by tapers or quarter-wave transformers with round waveguide.

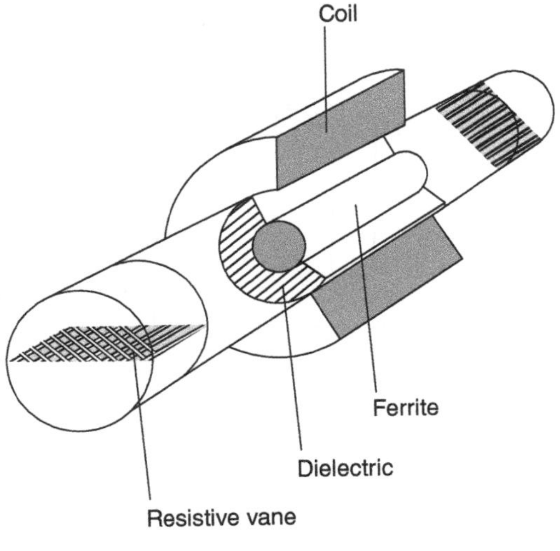

Figure 2.8 Schematic diagram of Faraday rotator isolator. *Source:* After Hogan (1952).

Resistance vanes are inserted into the round waveguide sections in a plane perpendicular to the electric fields of the input and output rectangular waveguides. A signal incident at the input port will be perpendicular to the first resistance card, and after a clockwise rotation through 45°, will also be perpendicularly polarized with respect to that at the output port. It will, therefore, be transmitted without attenuation through the isolator. In the reverse direction, a signal at the output, vane is likewise perpendicular to that card, but after rotation through 45° in a clockwise sense, will now be in the plane of the input one where it is attenuated. The wavelength in the transformer section is approximately the geometric mean of that of the rectangular waveguide and that of the isotropic round waveguide containing the ferrite rod. This device may also be used as an amplitude modulator by suitably varying the direct magnetic field.

2.6 Four-port Faraday Rotation Circulator

Another important application of a Faraday section is in connection with the realization of a four-port Faraday rotation circulator. A schematic diagram of this device is illustrated in Figure 2.9. In this device, power entering port 1 emerges as port 2, and so on in a cyclic manner. The physical arrangement is similar to the Faraday rotator isolator except that the sections containing the resistance vanes are replaced by two-mode transducers. Such transducers allow orthogonal linearly polarized waves to be applied to the round waveguide

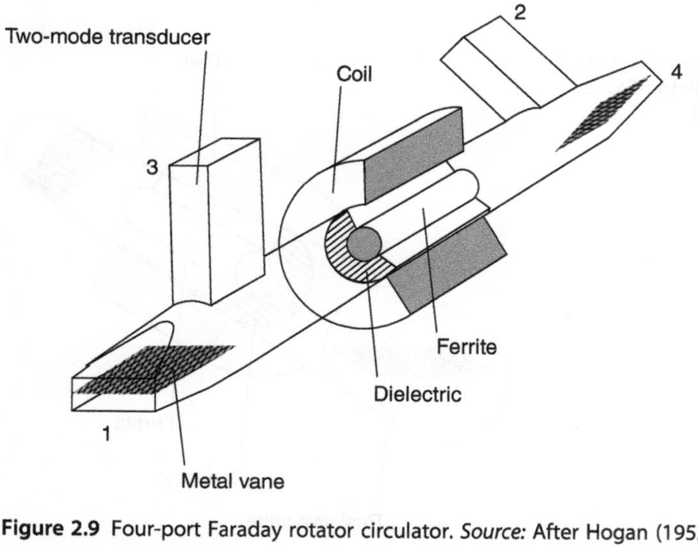

Figure 2.9 Four-port Faraday rotator circulator. *Source:* After Hogan (1952).

section. The Faraday rotator is again a 45° section. A wave entering port 1 with its electric field vertically polarized is rotated clockwise by 45° by the ferrite rotator and emerges at port 2. A wave entering port 2 is also rotated clockwise so that its electric field is now horizontally polarized at the input of the first two-mode transducer. It therefore emerges at port 3. Similarly, transmission occurs from port 3 to port 4, port 4 to port 1, and so on. This device may also be used as a switch by reversing the direction of the direct magnetic field.

2.7 Nonreciprocal Faraday Rotation-type Phase Shifter

It has already been noted in connection with Eq. (2.23) that if the input excitation to the Faraday rotation section corresponds to one of its normal modes, no Faraday rotation will occur. Instead, the wave travels in the same normal mode through the rotator section, but phase-shifted through either $\beta_+ z$ or $\beta_- z$ radians. This principle can be used to design nonreciprocal or reciprocal ferrite phase shifters in conjunction with reciprocal or nonreciprocal quarter-wave plates. One such nonreciprocal device will now be described. A nonreciprocal phase shifter is illustrated in Figure 2.10. The arrangement uses two reciprocal quarter-wave plates at either end of the Faraday rotation section. The first quarter-wave plate converts a linearly polarized input wave into a positive circularly polarized wave at the input of the rotator section. This wave is then phase-shifted through $\beta_+ z$ radians in the rotator section.

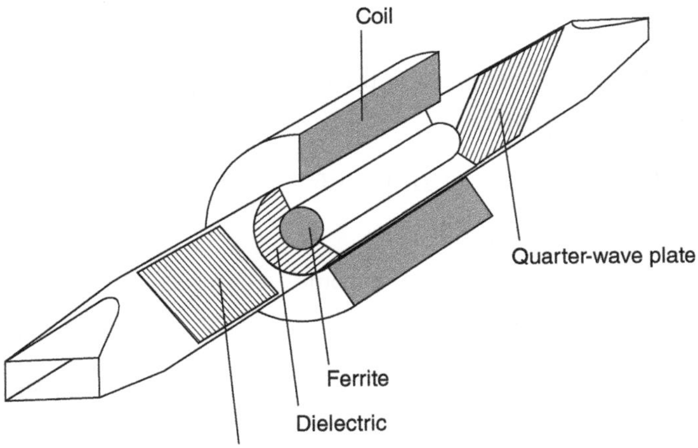

Figure 2.10 Nonreciprocal Faraday rotator phase shifter.

The phase-shifted circularly polarized wave is reconverted to a linearly polarized wave at the output by the second quarter-wave plate. In the reverse direction of propagation the circularly polarized wave is in the opposite sense and the wave is therefore phase-shifted through $\beta_{-}z$. The arrangement therefore behaves as a nonreciprocal phase shifter. A forward wave can of course also be switched from $\beta_{+}z$ to $\beta_{-}z$ by reversing the dc magnetic field on the rotator section.

2.8 Coupled Wave Theory of Faraday Rotation Section

Treating the rotator section as a four-port nonreciprocal direction coupler indicates that the entries of its scattering matrix with an input at port 1 may be expressed in terms of the properties of the counterrotating reflection $\rho_{\pm}$ and transmission $\tau_{\pm}$ variables of the system by

$$S_{11} = \frac{\rho_{-} + \rho_{+}}{2} \tag{2.31a}$$

$$S_{12} = \frac{\rho_{-} - \rho_{+}}{-2j} \tag{2.31b}$$

$$S_{13} = \frac{\tau_{-} + \tau_{+}}{2} \tag{2.31c}$$

$$S_{14} = \frac{\tau_{-} - \tau_{+}}{2j} \tag{2.31d}$$

If the magnetized line is matched to the demagnetized one by a stepped impedance transformer, the reflection coefficients $\rho_{\pm}$ are given by

$$\rho_{\pm} = \frac{-\beta_0 + \beta_{\pm}}{\beta_0 + \beta_{\pm}} \tag{2.32}$$

The transmission coefficients $\tau_{\pm}$ are given in the usual way by

$$\tau_{\pm} = \sqrt{1 - \rho_{\pm}\rho_{\pm}^{*}}\exp(-j\beta_{\pm}\ell) \tag{2.33}$$

It is readily verified that Eqs. (2.31a)–(2.31d) satisfy the unitary condition:

$$S_{11}S_{11}^{*} + S_{12}S_{12}^{*} + S_{13}S_{13}^{*} + S_{14}S_{14}^{*} = 1 \tag{2.34}$$

For symmetric splitting Eqs. (2.31a)–(2.31d) become

$$S_{11} = 0 \tag{2.35a}$$

$$S_{12} \cong j\frac{\beta_{-} - \beta_{+}}{2\beta_0} \tag{2.35b}$$

$$S_{13} = \sqrt{1 - \left(\frac{\beta_{-} - \beta_{+}}{2\beta_0}\right)^2} \cos\left(\frac{\beta_{-} - \beta_{+}}{2}\ell\right) \exp(-j\beta_0\ell) \tag{2.35c}$$

$$S_{14} = \sqrt{1 - \left(\frac{\beta_- - \beta_+}{2\beta_0}\right)^2} \sin\left(\frac{\beta_- - \beta_+}{2}\ell\right) \exp(-j\beta_0\ell) \tag{2.35d}$$

which also satisfies the unitary condition.

This result suggests that in a four-port nonreciprocal network, matching port 1 is not sufficient to decouple port 2. This feature is of course well understood. In order to decouple port 2 from port 1 by at least 20 dB, it is necessary to have

$$\frac{\beta_- - \beta_+}{2\beta_0} \leq 0.10 \tag{2.36}$$

This result places an upper bound on the normalized splitting and a lower bound on the overall length of the device. If the splitting is not symmetric, Eqs. (2.35a) and (2.35b) do not apply and the outputs at port 3 and 4 combine as an elliptically polarized wave instead of as a linearly polarized one.

2.9 The Partially Ferrite-filled Circular Waveguide

The geometry of a practical Faraday rotation section usually consists of a ferrite rod inside a circular waveguide filled by a suitable dielectric material. The geometry considered here is shown in Figure 2.11 where a is the radius of the circular waveguide and b is the radius of the ferrite rod. Its mode spectrum has been extensively investigated by Waldron in a number of classic papers. Figure 2.12 depicts the split phase constants of one solution.

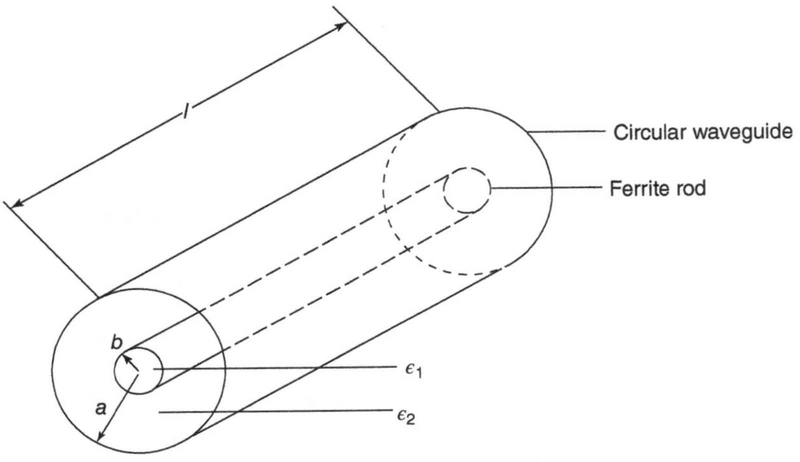

Figure 2.11 Circular waveguide containing a coaxial ferrite cylinder.

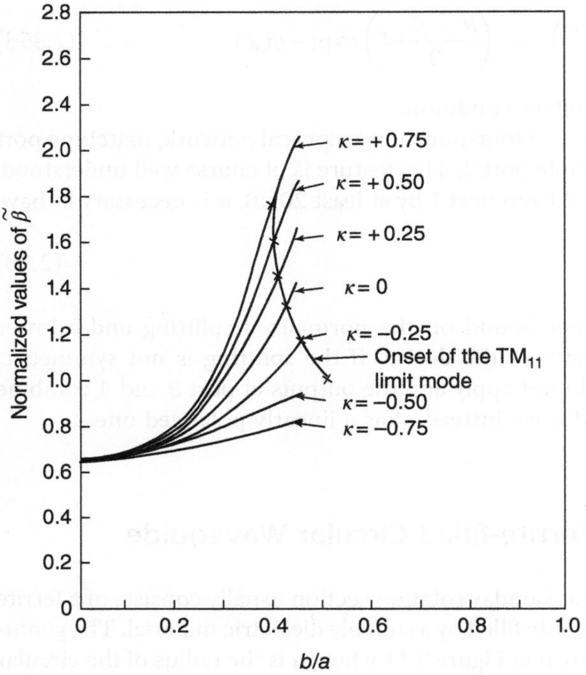

Figure 2.12 Normalized values of propagation constants $(\beta/k_0\sqrt{\varepsilon_d})$ for TM_{11} limit mode for $\varepsilon_r/\varepsilon_d = 5$, $a/\lambda_d = 0.40$ and parametric values of κ. *Source:* After Waldron (1958).

Bibliography

Allen, P.J. (1956). The turnstile circulator. *IRE Trans. Microw. Theory Tech.* **MTT-4**: 223–227.

Beust, W. and Johnston, E.G. (1960). High average power rotator. *Microw. J.* (May).

Casswell, D. (1964). Faraday rotation devices. *Microwaves* 24.

Chait, H. and Sakiotis, G. (1959). Broadband ferrite rotators using quadruply ridged circular waveguide. *IRE Trans. Microw. Theory Tech.* **MTT-7**: 38–41.

Clarricoats, P.J.B. (1959). A perturbation method for circular waveguides containing ferrites. *Proc. IEEE* **106** (Part B): 335–340.

Fay, C.E. and Comstock, R.L. (1965). Operation of the ferrite junction circulator. *IEEE Trans. Microw. Theory Tech.* **MTT-13**: 15–27.

Fox, A.G., Miller, S.E., and Weiss, M.T. (1955). Behaviour and applications of ferrites in the microwave region. *Bell. Syst. Tech.* **34**: 5–103.

Gamo, H. (1953). The Faraday rotation of waves in a circular waveguide. *J. Phys. Soc. Jpn.* **8**: 176–182.

Helszajn, J. (1968). A symmetrical design of a short Faraday rotator. *Microw. J.* (August): 35–39.

Helszajn, J. (1989). *Ferrite Phase Shifters and Control Devices*. New York: McGraw-Hill.

Hogan, C.L. (1952). The ferromagnetic Faraday effect at microwave frequencies and its applications – the microwave gyrator. *Bell. Syst. Tech.* **31**: 1–31.

Kales, M.L. (1953). Modes in waveguides containing ferrites. *J. Appl. Phys.* **24**: 604–608.

Meyer, M.A. and Goldberg, H.B. (1955). Applications of the turnstile junction. *IRE Trans. Microw. Theory Tech.* **MTT-3**: 40–45.

Montgomery, C.G., Dicke, R.H., and Purcell, E.M. (1948). Principles of microwave circuits. *Radiat. Lab. Ser.* **8**: 459.

Ohm, O.E. (1956). A broadband microwave circulator. *IRE Trans. Microw. Theory Tech.* **MTT-4**: 210.

Ragan, G.L. (1948). Microwave transmission circuits. *Radiat. Lab Ser.* **9**: 375.

Rizzi, P.A. (1967). High power ferrite circulators. *IRE Trans. Microw. Theory Tech.* **MTT-5**: 210.

Rowen, J.H. (1953). Ferrites in microwave applications. *Bell. Syst. Tech.* **32**: 1333–1369.

Scharfman, H. (1956). Three new ferrite phase shifters. *Proc. IRE* **44**: 1456–1459.

Schaug-Patterson, T. (1958). Novel design of a 3-port circulator. Norwegian Defence Research (January).

Suhl, H. and Walker, L.R. (1954). Topics in guided wave propagation through gyromagnetic media-part I. *Bell. Syst. Tech.* **33**: 579–659.

Van Trier, A.A. (1953). Guided electromagnetic waves in anisotropic media. *Appl. Sci. Res.* **3**: 305–371.

Vartanian, P.H., Melchor, J.L., and Ayres, W.P. (1956). A broadband ferrite isolator. *IRE Trans. Microw. Theory Tech.* **MTT-4**: 8.

Waldron, R.A. (1958). Electromagnetic wave propagation in cylindrical waveguides containing gyromagnetic media. *J. Br. Inst. Radio Eng.* **18**: 597–612.

3

Circular Polarization in Parallel Plate Waveguides

Joseph Helszajn

Heriot Watt University, Edinburgh, UK

A circular polarized spinning electron under the influence of a direct magnetic field and an alternating radio frequency magnetic field display two different scalar permeabilities depending upon whether the alternating radio frequency field is circularly polarized in the same sense as that of the electromagnet spin or not. This property of a magnetized ferrite medium is the basis of some of the most important nonreciprocal devices described to date. Identifying such planes of circular polarization in practical microwave circuits or establishing them by one mean or another is therefore an essential part of the development of practical ferrite phase shifters. Situations in which the rotation of these waves is different in the two directions of propagation are of course of special interest. Such polarization is defined by two equal vectors in space quadrature with one or the other in time quadrature. Counterrotating magnetic fields occur in fact naturally on either side of the symmetry plane of an ordinary rectangular waveguide propagating the dominant TE mode: at the interface and everywhere outside two different dielectric regions. Furthermore, in each instance, the hand of rotation is interchanged if the direction of propagation is reversed. The subject of circular polarization in round waveguides is the topic of Chapters 5, 7, and 8.

3.1 Circular Polarization in Rectangular Waveguide

Natural regions of circular polarization may be found in a rectangular waveguide propagating the dominant TE_{10} mode. This feature is particularly important as a number of nonreciprocal ferrite devices rely on this type of polarization

Microwave Polarizers, Power Dividers, Phase Shifters, Circulators, and Switches, First Edition. Joseph Helszajn.

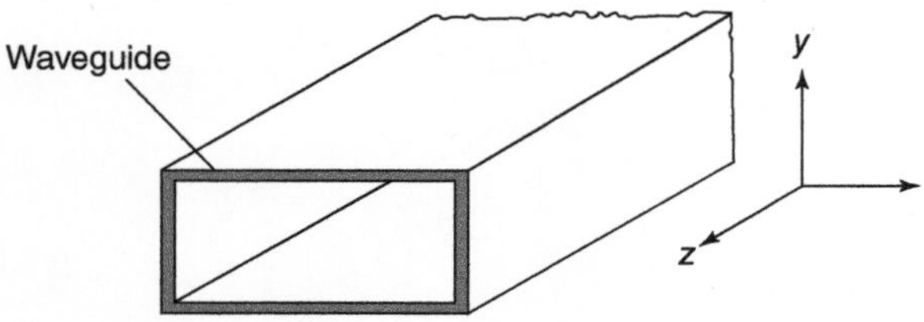

Figure 3.1 Schematic diagram of a rectangular waveguide.

for their operation. It can be demonstrated quite simply by first forming the three field components for the waveguide in Figure 3.1.

$$H_z = \cos\left(\frac{\pi x}{a}\right)\cdot \exp\cdot(-j\beta_z z) \tag{3.1}$$

$$H_x = j\left(\frac{\lambda_c}{\lambda_g}\right)\sin\left(\frac{\pi x}{a}\right)\cdot \exp\cdot(-j\beta_z z) \tag{3.2}$$

$$E_y = -j\left(\frac{\lambda_c}{\lambda_0}\right)\sqrt{\frac{\mu_0}{\varepsilon_0}}\sin\left(\frac{\pi x}{a}\right)\cdot \exp\cdot(-j\beta_z z) \tag{3.3}$$

where propagation is assumed along the positive z direction (Figure 3.2). If it is in the negative z direction the terms involving the λ_g are reversed (Figure 3.3).

In the above equations

$$\lambda_c = 2a \tag{3.4}$$

$$\left(\frac{2\pi}{\lambda_g}\right)^2 = \left(\frac{2\pi}{\lambda_0}\right)^2 - \left(\frac{2\pi}{\lambda_c}\right)^2 \tag{3.5}$$

$$\beta_z = \frac{2\pi}{\lambda_g} \tag{3.6}$$

It is observed from these equations that H_x and H_z are in time space quadrature. If a region can now be located where their amplitudes are also equal it would exhibit circular polarization. This condition is in fact met on either side of the center line of the waveguide provided.

$$\tan\left(\frac{\pi x}{a}\right) = \frac{\lambda_g}{\lambda_c} \tag{3.7}$$

The polarizations along such a waveguide are in opposite directions on either side of the center line of the waveguide. Furthermore, their hands are reversed, provided propagation is taken along the negative z-axis instead of the positive one.

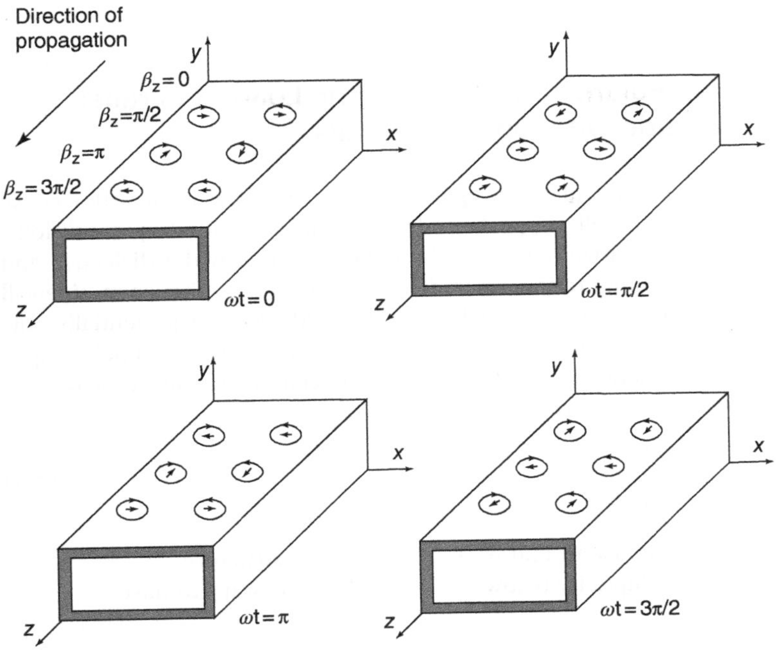

Figure 3.2 Planes of circular polarization in rectangular waveguide for forward direction of propagation.

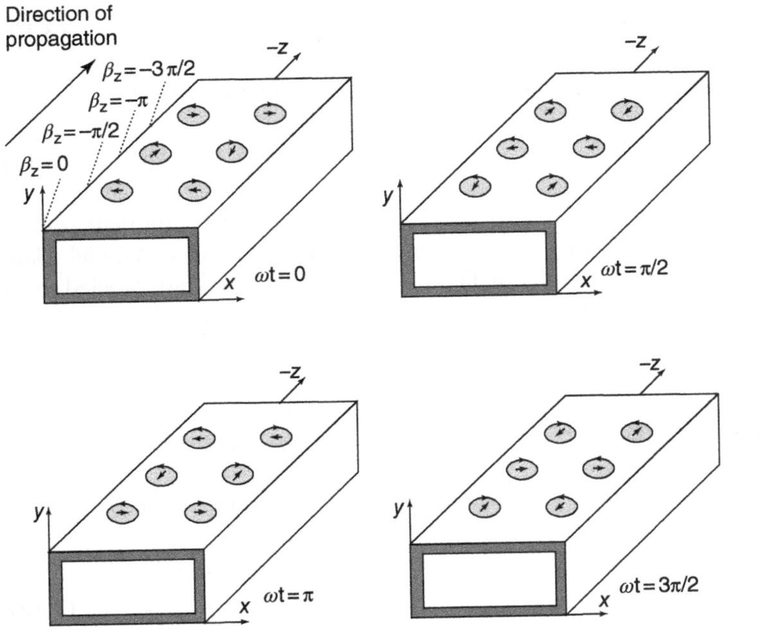

Figure 3.3 Planes of circular polarization in rectangular waveguide for backward direction of propagation.

3.2 Circular Polarization in Dielectric Loaded Parallel Plate Waveguide with Open Sidewalls

Dielectric loaded parallel plate waveguides with open sidewalls or with electric or magnetic sidewalls all support, under appropriate boundary conditions, planes of circular polarization at the boundaries between the dielectric and air regions. The configuration treated in this section is the open sidewall arrangement in Figure 3.4. In this situation the fields decay exponentially outside the dielectric region and the ratio of the transverse to longitudinal components of the r.f. magnetic field at the interface between the two dielectric regions and everywhere outside is readily given by

$$\frac{H_x}{H_z} = \frac{-j}{\sqrt{1-(k_0^2/\beta_z^2)}} \tag{3.8}$$

k_0 is the free space constant and β_z is that along the structure.

To maintain the ellipticity below 1.05 (say) it is necessary to have

$$\frac{\beta_z}{k_0} \geq 3 \tag{3.9}$$

Forming β_z/k_0 in the 2–4 GHz band for a dielectric slab with $\varepsilon_r = 9$, using the data given later in this chapter, gives

$$1.8 \lesssim \frac{\beta_z}{k_0} \lesssim 2.2 \tag{3.10}$$

and the corresponding ellipticity over the same frequency interval is

$$1.20 \geq \left|\frac{H_x}{H_z}\right| \geq 1.12 \tag{3.11}$$

The derivation of this result starts by developing the field components of the TE family of modes in the three regions of dielectric loaded parallel plate waveguide with open sidewalls in Figure 3.4. The solutions are labeled even or odd according to whether an electric or magnetic wall can be introduced along the place of symmetry at $x = a$. The dominant mode in such a waveguide is the so-called even one with no low frequency cutoff condition. The derivation proceeds from first principles in order to familiarize the reader with this class of boundary value problem, which will be met again in a number of related situations.

Maxwell's equations in differential form for a charge-free region ($\rho = 0$) are given by

$$\nabla \times \bar{E} = -\mu_0 \frac{\delta \bar{H}}{\delta t} \tag{3.12}$$

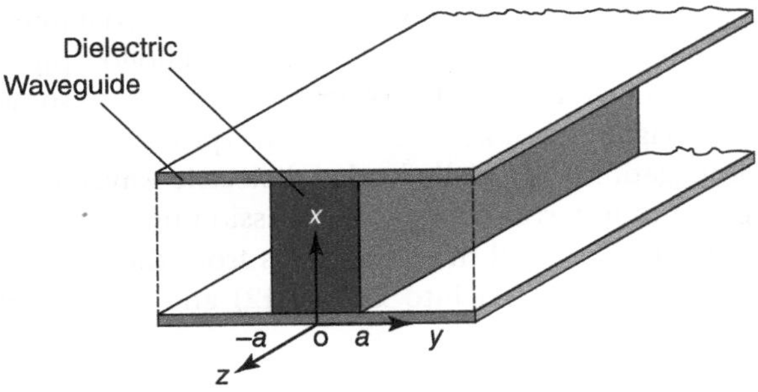

Figure 3.4 Schematic diagram of dielectric loaded parallel plate waveguide with open sidewalls.

$$\nabla \times \bar{H} = \varepsilon_0 \frac{\delta \bar{E}}{\delta t} \tag{3.13}$$

$$\nabla \cdot \bar{D} = 0 \tag{3.14}$$

$$\nabla \cdot \bar{B} = 0 \tag{3.15}$$

H is the magnetic field (A m^{-1}), E is the electric field (V m^{-1}), μ_0 is the free space permeability ($4\pi \times 10^{-7}$ H m^{-1}), ε_0 is the free space permittivity ($1/36\pi \times 10^{-11}$ F m^{-1}), $\bar{D} = \varepsilon_0 \bar{E}$, and $\bar{B} = \mu_0 \bar{H}$.

In transmission line theory it is usual to assume that all the fields vary in the positive z direction as

$$\exp(-\gamma z) \tag{3.16}$$

γ is a complex quantity known as the propagation constant

$$\gamma = \alpha + j\beta \tag{3.17}$$

α is the attenuation constant per unit length (nepers m^{-1}), and β is the phase constant (rad m^{-1}).

It is also usual to describe the time variation as

$$\exp(j\omega t) \tag{3.18}$$

In what follows, it is therefore only necessary to determine the variation of the fields in the transverse x–y plane.

Solutions to Maxwell's equations may be catalogued according to whether

$$E_z = H_z = 0, \quad \text{TEM} \tag{3.19}$$

$$E_z \neq 0, \quad H_z = 0, \quad \text{TM} \tag{3.20}$$

$$E_z = 0, \quad H_z \neq 0, \quad \text{TE} \tag{3.21}$$

$$E_z \neq 0, \quad H_z \neq 0, \quad \text{HE} \tag{3.22}$$

The first of these solutions is encountered in two wire systems or coaxial lines, the second and third class of solutions are met in the description of hollow pipes having rectangular, triangular, circular, and elliptical cross-sections and in some parallel plate waveguides with magnetic sidewalls. The fourth type of solution is met in the description of propagation in partially loaded dielectric waveguides or open dielectric waveguides and in ferrite-loaded transmission lines.

If E_z or H_z exist, it is possible to derive all the other fields from them.

Introducing Eqs. (3.16), (3.18), and (3.21) into Eqs. (3.12) and (3.13) and taking

$$E_x = E_z = H_y = 0 \tag{3.23}$$

$$\frac{\delta}{\delta y} = 0 \tag{3.24}$$

leads to

$$\begin{bmatrix} i_x & i_y & i_z \\ \dfrac{\delta}{\delta x} & 0 & -\gamma \\ 0 & E_y & 0 \end{bmatrix} = -j\omega\mu_0 \begin{bmatrix} H_x \\ 0 \\ H_z \end{bmatrix} \tag{3.25}$$

or

$$\gamma E_y = -j\omega\mu_0 H_x \tag{3.26}$$

$$\frac{\delta E_y}{\delta x} = -j\omega\mu_0 H_z \tag{3.27}$$

Likewise

$$\begin{bmatrix} i_x & i_y & i_z \\ \dfrac{\delta}{\delta x} & 0 & -\gamma \\ H_x & 0 & H_z \end{bmatrix} = j\omega\varepsilon_0 \begin{bmatrix} 0 \\ E_y \\ 0 \end{bmatrix} \tag{3.28}$$

gives

$$\frac{\delta H_z}{\delta x} + \gamma H_x = -j\omega\varepsilon_0 E_y \tag{3.29}$$

Making use of the two divergence relationships in Eqs. (3.14) and (3.15) yields

$$\frac{\delta E_y}{\delta y} = 0 \tag{3.30}$$

and

$$\frac{\delta H_x}{\delta x - \gamma H_z = 0} \tag{3.31}$$

in keeping with $\delta/\delta y = 0$ and $E_x = E_z = H_y = 0$.

H_x and E_y are now written in terms of H_z using the appropriate curl equations.

$$H_x = \frac{-\gamma}{\gamma^2 + \omega^2 \mu_0 \varepsilon_0} \frac{\delta H_z}{\delta x} \tag{3.32}$$

and

$$E_y = \frac{-j\mu_0 \omega}{\lambda} H_x \tag{3.33}$$

H_x and E_y may now be formed once H_z has been determined. This is done by satisfying the wave equation using the second divergence equation in Eq. (3.31).

The result is

$$\frac{\delta^2 H_z}{\delta x^2} + \left(\gamma^2 + \omega^2 \mu_0 \varepsilon_0\right) H_z = 0 \tag{3.34}$$

One solution to this equation in the dielectric region with

$$\gamma \rightarrow j\beta_z \tag{3.35}$$

$$\varepsilon_0 \longrightarrow \varepsilon_0 \varepsilon_1 \tag{3.36}$$

is

$$H_z = A \sin(k_1 x) \cdot \exp(-j\beta_z z) \tag{3.37}$$

which satisfies the wave equation with

$$-k_1^2 + \left(-\beta_z^2 + \omega^2 \mu_0 \varepsilon_0 \varepsilon_1\right) = 0 \tag{3.38}$$

and the magnetic wall boundary condition at the symmetry plane of the dielectric region for the coordinate system employed in Figure 3.4. The other field components in region 1 are now readily constructed in terms of H_z as

$$H_x = \frac{-j\beta_z}{k_1} A \cos(k_1 x) \cdot \exp(-j\beta_z z) \tag{3.39}$$

$$E_y = \frac{j\mu_0 \omega}{k_1} A \cos(k_1 x) \cdot \exp(-j\beta_z z) \tag{3.40}$$

A suitable decaying solution in region 2 in keeping with the open wall boundary condition adopted for this region at $x = -\infty$ is

$$H_z = B \cdot \exp[k_2(a + x) - j\beta_z z] \tag{3.41}$$

which satisfies the wave equation with

$$k_2^2 + \left(-\beta_z^2 + \omega^2 \mu_0 \varepsilon_0\right) = 0 \tag{3.42}$$

The constant B is determined by noting that H_z must be continuous at the boundary between the two regions

$$B = -A\sin(k_1 a) \tag{3.43}$$

The complete solution in region 2 is therefore described by the wave equation in Eq. (3.42) and by

$$H_z = -A\sin(k_1 a)\cdot\exp[k_2(a+x) - j\beta_z z] \tag{3.44}$$

$$H_x = \frac{-j\beta_z}{k_2} A\sin(k_1 a)\cdot\exp[k_2(a+x) - j\beta_z z] \tag{3.45}$$

$$E_y = \frac{j\mu_0\omega}{k_2} A\sin(k_1 a)\cdot\exp[k_2(a+x) - j\beta_z z] \tag{3.46}$$

The solution in region 3 has the same form as the one in 2 but with the sign of H_z reversed:

$$H_z = A\sin(k_1 a)\cdot\exp[k_2(a-x) - j\beta_z z] \tag{3.47}$$

$$H_x = \frac{-j\beta_z}{k_2} A\sin(k_1 a)\cdot\exp[k_2(a-x) - j\beta_z z] \tag{3.48}$$

$$E_y = \frac{j\mu_0\omega}{k_2} A\sin(k_1 a)\cdot\exp[k_2(a-x) - j\beta_z z] \tag{3.49}$$

and satisfies the same wave equation as the one in region 2.

The magnetic field is therefore elliptically polarized with one hand of rotation everywhere in region 2

$$\frac{H_x}{H_z} = \frac{-j\beta_z}{k_2} \tag{3.50}$$

and the one in region 3 is elliptically polarized with the other hand of rotation as asserted

$$\frac{H_x}{H_z} = \frac{-j\beta_z}{k_2} \tag{3.51}$$

For completion, it is now necessary to evaluate β_z and the field patterns of the structure. This may be done by first noting that the propagation constant β_z must be the same in each region

$$-k_2^2 + \omega^2\mu_0\varepsilon_0 = -k_1^2 + \omega^2\mu_0\varepsilon_0\varepsilon_1 \tag{3.52}$$

and furthermore noting that the electric field E_y must be continuous across the two regions. Applying this boundary condition gives

$$k_2 = k_1\tan(k_1 a) \tag{3.53}$$

The preceding two relationships may now be employed to evaluate k_1a and k_2a for parametric values of ε_1 and $\omega^2\mu_0\varepsilon_0$ and either separation constant may be employed with the appropriate wave equation to determine β_z. A knowledge of these three parameters is sufficient to construct the field patterns of the waveguide. Figure 3.5 depicts one result and Figure 3.6 illustrates the relationship between β_z/k_2 and $\sqrt{\varepsilon_1}(ka)$.

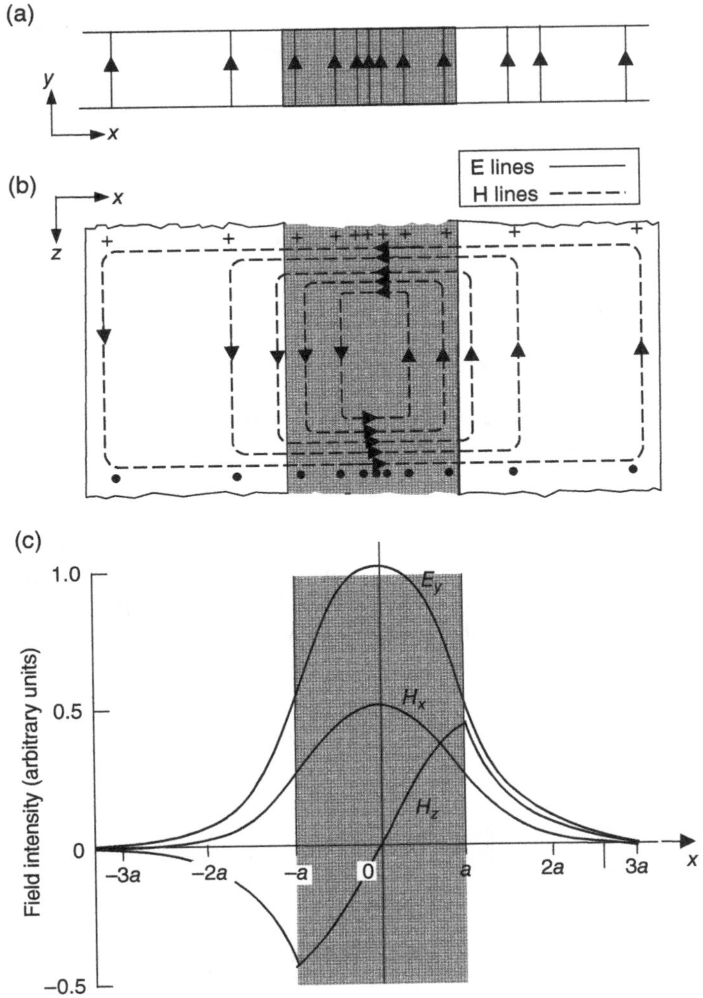

Figure 3.5 (a–c) Electric and magnetic fields of dielectric-loaded parallel plate waveguide with open sidewalls (dominant even mode solution) (Cohn 1959).

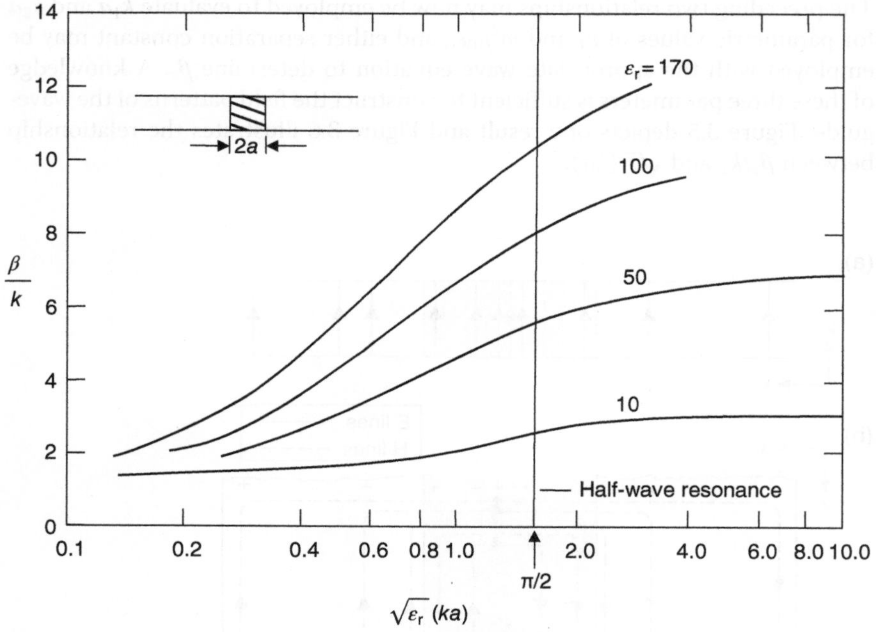

Figure 3.6 Relationship between β/k and $\sqrt{\varepsilon_1}(k_1a)$ (Anderson and Hines 1961).

The derivation of the second family of TE solutions for which the introduction of an electric wall at the plane of symmetry leaves the solution unperturbed is outside the remit of this chapter. However, Figure 3.7 depicts one result. It is in fact the next higher-order mode of this class of waveguide.

(a)

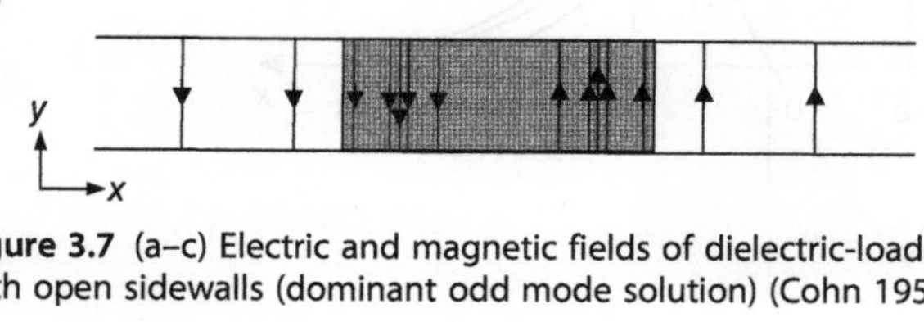

Figure 3.7 (a–c) Electric and magnetic fields of dielectric-loaded parallel plate waveguide with open sidewalls (dominant odd mode solution) (Cohn 1959).

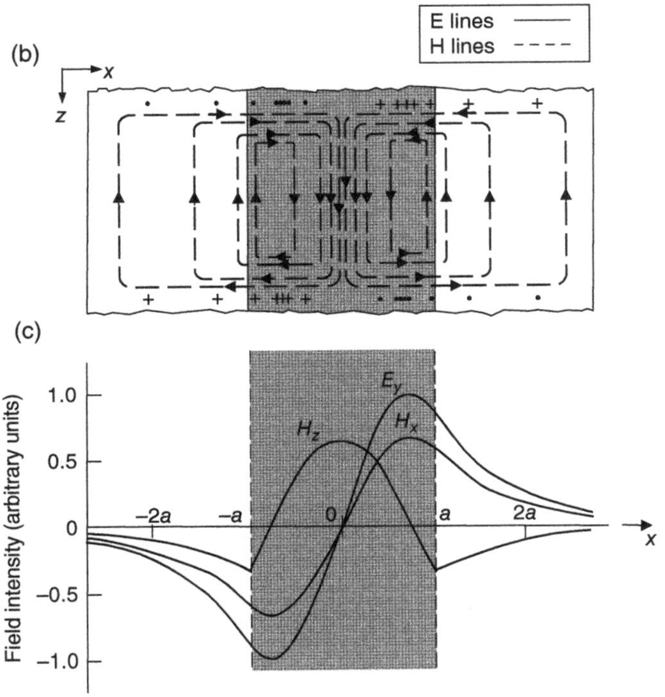

Figure 3.7 (Continued)

Bibliography

Anderson, W.W. and Hines, M.E. (1961). Wide-band resonance isolator. *IRE Trans. MTT* **MTT-9**: 63–67.

Button, K.J. (1958). Theory of non-reciprocal ferrite phase shifters in dielectric loaded coaxial line. *J. Appl. Phys.* **29**: 998.

Cohn, M. (1959). Propagation in a dielectric-loaded parallel plane waveguide. *IRE Trans. MTT* **MTT-7**: 202–208.

Duncan, B.J., Swern, L., Tomiyasu, K., and Hannwacker, J. (1957). Design considerations for broadband ferrite coaxial line isolators. *Proc. IRE* **45**: 483–490.

Fleri, D. and Hanley, G. (1959). Non-reciprocity in dielectric loaded TEM mode transmission lines. *IRE Trans. MTT* **MTT-7**: 23–27.

Jordan, A.R. (1962). Some effects of dielectric loading on ferrite phase shifters in rectangular waveguides. *IRE Trans. MTT* **MTT-10**: 83–84.

Soohoo, R.F. (1961). Theory of dielectric loaded and tapered field ferrite devices. *IRE Trans. MTT* **MTT-9**: 220–224.

Sucher, M. and Carlin, H.J. (1975). Coaxial line non-reciprocal phase shifters. *J. Appl. Phys.* **28**: 921.

Vartanian, P.H., Ayres, W.P., and Helgessen, A.L. (1958). Propagation in dielectric slab loaded rectangular waveguide. *IRE Trans. MTT* **MTT-6**: 215–222.

4

Reciprocal Quarter-wave Plates in Circular Waveguides

Joseph Helszajn

Heriot Watt University, Edinburgh, UK

Standard circular waveguide propagating the dominant TE mode supports degenerate counterrotating circularly polarized magnetic fields along its axis which may be split by the introduction of a suitably magnetized ferrite material, giving rise to so-called Faraday rotation of the polarization of the mode. However, if the incident magnetic field is clockwise or anticlockwise polarized, then it will be either phase advanced or retarded. Mode transducers or quarter-wave plates are therefore important building blocks in the design of ferrite phase-shifters and control devices. The purpose of this chapter is to describe the principle of one class of reciprocal quarter-wave plate, which may be used to implement a number of different control devices. Such plates convert a linearly polarized wave to either clockwise or anticlockwise circularly polarized ones. Its operation may be phenomenologically understood by decomposing the incident wave into equal orthogonal components and phase-shifting one or the other of them by 90° with respect to the other. It may also be interpreted by using coupled wave theory and this is the approach used here. The theory of nonreciprocal quarter-wave plates is left for a later chapter.

An equivalent waveguide model is separately employed to form the effective dielectric constants of the two polarizations from the experimental knowledge of the waveguide and free space wavelengths and the radius of the circular waveguide. This model allows the bandwidth of the plate to be estimated and also permits the results to be extrapolated to other waveguide bands. Some remarks about the optimum length of quarter-wave plates are included for completeness. The matching problem of different inhomogeneous waveguide sections met, for instance, in the design of Faraday rotation devices, is facilitated if equivalent waveguide models are defined for each different waveguide section.

Microwave Polarizers, Power Dividers, Phase Shifters, Circulators, and Switches, First Edition. Joseph Helszajn.

The theory of the half-wave plate and the rotary vane phase-shifter are included for completeness.

4.1 Quarter-wave Plate

The notion of horizontal or vertical linear polarization is well established. However, such polarizations may be regarded as limiting cases of elliptical polarization. Clockwise and anticlockwise circular polarizations are the other limiting forms. Many systems and microwave components, as already noted, rely on this type of polarization. One common type of birefringence that is suitable for the design of a circular polarizer or quarter-wave plate consists of a sheet of dielectric material at $\pm\pi/4$ rad in the x and y axis of a circular waveguide as indicated in Figure 4.1.

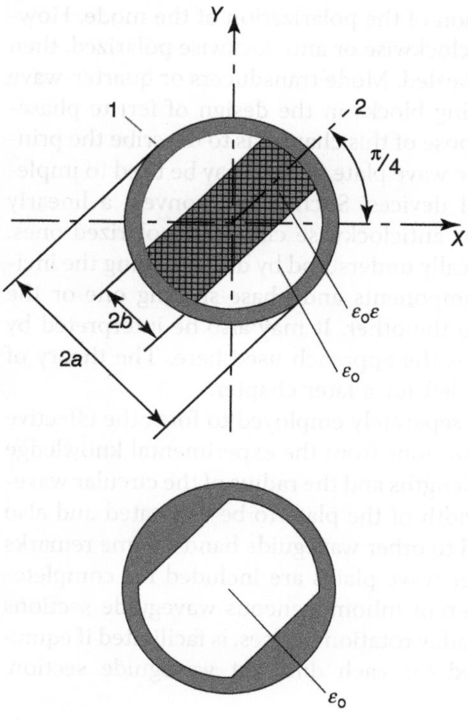

Figure 4.1 Schematic diagram of quarter-wave plates using partially filled circular waveguides.

If a vertically polarized wave is introduced into such a waveguide, it will have equal components normal and tangential to the dielectric vane which will propagate with different velocities β_n and β_t along the waveguide. The phase constant of the component normal to the dielectric sheet remains essentially unperturbed while that tangential to it increases. The output wave will be circularly polarized, provided its length ℓ is adjusted such that

$$(\beta_t - \beta_n)\ell = \pm\frac{\pi}{2} \tag{4.1}$$

The quarter-wave plate is a four-port network for which a scattering matrix is readily obtained by inspection from the schematic diagram in Figure 4.2.

The result for the port nomenclature in Figure 4.2 is

$$\bar{S} = \begin{bmatrix} 0 & 0 & 1 & 0 \\ 0 & 0 & 0 & -j \\ 1 & 0 & 0 & 0 \\ 0 & -j & 0 & 0 \end{bmatrix} \tag{4.2}$$

At the output, port 3 corresponds to port 1 and port 4 to port 2. Ports 1 and 2 form a right-banded system with the direction of propagation.

In obtaining this matrix it has been assumed that (i) the network is matched so that the main diagonal is zero, (ii) it is reciprocal so that the matrix is symmetrical about the main diagonal, (iii) all ports are decoupled except ports 1 and 3, and 2 and 4, and (iv) there is a $\pi/4$ differential phase-shift between ports 3 and 4.

If a linearly polarized wave is now incident on the device with its polarization at 45° to ports 1 and 2, the outgoing waves are obtained by forming the following input/output relationship:

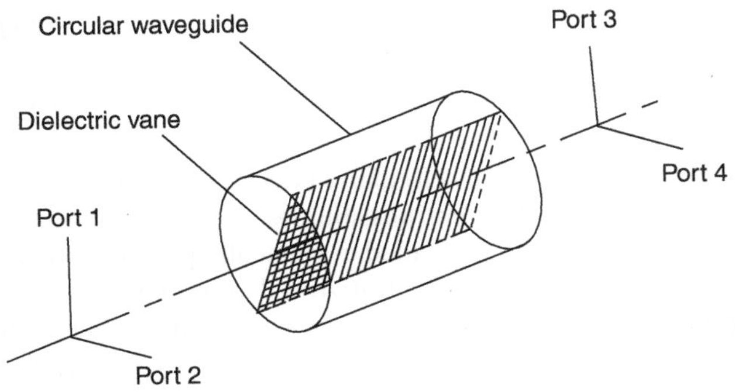

Figure 4.2 Schematic diagram of quarter-wave plate showing port nomenclature.

$$\begin{bmatrix} b_1 \\ b_2 \\ b_3 \\ b_4 \end{bmatrix} = \begin{bmatrix} 0 & 0 & 1 & 0 \\ 0 & 0 & 0 & -j \\ 1 & 0 & 0 & 0 \\ 0 & -j & 0 & 0 \end{bmatrix} \begin{bmatrix} 1/\sqrt{2} \\ 1/\sqrt{2} \\ 0 \\ 0 \end{bmatrix} \tag{4.3}$$

The results is

$$b_1 = 0 \tag{4.4}$$

$$b_2 = 0 \tag{4.5}$$

$$b_3 = \frac{1}{\sqrt{2}} \tag{4.6}$$

$$b_4 = \frac{-j}{\sqrt{2}} \tag{4.7}$$

If the dielectric vane is rotated by 90° with respect to its position in Figure 4.2, S_{13} and S_{14} are interchanged and the input/output relation of the network becomes:

$$\begin{bmatrix} b_1 \\ b_2 \\ b_3 \\ b_4 \end{bmatrix} = \begin{bmatrix} 0 & 0 & -j & 0 \\ 0 & 0 & 0 & 1 \\ -j & 0 & 0 & 0 \\ 0 & 1 & 0 & 0 \end{bmatrix} \begin{bmatrix} 1/\sqrt{2} \\ 1/\sqrt{2} \\ 0 \\ 0 \end{bmatrix} \tag{4.8}$$

The results is

$$b_1 = 0 \tag{4.9}$$

$$b_2 = 0 \tag{4.10}$$

$$b_3 = \frac{-j}{\sqrt{2}} \tag{4.11}$$

$$b_4 = \frac{1}{\sqrt{2}} \tag{4.12}$$

and the output wave at ports 3 and 4 is now circularly polarized in the opposite direction. Table 4.1 summarizes the different possibilities.

If two similar quarter-wave plates are cascaded, a vertically polarized wave incident on the first plate is converted to a horizontal one at the output plate, i.e. a linearly polarized wave is rotated by twice the angle between the input and the plate orientations. However, if one plate of each type is connected in tandem the polarization at the output plate is the same as that at the input.

Table 4.1 Input and output polarizations with quarter-wave plates.

Device	Input	Output polarization
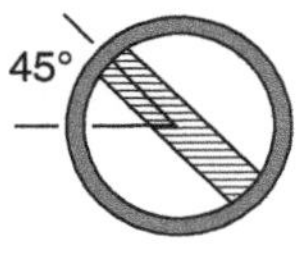	Linear ↑	+90° ACP
	Linear ↓	+90° ACP
	Linear →	+90° CCP
	Linear ←	+90° CCP
	CCP +90°	≡ ↑ Linear
	ACP +90°	≡ → Linear

Device	Input	Output polarization
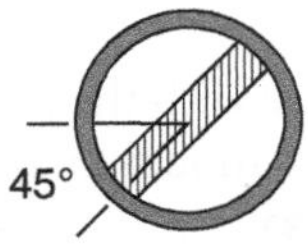	Linear ↑	+90° CCP
	Linear ↓	+90° CCP
	Linear →	+90° ACP
	Linear ←	+90° ACP
	CCP +90°	≡ ← Linear
	ACP +90°	≡ ↓ Linear

4.2 Coupled Mode Theory of Quarter-wave Plate

In the coupled wave model of the quarter-wave plate to be described, the port nomenclature employed in Figure 4.2 is rotated by 45° as illustrated in Figure 4.3 and two coupled waves each polarized at 45° to the dielectric sheet are defined at both the input and output terminals of the device. The scattering matrix is now formed at these new rotated ports in terms of two normal modes; one consisting of two equal in-phase waves in-space quadrature with the phase constant of the wave polarized normal to the dielectric sheet and the other

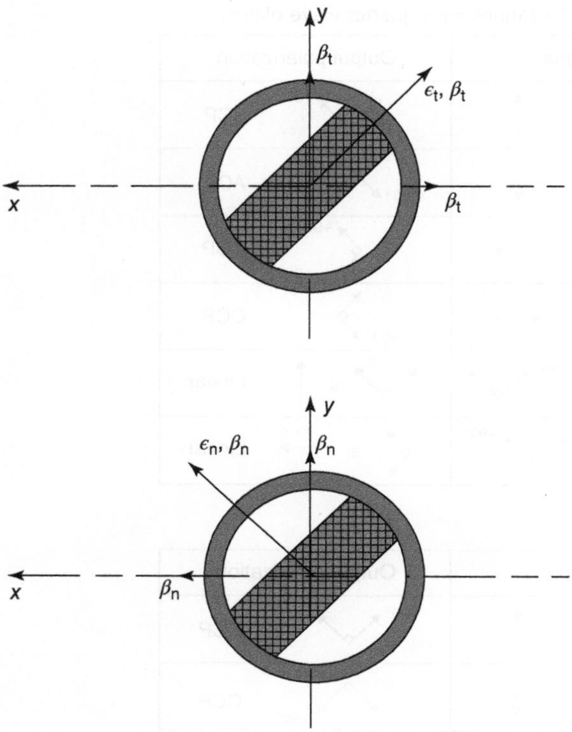

Figure 4.3 Construction of normal modes of a quarter-wave plate.

consisting of two equal out-of-phase waves in-space quadrature with the phase constant of the dielectric sheet. The nature of these normal modes may be derived by decomposing a vertically polarized wave at port 1 (say) into components normal and parallel to the dielectric sheet and thereafter individually decomposing each of them back along the coupled ports (see Figure 4.3).

The scattering matrix is now constructed in terms of the phase constants and reflection coefficients of these in-phase and out-of-phase in-space quadrature normal modes by taking the components of each one at a time as the input waves of the quarter-wave plate and constructing the output ones. It is readily appreciated that the quarter-wave plate has fourfold symmetry so that its scattering matrix may be written as

$$\bar{S} = \begin{bmatrix} S_{11} & S_{12} & S_{13} & S_{14} \\ S_{12} & S_{11} & S_{14} & S_{13} \\ S_{13} & S_{14} & S_{11} & S_{12} \\ S_{14} & S_{13} & S_{12} & S_{11} \end{bmatrix} \tag{4.13}$$

The above scattering matrix assumes that the device is reciprocal and symmetrical, but no assumption is made about the boundary conditions of the quarter-wave plate.

For the in-phase in-space quadrature normal mode with propagation constant β_t, the input/output relationship of the quarter-wave plate is

$$\begin{bmatrix} b_1 \\ b_2 \\ b_3 \\ b_4 \end{bmatrix} = \begin{bmatrix} S_{11} & S_{12} & S_{13} & S_{14} \\ S_{12} & S_{11} & S_{14} & S_{13} \\ S_{13} & S_{14} & S_{11} & S_{12} \\ S_{14} & S_{13} & S_{12} & S_{11} \end{bmatrix} \begin{bmatrix} 1/2 \\ 1/2 \\ 0 \\ 0 \end{bmatrix} \tag{4.14}$$

Expanding the above matrix relationship gives

$$b_1 = \frac{S_{11} + S_{12}}{2} \tag{4.15}$$

$$b_2 = \frac{S_{11} + S_{12}}{2} \tag{4.16}$$

$$b_3 = \frac{S_{13} + S_{14}}{2} \tag{4.17}$$

$$b_4 = \frac{S_{13} + S_{14}}{2} \tag{4.18}$$

In-phase in-space quadrature mode reflection ρ_t and transmission τ_t coefficients may now be defined for each waveguide or polarization as

$$\rho_t = \frac{b_1}{a_1} = \frac{b_2}{a_2} = S_{11} + S_{12} \tag{4.19}$$

$$\tau_t = \frac{b_3}{a_1} = \frac{b_4}{a_2} = S_{13} + S_{14} \tag{4.20}$$

Since there is no coupling between the orthogonal polarizations for this set of incident waves, the coupled waveguides may be replaced by a single waveguide with an in-phase in-space quadrature field pattern with parameters τ_t and ρ_t and propagation constant β_t.

For the out-of-phase in-space quadrature mode excitation with propagation constant β_t, the input/output relation of the network becomes

$$\begin{bmatrix} b_1 \\ b_2 \\ b_3 \\ b_4 \end{bmatrix} = \begin{bmatrix} S_{11} & S_{12} & S_{13} & S_{14} \\ S_{12} & S_{11} & S_{14} & S_{13} \\ S_{13} & S_{14} & S_{11} & S_{12} \\ S_{14} & S_{13} & S_{12} & S_{11} \end{bmatrix} \begin{bmatrix} 1/2 \\ -1/2 \\ 0 \\ 0 \end{bmatrix} \tag{4.21}$$

Thus,

$$b_1 = \frac{S_{11} - S_{12}}{2} \tag{4.22}$$

$$b_2 = \frac{-S_{11} + S_{12}}{2} \tag{4.23}$$

$$b_3 = \frac{S_{13} - S_{14}}{2} \tag{4.24}$$

$$b_4 = \frac{-S_{13} + S_{14}}{2} \tag{4.25}$$

Out-of-phase in-space quadrature reflection ρ_n and transmission τ_n coefficients for each waveguide are in this case defined by

$$\rho_n = \frac{b_1}{a_1} = \frac{b_2}{a_2} = S_{11} - S_{12} \tag{4.26}$$

$$\tau_n = \frac{b_3}{a_1} = \frac{b_4}{a_2} = S_{13} - S_{14} \tag{4.27}$$

The reflection and transmission coefficients are again identical for each polarization so that the four-port network may be once more replaced by a two-port one but with constitutive parameters ρ_n, τ_n, and β_n. Taking linear combinations of these two solutions gives

$$S_{11} = \frac{\rho_t + \rho_n}{2} \tag{4.28}$$

$$S_{12} = \frac{\rho_t - \rho_n}{2} \tag{4.29}$$

$$S_{13} = \frac{\tau_t + \tau_n}{2} \tag{4.30}$$

$$S_{14} = \frac{\tau_t - \tau_n}{2} \tag{4.31}$$

The entries of the scattering matrix are therefore linear combinations of ρ_n, ρ_t, τ_n, and τ_t. A knowledge of these quantities is therefore sufficient to characterize the quarter-wave plate.

It is apparent from Eqs. (4.28) and (4.29) that an ideal quarter-wave plate requires

$$\rho_t = \rho_n = 0 \tag{4.32}$$

It is thus necessary to separately match both normal modes of the system.

If the quarter-wave plate is matched to the input and output waveguides with phase-constants β_0 by a stepped impedance transformer, the reflection coefficients $\rho_{n,t}$ may be expressed in terms of ρ_0, ρ_n, and β_t as

$$\rho_t = \frac{-\beta_0 + \beta_t}{\beta_0 + \beta_t} \tag{4.33}$$

$$\rho_n = \frac{-\beta_0 + \beta_n}{\beta_0 + \beta_n} \tag{4.34}$$

The transmission variables $\tau_{t,n}$ may be defined in terms of the propagation constants $\beta_{t,n}$ and the reflection coefficients $\rho_{t,n}$ by

$$\tau_t = \left(1 - \rho_t \rho_t^*\right)^{1/2} \exp\left(-j\beta_p\right) l \tag{4.35}$$

$$\tau_n = \left(1 - \rho_n \rho_n^*\right)^{1/2} \exp(-j\beta_n) l \tag{4.36}$$

It is readily verified that Eqs. (4.28)–(4.31) satisfy the unitary condition:

$$S_{11}S_{11}^* + S_{12}S_{12}^* + S_{13}S_{13}^* + S_{14}S_{14}^* = 1 \tag{4.37}$$

For symmetric splitting Eqs. (4.28)–(4.31) become

$$S_{11} \approx 0 \tag{4.38}$$

$$S_{12} \approx j\left(\frac{\beta_t - \beta_n}{2\beta_0}\right) \tag{4.39}$$

$$S_{13} \approx \left[1 - \left(\frac{\beta_t - \beta_n}{2\beta_0}\right)^2\right]^{\frac{1}{2}} \cos\left(\frac{\beta_t - \beta_n}{2} l\right) \exp(-j\beta_0) l \tag{4.40}$$

$$S_{14} \approx j\left[1 - \left(\frac{\beta_t - \beta_n}{2\beta_0}\right)^2\right]^{\frac{1}{2}} \sin\left(\frac{\beta_t - \beta_n}{2} l\right) \exp(-j\beta_0) l \tag{4.41}$$

which also satisfies the unitary condition.

This result suggests that in a four-port reciprocal network, matching port 1 is not sufficient to decouple port 2. This feature is of course well understood. In order to decouple port 2 from port 1 by at least 20 dB it is necessary to have

$$\frac{\beta_t - \beta_n}{2\beta_0} \le 0.10 \tag{4.42}$$

This condition places an upper bound on the normalized splitting of the normal-mode phase constants and a lower bound on the overall length of the device.

The wave at ports 2 and 3 is circularly polarized, provided

$$\frac{(\beta_t - \beta_n) l}{2} = \frac{\pi}{4} \tag{4.43}$$

in keeping with Eq. (4.1).

If the splitting is not symmetric, Eqs. (4.4) and (4.41) do not apply and the outputs at ports 3 and 4 combine as an elliptically polarized wave instead of a circularly polarized one.

4.3 Effective Waveguide Model of Quarter-wave Plate

The description of an inhomogenous waveguide partially filled by some dielectric material is often facilitated by forming an equivalent waveguide model consisting of a homogenous waveguide having the same cross-section but fully filled by an effective relative dielectric constant. Such a model is useful for matching purposes and for estimating the frequency characteristic of the inhomogenous waveguide. Figure 4.4 illustrates this equivalence for the two orientations of the quarter-wave plate in Figure 4.1.

The relationship between the waveguide wavelength ($\lambda_{g,n,p}$), the free space wavelength (λ_0), the radius of the waveguide R, and the effective relative dielectric constants ($\varepsilon_{n,p}$) of the two equivalent waveguides is defined in the absence of fringing by

$$\left(\frac{2\pi}{\lambda_{g,n,p}}\right)^2 = \left(\frac{2\pi}{\lambda_0}\right)^2 \varepsilon_{n,p} - \left(\frac{1.84}{R}\right)^2 \tag{4.44}$$

The equivalent waveguide model employed here is of course also appropriate for use with the three different definitions of impedance in circular waveguide. It assumes, however, that the radius of the equivalent waveguide is the same for the two waveguides. The equivalence between the two is therefore exact at a single frequency. In order to cater for frequency effects, it is strictly

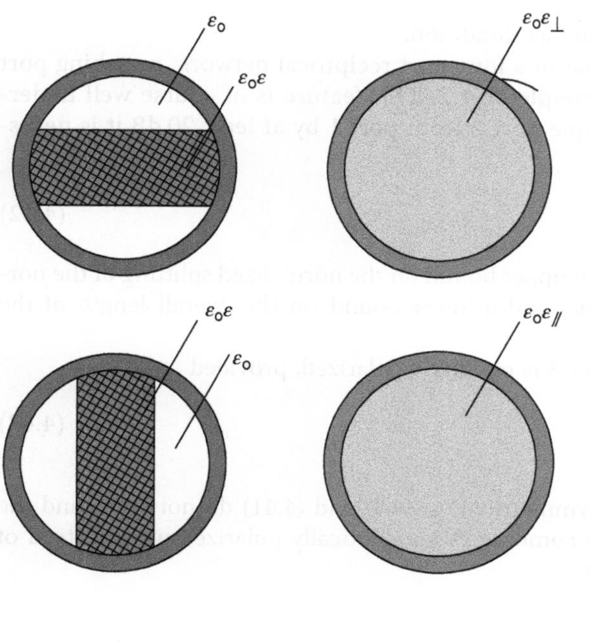

Figure 4.4 Equivalence between inhomogenous and homogenous circular waveguides.

necessary to adopt an equivalent radius as well as an effective dielectric constant.

The effective relative dielectric constants can be determined in terms of the phase constant (β_0) of the matching waveguides and the required differential phase-shift. Solving Eq. (4.44) for ε_t and ε_n leads to

$$\varepsilon_t = \frac{\beta_0^2 + k_c^2 + (\Delta\beta)^2 + 2\beta_0(\Delta\beta)}{k_0^2} \tag{4.45}$$

$$\varepsilon_n = \frac{\beta_0^2 + k_c^2 + (\Delta\beta)^2 - 2\beta_0(\Delta\beta)}{k_0^2} \tag{4.46}$$

4.4 Phase Constants of Quarter-wave Plate Using the Cavity Method

If an equivalent waveguide model is also adopted for the rotator section, k_0, k_c, and β_0 of the quarter-wave plate in Eqs. (4.45) and (4.46) are fixed by the corresponding parameters of the rotator section. It now only remains to determine the relationship between the effective dielectric constants of the two polarizations and the details of the quarter-wave plate. This may be done by incorporating the inhomogeneous waveguide into a half-wave long undercoupled cavity resonator and forming an experimental relationship between its resonance frequency, the length of the cavity, and the transverse parameters of the plate. The definition of such a resonator is indicated in Figure 4.5.

The orthogonal phase constants from which the effective dielectric constants may be evaluated are given in terms of the length $l_{t,n}$ of the resonator by

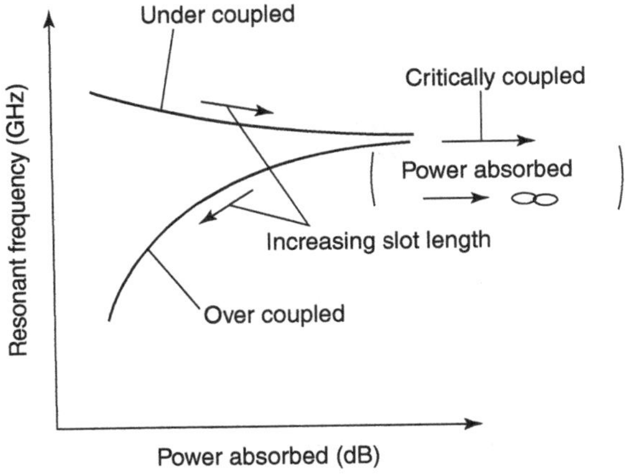

Figure 4.5 Sketch of power absorbed versus resonant frequency for undercoupled critically coupler and overcoupled cavity resonator.

$$\beta_{t,n} = \frac{2\pi}{\lambda_{g,t,n}} = \frac{\pi}{l_{t,n}} \tag{4.47}$$

To cater for possible dispersion effects this relationship should, strictly speaking, be formed at the design frequency of the device. Figures 4.6 and 4.7 display the two effective dielectric constants of the polarizer versus that of the plate for parametric value of H/R. This result may be used for design as outlined below.

Taking the following entries based on the equivalent waveguide model in Figure 4.4 or a Faraday rotator section described in Chapter 2 as an example,

$$k_0 = 0.1962\,\text{rad}\,\text{mm}^{-1}$$

$$k_c = 0.2968\,\text{rad}\,\text{mm}^{-1}$$

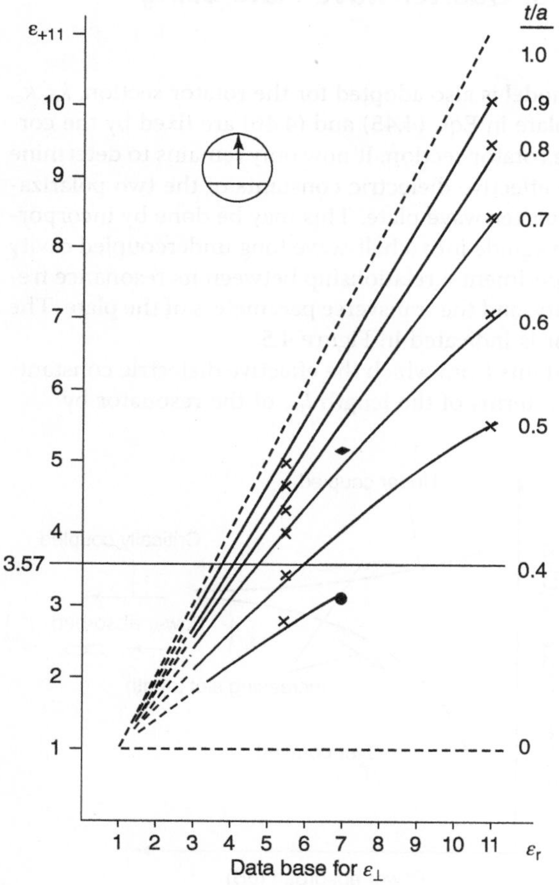

Figure 4.6 Effective dielectric constants of round waveguide with polarization perpendicular to dielectric vane versus ε_r for parametric values of *H/R*.

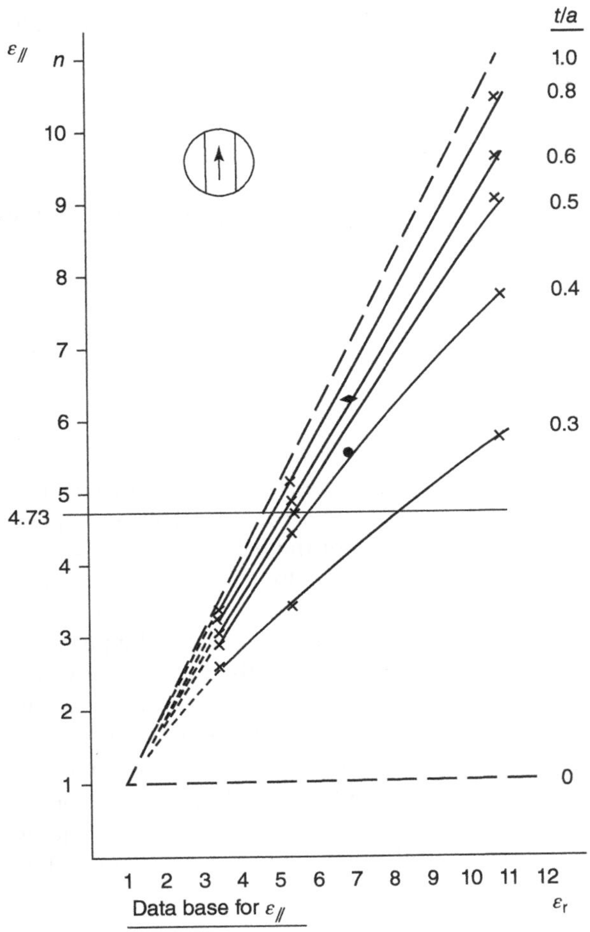

Figure 4.7 Effective dielectric constants of round waveguide with polarization parallel to dielectric vane versus ε_r for parametric values of *H/R*.

$$\beta_0 = 0.3317\,\text{rad}\,\text{mm}^{-1}$$

$$\Delta\beta = 0.0663\,\text{rad}\,\text{mm}^{-1}$$

$$l_0 = 23.6660\,\text{mm}$$

gives

$$\varepsilon_t = 5.7750$$

$$\varepsilon_n = 4.6324$$

The required variables ε_r and H/R are now determined graphically as indicated in Figures 4.6 and 4.7:

$$\varepsilon_r = 6.1$$

$$\frac{H}{R} = 0.40$$

The effective dielectric constant of the equivalent waveguide model of the rotor section with k_0, k_c, and β_0 noted above is

$$\varepsilon_{eff} = 5.1451$$

4.5 Variable Rotor Power Divider

One application of the half-wave plate is in the construction of a mechanical variable coupler or power divider. It consists of a rotatable half-wave plate between two-mode transducers. The operation of the half-wave plate is illustrated in Figure 4.8. The two-mode transducer is a four-port network, which connects orthogonal TE_{11} modes in a round waveguide to rectangular waveguides propagating the dominant TE_{10} mode. A vertically polarized input wave in the round waveguide of this network is emergent in the rectangular waveguide which support the same polarization; a horizontally polarized wave in the round waveguide is reflected by a suitably located septum to the other rectangular waveguide. The operation of this latter transmission path may be understood by recognizing that it is always possible to achieve perfect transmission between any two ports of a three-port lossless network by terminating the third one by a suitable short-circuit termination. The network is of course

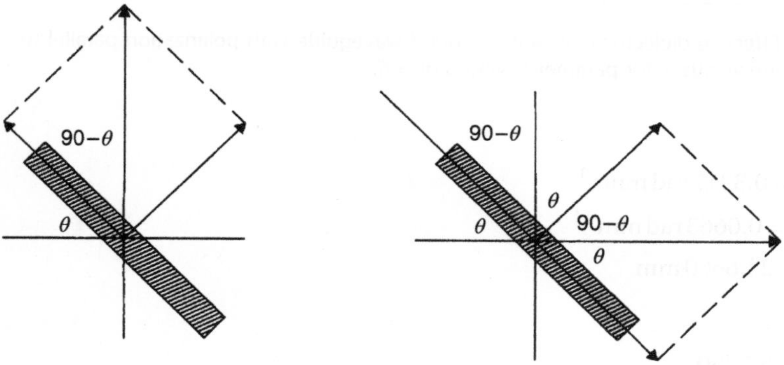

Figure 4.8 Input and output waves on half-wave plate with $\theta = \pi/4$.

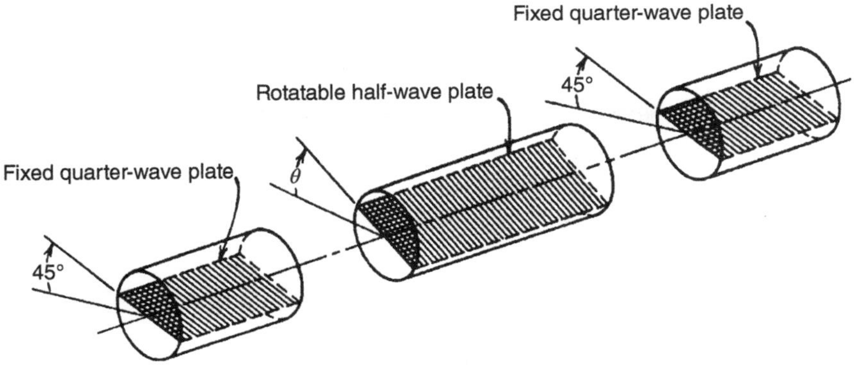

Figure 4.9 Rotary vane phase-shifter.

reciprocal, so that the input wave at the rectangular waveguide is emergent at the corresponding ports in the circular waveguide. The operation of the overall network is noted without further ado by recalling that the polarization of a linearly polarized wave is rotated in traversing the plate by twice the angle of the plate orientation. A standard variable rotor phase shifter is depicted in Figure 4.9. The first quarter-wave plate converts a vertically polarized wave into a circularly polarized wave which is phase shifted through an angle 2θ; the second converts the latter wave back to linear polarization.

Bibliography

Ayres, W.P. (1957). Broadband quarter-wave plates. *IRE Trans. Microw. Theory Tech.* **MTT-5**: 258–261.

Fox, A.G. (1947). An adjustable waveguide phase changer. *Proc. IRE* **35**: 1489–1498.

Helszajn, J. (1964). A novel ferrite quarter-wave plate. *Radio Electron. Eng.* **27**: 455–458.

Helszajn, J. (1985). Design of a quarter-wave plate using a cavity resonator technique. *Proc. IEE Microw. Antennas Propag.* **134** (Part H): 139–144.

Monaghan, S.R. and Mohr, M.C. (1969). Polarization insensitive phase-shifter for use in phase-array antennas. *Microw. J.* **12**: 75–80.

Pyle, J.R. (1964). Circular polarizers of fixed bandwidth. *IEEE Trans. Microw. Theory Tech.* **MTT-12**: 567–568.

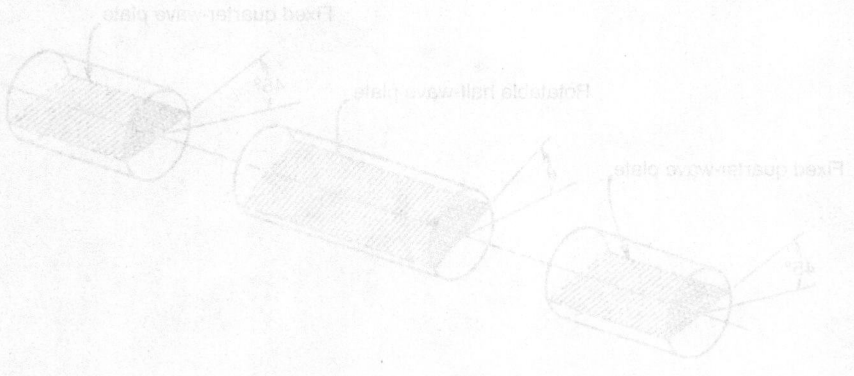

Figure 4.9 Rotary vane phase shifter.

reciprocal, so that the input wave at the rectangular waveguide is emergent at the corresponding ports to the circular waveguide. The operation of the overall network is noted without further ado by recalling that the polarization of a linearly polarized wave is rotated in traversing the plate by twice the angle of the plate orientation. A standard variable rotor phase shifter is depicted in Figure 4.9. The first quarter-wave plate converts a vertically polarized wave into a circularly polarized wave which is phase shifted through an angle 2θ; the second converts the latter wave back to linear polarization.

Bibliography

Ayres, W.P. (1957). Broadband quarter-wave plates. *IRE Trans. Microw. Theory Tech.* MTT-5: 258–261.

Fox, A.G. (1947). An adjustable waveguide phase changer. *Proc. IRE* 35: 1489–1498.

Helszajn, J. (1964). A novel ferrite quarter-wave plate. *Radio Electron. Eng.* 27: 455–458.

Helszajn, J. (1985). Design of a quarter-wave plate using a cavity resonator technique. *Proc. IEE Microw. Antennas Propag.* 132 (Part H): 139–144.

Monaghan, S.K. and Molly, M.C. (1969). Polarization insensitive phase shifter for use in phase array antennas. *Microw. J.* 12: 75–80.

Pyle, J.R. (1964). Circular polarization of fixed bandwidth. *IEEE Trans. Microw. Theory Tech.* MTT-12: 557–558.

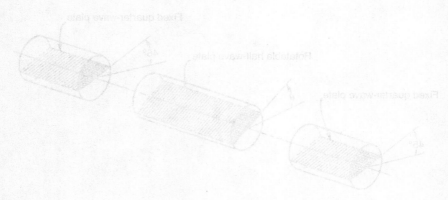

5

Nonreciprocal Ferrite Quarter-wave Plates

Joseph Helszajn

Heriot Watt University, Edinburgh, UK

5.1 Introduction

A nonreciprocal phase-shifter may be constructed by placing a Faraday rotation section between two reciprocal quarter-wave plates. Reciprocal ferrite phase-shifters may, however, be realized by utilizing nonreciprocal, instead of reciprocal, quarter-wave plates. The purpose of this chapter is to describe such as a plate based on the birefringence displayed by a magnetic insulator with the direct magnetic field intensity perpendicular to the direction of propagation. This type of waveguide is characterized by orthogonal linearly polarized normal modes and counterrotating circularly polarized coupled modes. It is therefore birefringent for orthogonal modes. Another type of plate makes use of a suitably longitudinally magnetized ferrite-filled elliptical waveguide in which the normal modes are counterrotating circularly polarized and the coupled modes are orthogonal linearly polarized. These two possible quarter-wave plates are therefore dual. The chapter includes a description of a number of reciprocal and nonreciprocal phase-shifters utilizing Faraday rotation and birefringent sections between reciprocal and nonreciprocal quarter-wave plates. A quarter-wave plate using a hexagonal ferrite with planar anisotropy is separately described.

5.2 Birefringence

The situation in which the direction of the direct magnetic field intensity is along the y-coordinate and that of propagation is in the z-direction proceeds by forming the wave equation with

Microwave Polarizers, Power Dividers, Phase Shifters, Circulators, and Switches,
First Edition. Joseph Helszajn.

$$\frac{\partial}{\partial z} = -\gamma \tag{5.1}$$

$$\frac{\partial}{\partial x} = \frac{\partial}{\partial y} = 0 \tag{5.2}$$

and

$$\bar{\mu} = \begin{pmatrix} \mu & 0 & -j\kappa \\ 0 & 1 & 0 \\ j\kappa & 0 & \mu \end{pmatrix} \tag{5.3}$$

The characteristic equation is here given by

$$\begin{pmatrix} \mu - \dfrac{\gamma^2}{\omega^2 \varepsilon_r \varepsilon_0 \mu_0} & 0 & -j\kappa \\ 0 & 1 - \dfrac{\gamma^2}{\omega^2 \varepsilon_r \varepsilon_0 \mu_0} & 0 \\ j\kappa & 0 & \mu \end{pmatrix} = 0 \tag{5.4}$$

A wave with its polarization parallel to the direct magnetic field propagates with a propagation constant given by

$$\beta_{11}^2 = \omega^2 \varepsilon_r \varepsilon_0 \mu_0 \tag{5.5}$$

and therefore exhibits a relative permeability

$$\mu_{11} = 1 \tag{5.6}$$

It is therefore unaffected by the electron spin.

Waves polarized perpendicular to the direction of the direct magnetic field have propagation constants given by

$$\beta_\perp^2 = \omega^2 \varepsilon_r \varepsilon_0 \mu_0 \mu_\perp \tag{5.7}$$

The effective permeability in this instance is described by

$$\mu_\perp = \mu_{\text{eff}} = \frac{\mu^2 - \kappa^2}{\mu} \tag{5.8}$$

The variation of μ_{eff} for a finite medium is indicated in Figure 5.1 as a function of the direct magnetic field.

μ_{11} is larger than $\mu_\perp$ so that the phase velocity associated with the latter permeability is less than that connected with the former one. Consequently, as in optics, the parallel wave is sometimes referred to as the ordinary wave and the perpendicular one as the extraordinary wave. The differential phase-shift per unit length between the two waves is

$$\beta_{11} - \beta_\perp = \omega \sqrt{\varepsilon_r \varepsilon_0 \mu_0} (1 - \mu_{\text{eff}}) \tag{5.9a}$$

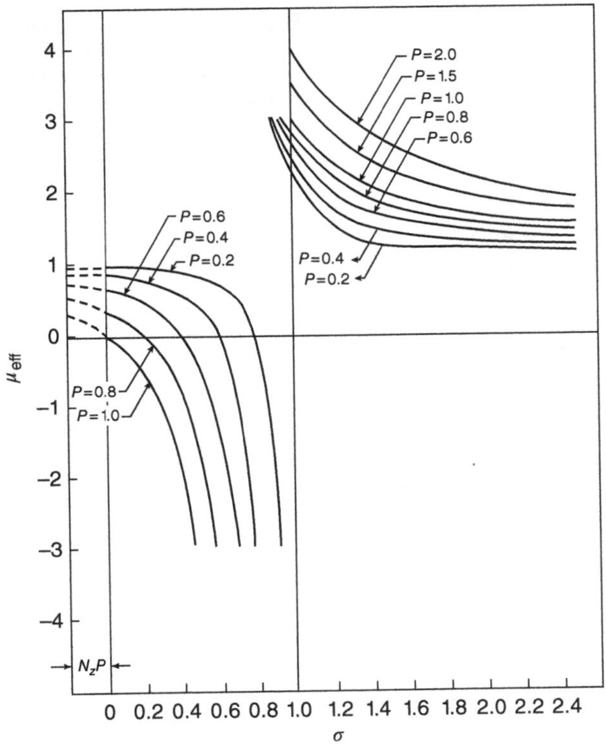

Figure 5.1 Variation of μ_{eff} as a function of the direct magnetic field intensity.

$$\beta_{11} - \beta_{\perp} \approx \frac{1}{2}\omega\sqrt{\varepsilon_r\varepsilon_0\mu_0}\kappa^2 \tag{5.9b}$$

provided the material is saturated. This result is derived by noting that for a saturated material

$$\mu = 1 \tag{5.10}$$

$$\kappa = \frac{\omega_m}{\omega} \tag{5.11}$$

One application of this birefringence is in the construction of nonreciprocal quarter- and half-wave plates.

5.3 Nonreciprocal Quarter-wave Plate Using the Birefringence Effect

The principle of one quarter-wave plate in square waveguide is illustrated in Figure 5.2. A horizontally polarized TE_{11} mode in this square waveguide has

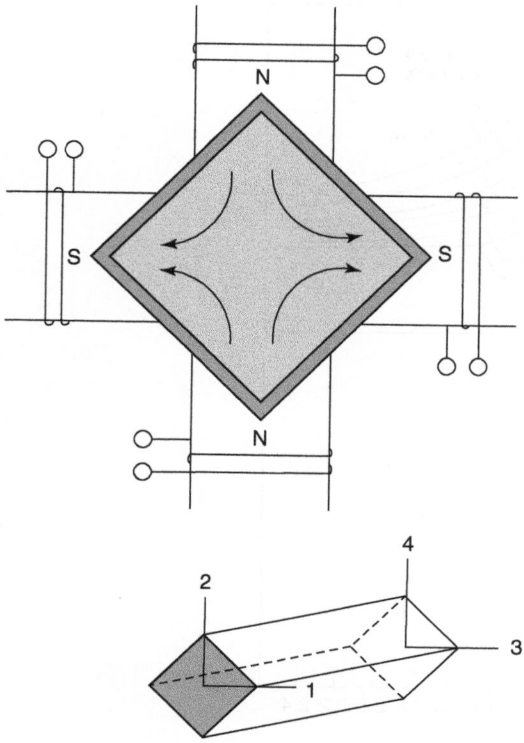

Figure 5.2 Schematic diagram of square birefringent waveguide showing quadrature magnetic coil arrangement.

its magnetic field parallel to the direct magnetic field and displays a relative permeability of unity. A similar vertically polarized mode in this same waveguide has a component of its magnetic field perpendicular to the direct magnetic field and therefore exhibits a relative permeability μ_{eff}. If both modes are established simultaneously in such a waveguide, then the two will be in time–space quadrature, provided

$$\frac{(\beta_{10}-\beta_{10})\ell}{2}=\frac{\pi}{4} \tag{5.12}$$

The port nomenclature of the arrangement in question is separately illustrated in Figure 5.2. The quadruple coil geometry is also shown in this diagram. The decomposition of a typical input wave at one port into orthogonal TE_{01} and TE_{10} modes is indicated in Figure 5.3. Figure 5.4 depicts the coil arrangement met in connection with a circular waveguide and Figure 5.5 depicts some possible geometries using ferrite tiles.

One semiempirical relationship due to Boyd for the differential phase per unit length of the square waveguide in Figure 5.2 is

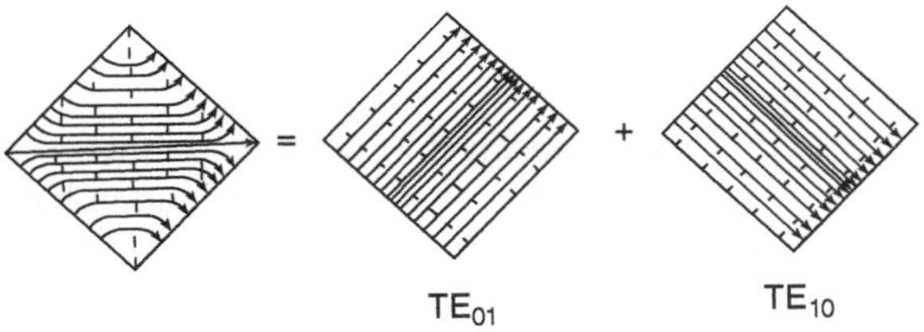

Figure 5.3 Linear combination of TE_{10} and TE_{01} in square waveguide.

Figure 5.4 Schematic diagram of round birefringent waveguide showing quadrature magnetic coil arrangement.

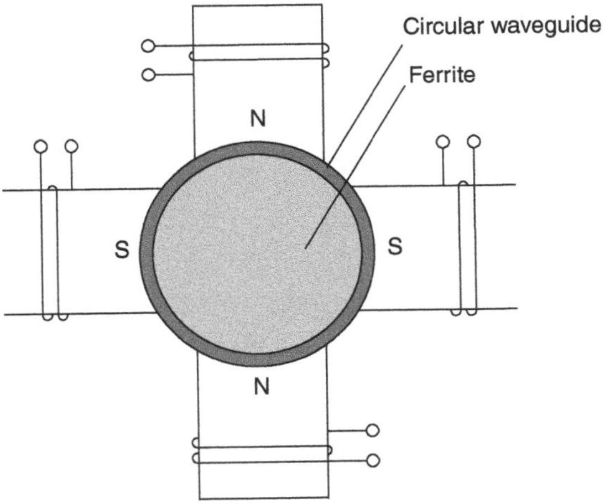

Figure 5.5 Schematic diagrams of round birefringent waveguides using ferrite slabs.

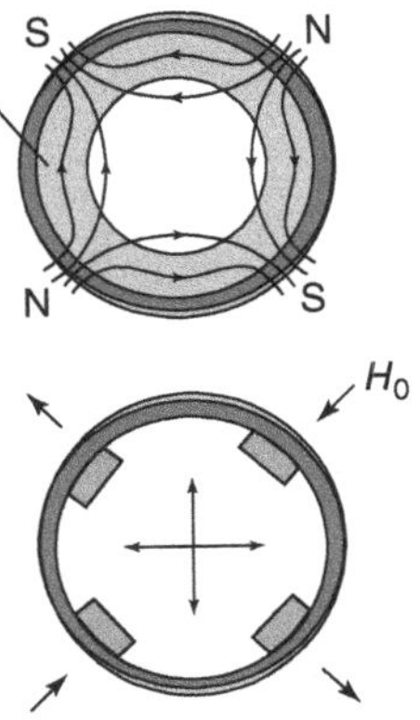

$$\frac{\Delta\phi}{\phi_0} \approx \frac{4f_c}{\pi f}\frac{\kappa}{\mu} \tag{5.13}$$

where f_c is the cutoff frequency of the waveguide, f is the operating frequency, κ and μ are the elements of the permeability tensor, and ϕ_0 is the insertion phase of a uniform transmission line embedded in a ferrite medium.

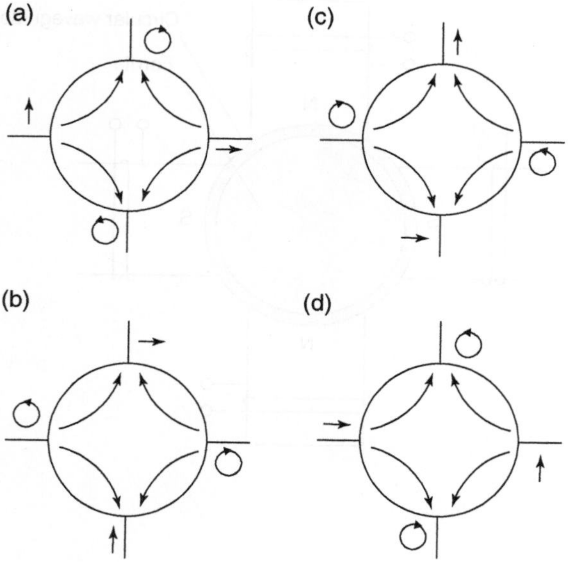

Figure 5.6 (a–d) Equivalent circuits of nonreciprocal quarter-wave plate.

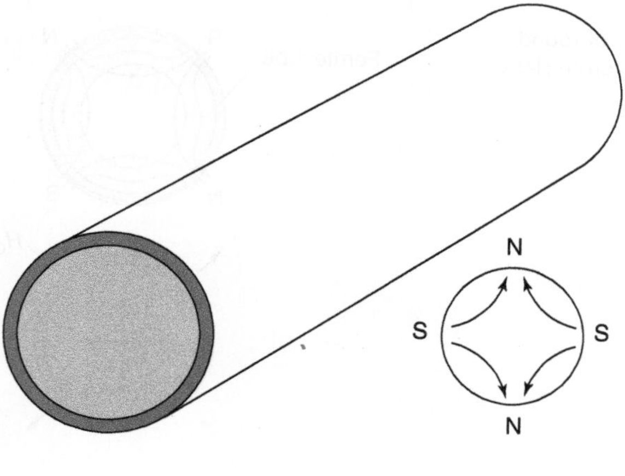

5.4 Circulator Representation of Nonreciprocal Quarter-wave Plates

One possible equivalent circuit of the birefringent or nonreciprocal quarter-wave plate is a four-port circulator. A vertically polarized signal, say, at the input port 1 of this device is converted at an adjacent port 3 into a circularly polarized signal with a hand of polarization whose direction of rotation is dependent upon that of the direct magnetic field; any reflection of such an incident wave at this port is converted to a horizontally polarized wave at the third port 3; any

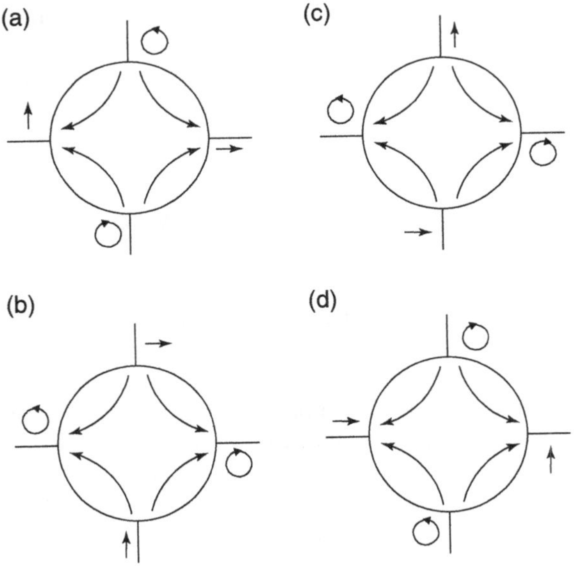

Figure 5.7 (a–d) Equivalent circuits of nonreciprocal quarter-wave plate (direct magnetic field conditions reversed).

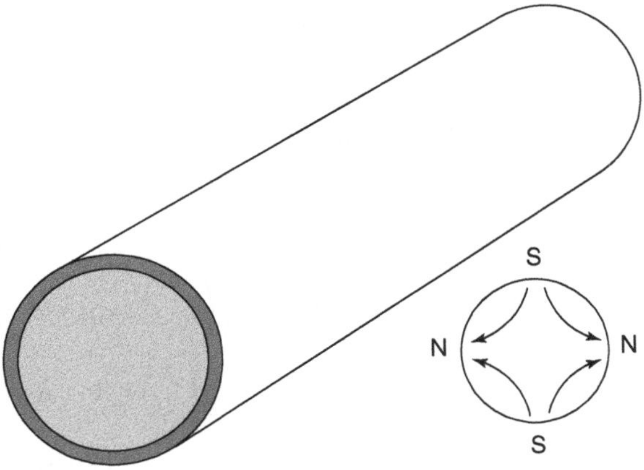

reflection of a horizontally polarized wave incident on this port is converted to a circularly polarized wave at the fourth port 4 with the opposite hand of rotation to that at port 2; the cycle is complete by noting that a signal with this polarization at port 4 is coupled or reconverted to a vertically polarized wave at port 1. Figure 5.6a depicts one possible equivalent circuit of this device.

Any of the four polarizations may of course be taken as an input port. Figure 5.6b–d illustrates the nature of the polarizations at ports 2–4 with that at port 1 circularly polarized in an anticlockwise direction, respectively. Figure 5.7a–d depicts the appropriate terminal conditions for the direct magnetic field conditions reversed.

Scrutiny of these diagrams indicates that this circulator does not have the fourfold symmetry associated with many conventional circulators. In obtaining these diagrams, note is made of the fact that the direct magnetic field is reversed with respect to the direction of the propagation after each transition through the network. Since the device is a circulator, it may not be amiss to invoke the notion of a gyrator impedance in its description.

5.5 Coupled and Normal Modes in Magnetized Ferrite Medium

Propagation in infinite space or a circular waveguide may be described either in terms of two degenerate orthogonal linearly polarized waves or in terms of two degenerate counterrotating circularly polarized waves. If the degeneracy between either descriptions is somehow removed, as in a ferrite medium, then the variables whose degeneracy is removed are known as the normal modes of the structure and the others are known as the coupled ones. If the direct magnetic field in a ferrite medium coincides with the direction of propagation, then the degeneracy between the circular polarized waves is removed and these are the normal modes; the orthogonal waves are then the coupled modes. If the degeneracy is removed between the orthogonal waves, as in the case of a ferrite medium with the direct field normal to the direction of propagation, then the orthogonal waves are the normal modes of the system and the circularly polarized ones are the coupled modes. The former situation leads to the phenomena of Faraday rotation, applicable to the description of nonreciprocal quarter-wave plates, as will now be demonstrated.

Decomposing the vertically and horizontally linearly polarized waves into counterrotating circular ones, assuming that they are equal in amplitude, gives

$$h_{\mathrm{v}} = \frac{1}{2}\begin{pmatrix} 1 \\ j \end{pmatrix} \exp(-j\beta_{\mathrm{v}} z) + \frac{1}{2}\begin{pmatrix} 1 \\ -j \end{pmatrix} \exp(-j\beta_{\mathrm{v}} z) \tag{5.14a}$$

$$h_{\mathrm{h}} = \frac{1}{2}\begin{pmatrix} -j \\ 1 \end{pmatrix} \exp(-j\beta_{\mathrm{h}} z) + \frac{1}{2}\begin{pmatrix} j \\ 1 \end{pmatrix} \exp(-j\beta_{\mathrm{h}} z). \tag{5.14b}$$

Taking linear combinations of the clockwise and anticlockwise circularly polarized waves yields

$$h_+ = \begin{pmatrix} 1 \\ j \end{pmatrix} \cos\left(\frac{\beta_v - \beta_h}{2}\right)\ell \exp\left[-j\left(\frac{\beta_v + \beta_h}{2}\right)\ell\right] \tag{5.15a}$$

$$h_- = j\begin{pmatrix} 1 \\ -j \end{pmatrix} \sin\left(\frac{\beta_v - \beta_h}{2}\right)\ell \exp\left[-j\left(\frac{\beta_v + \beta_h}{2}\right)\ell\right] \tag{5.15b}$$

The circular variables are in this instance the periodic or coupled waves and travel with similar propagation constants whereas the normal modes propagate with equal amplitudes but different phase constants. This is the same dual situation as that encountered in the classic Faraday rotation case.

Adding the two circularly polarized waves gives

$$h_v = \exp(-j\beta_v\ell) \tag{5.16a}$$

$$h_h = \exp(-j\beta_h\ell) \tag{5.16b}$$

The wave is circularly polarized, provided

$$\frac{(\beta_v - \beta_h)\ell}{2} = \frac{\pi}{4} \tag{5.17}$$

5.6 Variable Phase-shifters Employing Birefringent, Faraday Rotation, and Dielectric Half-wave Plates

One classic type of reciprocal variable phase-shifter consists of a reciprocal half-wave plate between reciprocal quarter-wave plates. The first quarter-wave plate converts a linearly polarized wave into a circularly polarized one, which is phase-shifted through an angle 2θ by the half-wave plate and has its hand of polarization reversed by it; the second converts the latter wave back to linear polarization. The half-wave plate resembles the quarter-wave plate in construction but introduces a 180° instead of a 90° phase delay. This arrangement will produce a continuous change in phase 2θ in a linearly polarized wave if the half-wave plate is rotate about its axis θ, retarding or advancing it depending on the orientation of the quarter-wave plate, with respect to the polarization.

An electronic version of this mechanically controlled phase-shifter may be constructed by replacing the dielectric half-wave plate by a birefringent one employing a rotating magnetic field. It consists of a birefringent bit utilizing a ferrite section magnetized by a rotating magnetic field similar to the classical alternator. The magnetic field is established by quadrature currents in identical quadrature coils. This provides a continuously increasing phase-shift, which is linear in time to the same extent that the angular magnetic field is constant. The ensuing structure, however, is now nonreciprocal. If nonreciprocal birefringent quarter-wave plates are substituted for the reciprocal ones, then the overall network is once more reciprocal. Another possibility, of course, is a reciprocal section between nonreciprocal quarter-wave plates. A Faraday rotation section between two reciprocal quarter-wave plates or between nonreciprocal ones are two other possibilities. There are altogether three reciprocal and three nonreciprocal configurations, which are summarized in Figures 5.8 and 5.9.

Figure 5.10 illustrates one possible equivalent circuit of the nonreciprocal ferrite phase-shifter.

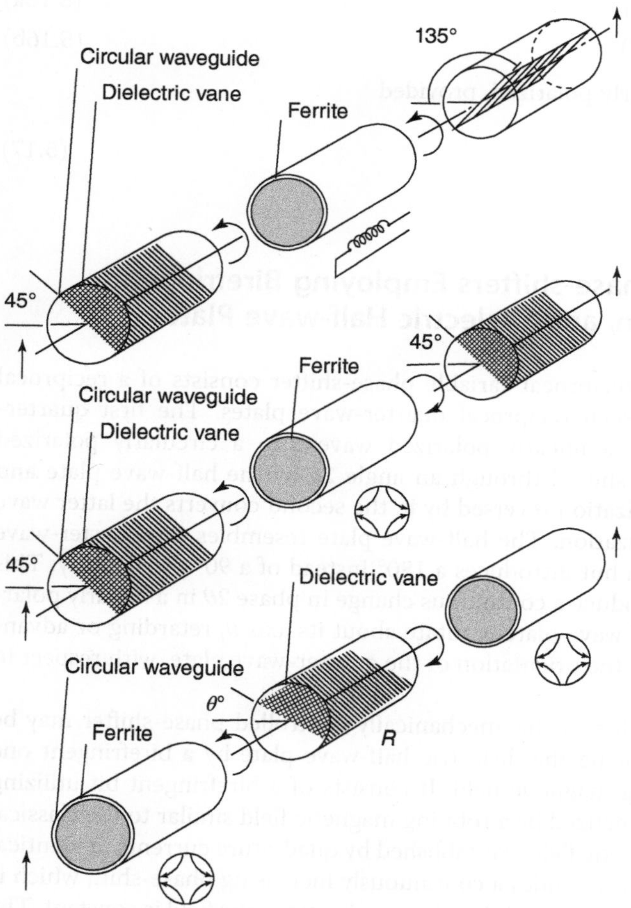

Figure 5.8 Nonreciprocal ferrite phase-shifter using reciprocal and nonreciprocal quarter-wave plates and birefringent bits.

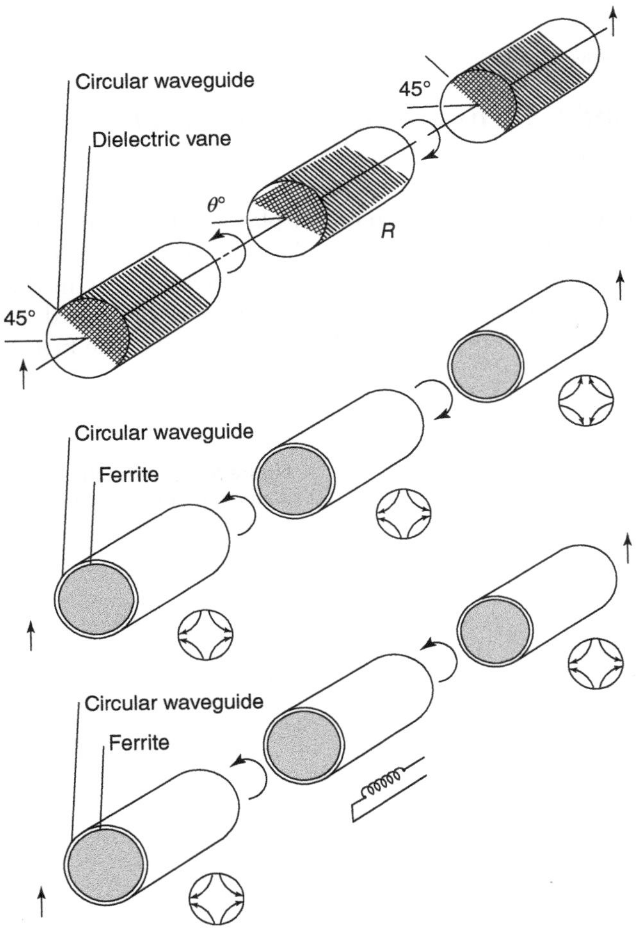

Figure 5.9 Reciprocal ferrite phase-shifter using reciprocal and nonreciprocal birefringent quarter-wave and half-wave birefringent bits.

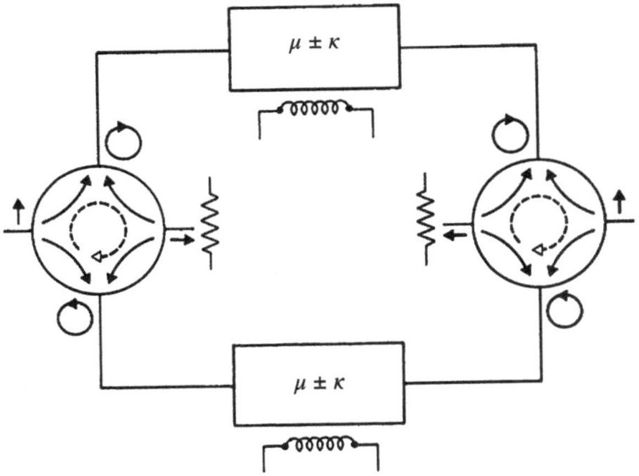

Figure 5.10 Equivalent circuit of nonreciprocal ferrite phase-shifter.

5.7 Circulators and Switches Using Nonreciprocal Quarter-wave Plates

The use of reciprocal and nonreciprocal quarter-wave plates in the construction of ferrite phase-shifters is only one of a host of applications of these devices.

One simple application of the nonreciprocal quarter-wave plate is in the construction of the switchable circulator polarized antenna, illustrated in Figure 5.11. Figures 5.12 and 5.13 depict two other arrangements. The first is

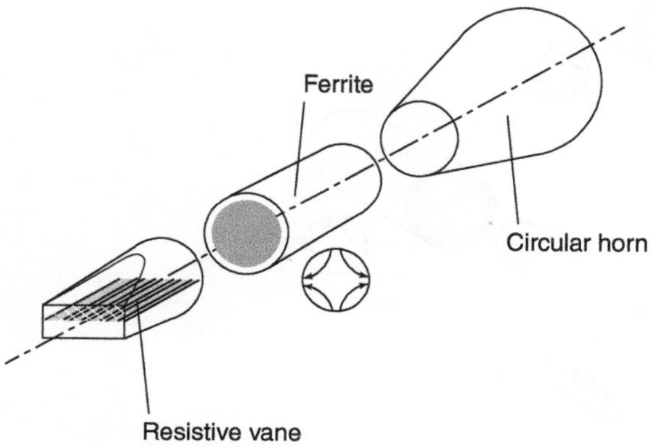

Figure 5.11 Schematic diagram of a switchable circular polarizer.

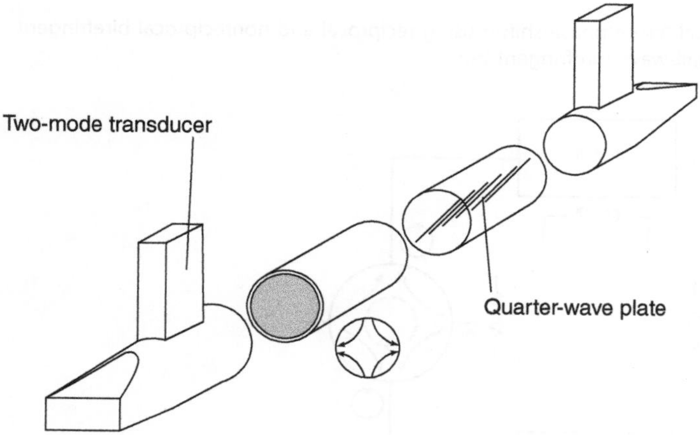

Figure 5.12 Circulator network using 2 two-mode transducers, a ferrite quarter-wave plate and a reciprocal quarter-wave plate.

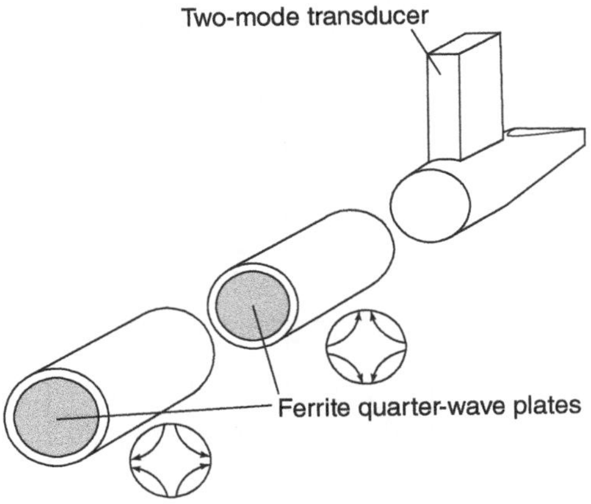

Figure 5.13 Two-mode transducer and 90° rotator.

a circulator type based on the use of 2 two-mode transducers, a ferrite and a dielectric quarter-wave plate. The other is a 90° rotator switch using two non-reciprocal quarter-wave plates and a two-mode transducer.

5.8 Nonreciprocal Circular Polarizer Using Elliptical Gyromagnetic Waveguide

Still another quarter-wave plate may be constructed by recourse to the elliptical gyromagnetic waveguide in Figure 5.14. Figure 5.15 shows the required solution.

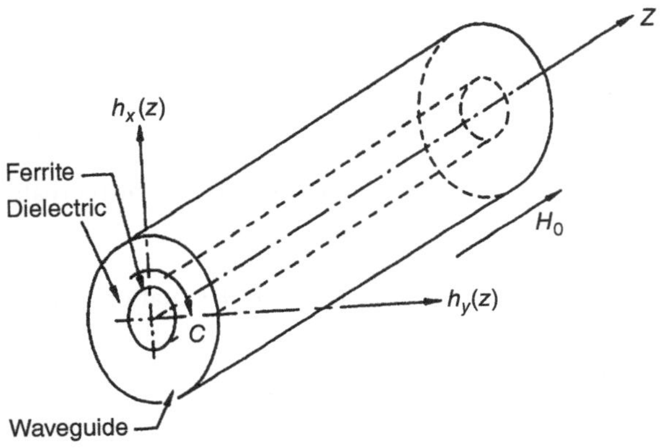

Figure 5.14 Elliptical Faraday rotator.

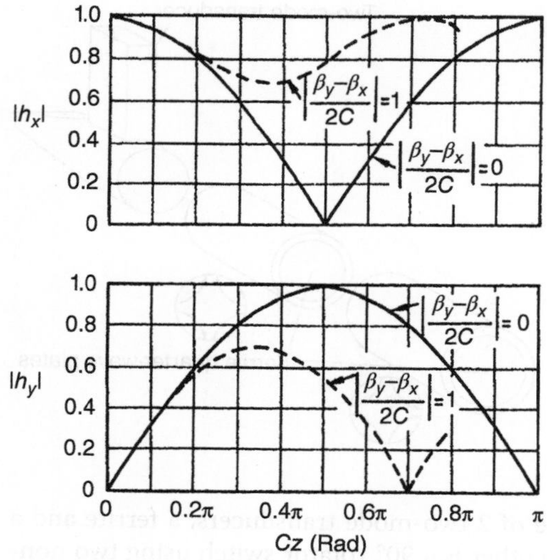

Figure 5.15 Wave amplitude in two polarizations in elliptical Faraday rotator for $(\beta_x - \beta_y)/c = 2$ and $(\beta_x - \beta_y)/c = 0$.

$$\frac{dE_1}{dz} = -\Gamma_1 E_1 + k_{21} E_2 \tag{5.18}$$

$$\frac{dE_2}{dz} = k_{12} E_1 - \Gamma_2 E_2 \tag{5.19}$$

where E_1 and E_2 are the complex wave amplitudes in the primary and secondary waveguides. k_{21} and k_{12} represent the perturbed transfer effects of the coupling mechanism. Γ_1 and Γ_2 are the perturbed propagation constants in the primary and secondary waveguides, i.e.

$$\Gamma_1 = \alpha_1 + j\beta_1 \quad \Gamma_2 = \alpha_2 + j\beta_2 \tag{5.20}$$

For reciprocal coupling conservation of energy requires that the coupling coefficient be imaginary:

$$k_{12} = k_{21} = jk \tag{5.21}$$

For nonreciprocal coupling it requires that the coupling coefficient be real:

$$k_{12} = -k_{21} = C \tag{5.22}$$

The solutions of Eqs. (5.18) and (5.19) for the reciprocal situation are of the form:

$$E_1 = A\exp(\Gamma_0 z) + B\exp(\Gamma_e z) \tag{5.23}$$

$$E_2 = C\exp(\Gamma_0 z) + D\exp(\Gamma_e z) \tag{5.24}$$

where

$$\Gamma_0 = -\left(\frac{\Gamma_1 + \Gamma_2}{2}\right) - jk\sqrt{\left[\frac{(\Gamma_1 - \Gamma_2)^2}{-4k^2} + 1\right]} \quad (5.25)$$

$$\Gamma_e = -\left(\frac{\Gamma_1 + \Gamma_2}{2}\right) + jk\sqrt{\left[\frac{(\Gamma_1 - \Gamma_2)^2}{-4k^2} + 1\right]} \quad (5.26)$$

Bibliography

Boyd, C.R. (1970). A dual-mode latching reciprocal ferrite phase shifter. *IEEE Trans. Microw. Theory Tech.* **MTT-18**: 1119–1124.

Boyd, C.R. (1975). Design of ferrite differential phase shift sections. *IEEE Transactions on Microwave Theory and Techniques Symposium*, Palo Alto, CA (12–14 May 1975), pp. 240–242.

Boyd, C.R. (1982). A 60 GHz dual-mode ferrite phase shifter. *IEEE Transactions on Microwave Theory and Techniques Symposium*, Dallas, TX (15–17 June 1982), pp. 257–259.

Fox, A.G. (1947). An adjustable waveguide phase changer. *Proc. IRE* **35**: 1489–1498.

Fox, A.G., Miller, S.E., and Weiss, M.T. (1955). Behaviour and applications of ferrites in the microwave region. *Bell Syst. Tech. J.* **34**: 5–103.

Glass, H.I. (1959). A short rugged ferrite half-wave plate for a single sideband modulator. *IRE Trans. Microw. Theory Tech.* **MTT-7**: 295.

Helszajn, J. (1964). A novel ferrite quarter-wave plate. *Radio Electron. Eng.* **27**: 455–458.

Helszajn, J. and McStay, J. (1968). Non-reciprocal circular polarizer using hexagonal ferrites with planar anisotropy. *Electron. Lett.* **4** (10): 184–185.

Hogan, C.L. (1952). The ferromagnetic Faraday effect microwave frequencies and its applications – the microwave gyrator. *Bell Syst. Tech. J.* **31**: 22–26.

Hord, W.E., Boyd, C.R., and Rosenbaum, F.J. (1968). Applications of reciprocal latching ferrite phase shifters to lightweight electronic scanned phased arrays. *Proc. IEEE* **56**: 1931–1939.

Karayianis, N. and Cacheris, J.C. (1956). Birefringence of ferrites in circular waveguide. *IRE Proc.* 1414–1421.

Monaghan, N.B. and Mohr, M.C. (1969). Polarization insensitive phase shifter for use in phased-array antennas. *Microw. J.* **12**: 75–80.

Roberts, R.G. (1970). An X-band reciprocal latching Faraday rotator. *IEEE Transactions on Microwave Theory and Techniques Symposium*, Newport Beach, CA (11–14 May 1970), pp. 341–345.

Sultan, N.B. (1971). Generalized theory of waveguide differential phase shift sections and applications to novel ferrite devices. *IEEE Trans. Microw. Theory Tech.* **MTT-19**: 348–357.

Turner, E.H. (1953). A new non-reciprocal waveguide medium using ferrites. *IRE Proc.* **41**: 937.

6

Ridge, Coaxial, and Stripline Phase-shifters[1]

Joseph Helszajn

Heriot Watt University, Edinburgh, UK

TEM transmission lines do not support natural planes of circular polarization but may do so together with a suitable dielectric insert. Two typical structures employing transverse direct fields which rely on the scalar permeabilities displayed by a magnetic insulator under the influence of counterrotating magnetic fields are the coaxial-line and ridge waveguide ones. Since neither of these exhibit natural regions where the alternating magnetic fields are circularly polarized, some means of establishing such polarizations is necessary. In each case this is done by loading the line by some suitable dielectric insert. The operation of the coaxial device may be understood by introducing an electric or magnetic wall at its symmetry plane and making a one-to-one equivalence between it and a ferrite-loaded parallel plate waveguide with similar walls. The symmetric *E*-plane ferrite-loaded ridge waveguide can also be reduced to an approximate parallel waveguide problem but the more general *H*-plane layout requires more advanced analytical methods. The possibility of realizing ridge or square coaxial lines is also noted. The stripline edge mode or field displacement phase-shifter is separately considered in Chapter 8. The approach employed in this chapter is practice orientated and is essentially descriptive in nature.

---

1 Reprinted with amendments (Helszajn, J. (1982). 3.7–4.2 GHz 90° coaxial ferrite differential phase shifter. *IEE Proc.* **129** (Part H): 199–202).

Microwave Polarizers, Power Dividers, Phase Shifters, Circulators, and Switches,
First Edition. Joseph Helszajn.

6.1 Differential Phase-shift, Phase Deviation, and Figure of Merit of Ferrite Phase-shifter

The practical specification of ferrite phase-shifters is dealt with here. Its description involves a compromise between conflicting parameters such as phase deviation over some frequency interval, peak and average power rating, etc. Some typical parameters that are of interest are summarized below as a preamble to investigating some practical devices. Two quantities that are of obvious interest are the phase deviation ($\Delta\theta$) with respect to a 90° bit and the differential phase-shift ($\Delta\phi$) per unit length defined, respectively, by

$$\frac{\Delta\theta}{(\pi/2)} \tag{6.1}$$

and

$$\frac{\Delta\phi}{L}\,\text{rad}\,\text{m}^{-1} \tag{6.2}$$

The figure of merit (F) of the device is separately defined in terms of the insertion loss (α) per unit length and the differential phase-shift per unit length by

$$F = \frac{\Delta\phi}{\alpha}\,\text{rad}\,\text{dB}^{-1} \tag{6.3}$$

For completeness, it is also necessary to spell out the normalized bandwidth (BW) of the device, which is given in terms of the bandwidth (Δf) and midband frequency (f_0) by

$$\frac{\Delta f}{f_0} \tag{6.4}$$

The insertion phase at midband ϕ_0 may also be of some interest and is defined below:

$$\frac{\Delta\phi}{\phi_0} \tag{6.5}$$

Hysteresis effects, temperature, switching speed, and voltage standing wave ratio (VSWR) or return loss are often other parameters of concern.

6.2 Coaxial Differential Phase-shifter

The nonreciprocal coaxial phasor to be described now relies for its operation on the two different scalar permeabilities associated with counterrotating magnetic fields in a suitably magnetized ferrite-loaded coaxial transmission line. Whereas a homogenous coaxial line does not support planes of circular polarization, a partially filled dielectric line does as is by now understood. This transmission line is illustrated in Figure 6.1. The direct magnetic field in this device is perpendicular to the direction of the propagation.

One approximate solution to the problem of the coaxial ferrite phase-shifter is obtained by forming a one-to-one correspondence between it and a parallel plate waveguide with magnetic sidewalls. This equivalence is achieved by introducing a magnetic wall at the plane of symmetry of the coaxial network and unwrapping it to form a parallel plate waveguide as indicated in Figure 6.2.

$$A = 2\pi\left(\frac{r_a + r_b}{2}\right) \tag{6.6}$$

$$B = r_a - r_b \tag{6.7}$$

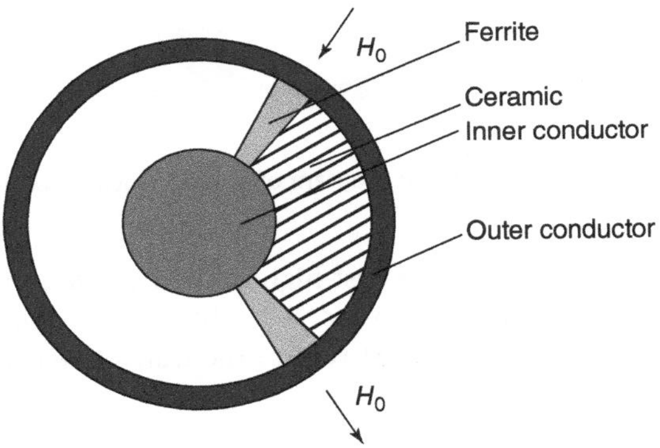

Figure 6.1 Schematic diagram of nonreciprocal coaxial transmission line.

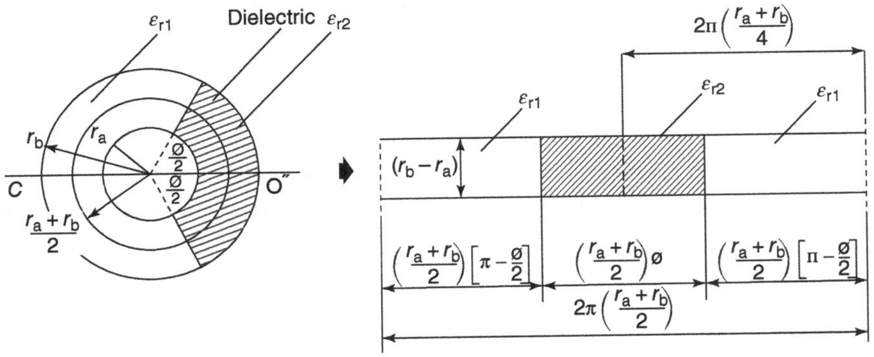

Figure 6.2 Equivalent parallel plate waveguide of partially dielectric-filled coaxial transmission line.

The solution of this parallel plate waveguide problem with decaying fields outside the dielectric region has been tackled in Chapter 3; the related ferrite-loaded arrangement has been dealt with in Chapter 12.

For completeness, it is observed that the introduction of a metal septum (electric wall) as indicated in Figure 6.3 makes the coaxial device in Figure 6.1 dual to the much-used waveguide structure. The influence of such sidewalls, on the split phase constants, is often minimal in practical devices.

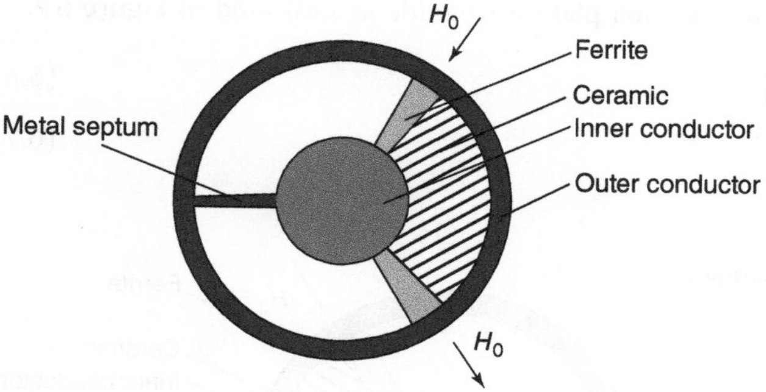

Figure 6.3 Schematic diagram of nonreciprocal coaxial transmission line loaded with metal septum.

The onset of the first higher-order mode in this class of waveguide has not been investigated so far. It is obtained, in the absence of the ferrite section, by evaluating the first root of the transverse resonance condition of the structure given below.

$$\frac{2\pi\sqrt{\varepsilon_r}}{\lambda_c}\tan\left[\left(\frac{2\pi\sqrt{\varepsilon_r}}{\lambda_c}\right)\left(\frac{r_a+r_b}{4}\pi\right)\right] = -\frac{2\pi}{\lambda_c}\tan\left[\left(\frac{2\pi}{\lambda_c}\right)\left(\frac{r_a+r_b}{4}\pi\right)\right] \tag{6.8}$$

This equation applies to a half dielectric-filled coaxial line derived by placing a magnetic wall at the symmetry plane of the circuit.

For $\varepsilon_r = 15$, the result is

$$\frac{2\pi}{\lambda_c}\sqrt{\varepsilon_r}\left(\frac{r_a+r_b}{4}\pi\right) \approx 2.906\,\text{rad} \tag{6.9}$$

The matching problem for this class of transmission line may be attended to by replacing the inhomogeneous line by the one completely filled by a material with a relative dielectric constant ε_{eff}.

The impedance Z_t and the wavelength λ_t of the inhomogeneous transformer region may be defined in terms of the effective dielectric constant, provided the propagation along the line is assumed quasi-TEM.

$$Z_t = \frac{Z_0}{\sqrt{\varepsilon_{eff}}} \tag{6.10}$$

$$\lambda_t = \frac{\lambda_0}{\sqrt{\varepsilon_{eff}}} \tag{6.11}$$

The effective dielectric constant of the inhomogeneous line is derived by making use of either conformal transformation between the coaxial line and the equivalent parallel plate waveguide indicated in Figure 6.4.

The derivation proceeds by calculating the capacitance of the partially dielectric-filled coaxial line. This is readily given as

$$C = \frac{\varepsilon_0}{\ln(r_b/r_a)_c}[(2\pi - \phi)\varepsilon_{r1} + \varepsilon_{r2}\phi] \text{ F m}^{-1} \tag{6.12}$$

The capacitance of the original homogenous line is now formed by writing $\phi = 0$ in the preceding equation:

$$C = \frac{2\pi\varepsilon_{r1}\varepsilon_0}{\ln(r_b/r_a)_c} \text{ F m}^{-1} \tag{6.13}$$

The required result is obtained by taking the ratio of the preceding two capacitances.

$$\varepsilon_{eff} = \frac{\varepsilon_{r1}(2\pi - \phi) + \varepsilon_{r2}\phi}{2\pi\varepsilon_{r1}} \tag{6.14}$$

ε_{r1} and ε_{r2} are the relative dielectric constants of the two dielectric regions, and ϕ is the angular angle defined in Figure 6.4.

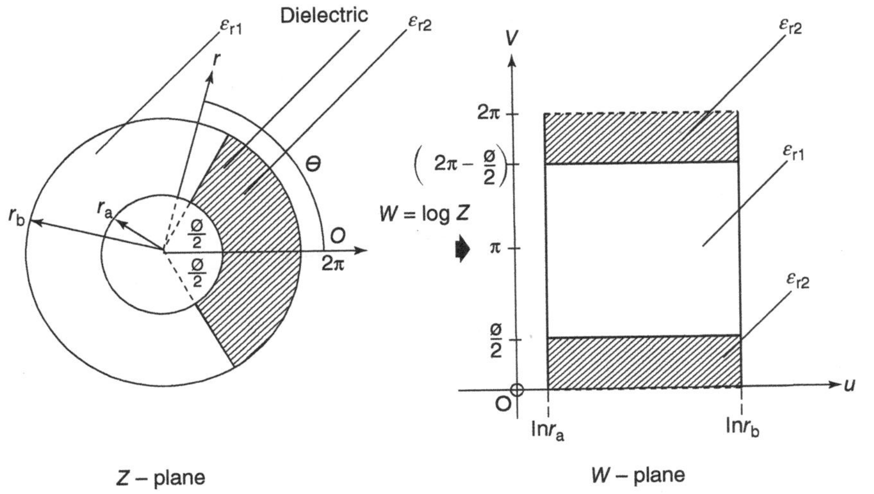

Figure 6.4 Equivalent parallel waveguides of partially filled coaxial line using conformal mappings.

In order to overcome difficulties in machining, it is sometimes desirable to have ε_r equal to π radians. This may be done by rearranging Eq. (6.9) as

$$\varepsilon_{r2} = \frac{\varepsilon_{r1} + [2\pi(\varepsilon_{eff} - 1)] + \phi}{\phi} \tag{6.15}$$

The conformal transformation employed here to describe the quasi-static capacitance or characteristic impedance of the line cannot be used to construct the transverse resonance condition in Eq. (6.3) because the corresponding width of the equivalent waveguide is 2π, which is not related in any way to the actual dimensions of the coaxial line. The capacitance or impedance of the line may, however, be derived from either circuit. This will now be demonstrated.

The capacitance C' of the equivalent waveguide in Figure 6.2 may be written as

$$C' = \frac{\varepsilon_0[1 + (r_b/r_a)]}{2[(r_b/r_a) - 1]_c}[\varepsilon_{r1}(2\pi - \phi) + \varepsilon_{r2}\phi]\ \mathrm{F\,m^{-1}} \tag{6.16}$$

For r_b/r_a small:

$$\ln\left(\frac{r_b}{r_a}\right)_c \approx \frac{2[(r_b/r_a) - 1]}{[1 + (r_b/r_a)]} \tag{6.17}$$

Making the use of this identity in Eq. (6.11) indicates that Eqs. (6.8) and (6.11) are equivalent for r_b/r_a small. For a 50 Ω line, r_b/r_a is 0.37 and the error is about 5.4%.

Figure 6.5 depicts the split phase constants of one 3700–4200 MHz device based on the discussion outlined in this section.

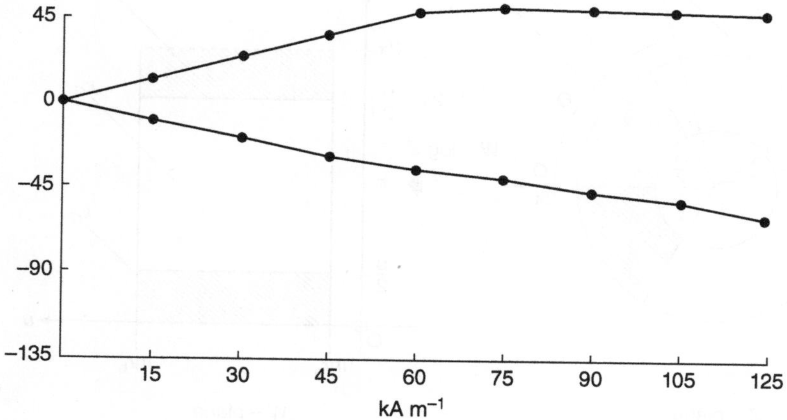

Figure 6.5 Phase constants versus direct magnetic for 100 mm × 2.99 mm × 2 mm long ferrite sheet (M_0 0.0800 T) (Helszajn 1982).

The experimental results are summarized below.

$$\frac{\Delta\theta}{(\pi/2)} \approx \pm\frac{2}{90}$$

$$\frac{\Delta\phi}{L} = \frac{(\pi/2)}{0.100}\,\text{rad}\,\text{m}^{-1}$$

$$F = \frac{(\pi/2)}{0.150}\,\text{rad}\,\text{dB}^{-1}$$

$$\frac{\Delta f}{f_0} = \frac{500}{3950}$$

$$\frac{\Delta\phi}{\phi_0} = \frac{900}{\phi_0}$$

$\Delta\theta$ is the phase deviation (rad), $\Delta\phi$ is the differential phase-shift (rad), ϕ_0 is the midband insertion phase (rad), α is the insertion loss (dB), L is the overall length (m), and f_0 and f are the center frequency and bandwidth, respectively (MHz).

The material employed in obtaining this data had a magnetization of 0.0800 T, a dielecric constant of 14.6, a spin-wave line-width of 23 kA m^{-1}, and a dielectric loss tangent of 0.0002. The dimensions of the ferrite sheets of the 90° section in the final device were 1.20 mm by 3.40 mm by 100mm. The direct magnetic field was about 65 kA m^{-1}. The relative dielectric constant of the dielectric insert was 15.0. To reduce the reluctance of the magnetic circuit the inner conductor of the

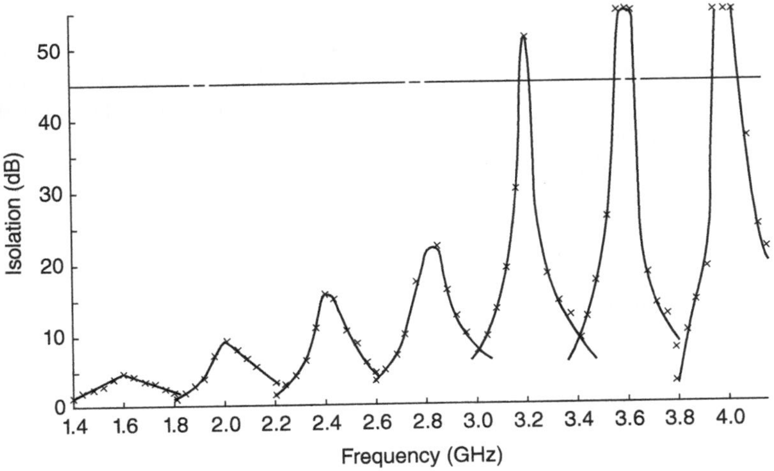

Figure 6.6 Ferrimagnetic resonance in a half dielectric-filled coaxial line loaded by garnet sheets (70 mm long by 0.80 mm × 2.0 mm) (Helszajn 1982).

coaxial line is made of magnetic steel and the conductor is silver plated to minimize the insertion of the device.

Higher-order modes occurred in the device at about 4.6 GHz in excellent agreement with the calculation. The quality of the circular polarization was investigated separately by biasing the phasor to ferrimagnetic resonance and measuring the forward and backward transmission losses between 1.5 and 4.2 GHz. Figure 6.6 indicates that the polarization is indeed nearly circular in the required 3.7–4.2 GHz band, and is nearly so over the full 2–4 GHz band.

6.3 Ridge Waveguide Differential Phase-shifter

A closely related geometry to that of the coaxial ferrite phase-shifter is that of the ridge waveguide one. The operation of this device again rests on the two different values of permeability exhibited by counterrotating magnetic fields in a magnetized ferrite medium. The field pattern of ridge waveguide does not, of course, have natural planes of circular polarization so that some means of establishing them is necessary. This is achieved here, in keeping with common practice, by again recognizing that such fields are nearly always displayed at the interface between two different dielectric regions and everywhere outside it. The details employed in Chapter 3 in connection with this problem apply here. It is furthermore assumed, for simplicity, that the introduction of thin *H* or *E* plane ferrite or garnet substrates in the vicinity of the dielectric wall does not in the first instance disturb the polarization in its neighborhood. Some possible topologies are illustrated in Figure 6.7.

Figure 6.8 depicts the experimental frequency response of a 90°bit over the 2–4 GHz band in WRD200 waveguide using the *H*-plane configuration. This result is obtained by normalizing the phase for one orientation of the applied field and recording the result for the reversed field.

The performance of the 90°bit may be summarized by

$$\frac{\Delta\theta}{(\pi/2)} \approx \pm\frac{3}{90}$$

$$\frac{\Delta\phi}{L} = \frac{(\pi/2)}{0.120}\,\text{rad}\,\text{m}^{-1}$$

$$F = \frac{(\pi/2)}{0.50}\,\text{rad}\,\text{dB}^{-1}$$

$$\frac{\Delta f}{f_0} = \frac{2000}{3000}$$

$$\frac{\Delta\phi}{\phi_0} = \frac{900}{860}$$

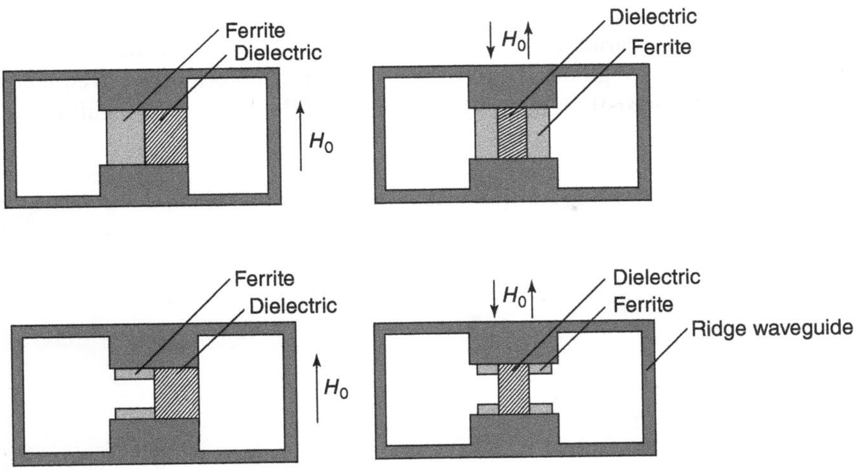

Figure 6.7 Schematic diagram of ridge waveguide phase-shifter.

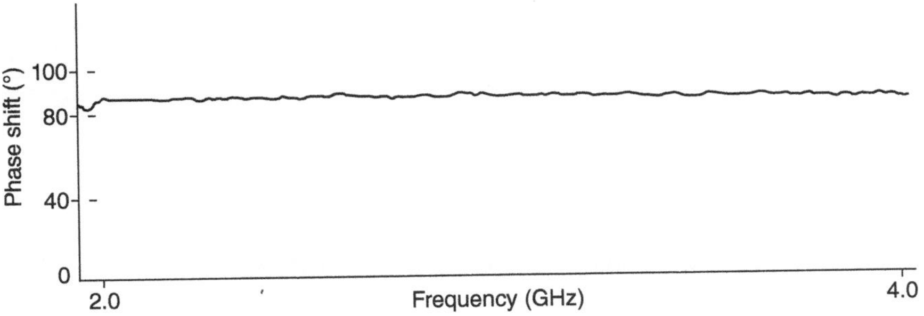

Figure 6.8 Differential phase-shift in WRD200 ridge waveguide (Hastings and Helszajn 1987).

The figure of merit F of this device reflects, in part, the use of a heavily doped garnet material with a spin-wave line-width of $\Delta H_k = 0.68$ kA m^{-1} to suppress spin-wave instability (or nonlinear loss), which can occur at high-peak power levels in ferrite devices. The other material details are described by $M_0 = 0.0800$ T, $\varepsilon_r = 14.7$, $\Delta H = 5.6$ kA m^{-1}, and tan $\delta \approx 0.002$. The direct magnetic field employed at the 90° phase state is about 52 kA m^{-1}.

Two features of note already remarked upon that may have some bearing on this excellent result are that the phase constant of the unperturbed parallel-plate waveguide is (for the parameters employed) quasi-TEM over the full 2–4 GHz frequency interval and that the polarizations of the magnetic fields are nearly circularly polarized over the same interval; another property of this geometry is that its insertion phase (860° at midband) is quite large compared to the phase deviation of the magnetized bit (±45°) so that perturbation conditions may be assumed to prevail. Faraday rotation in free space or under perturbation conditions in circular waveguides is another device that exhibits a similar frequency characteristic. Figure 6.9 illustrates one practical differential phase-shift circulator using ridges.

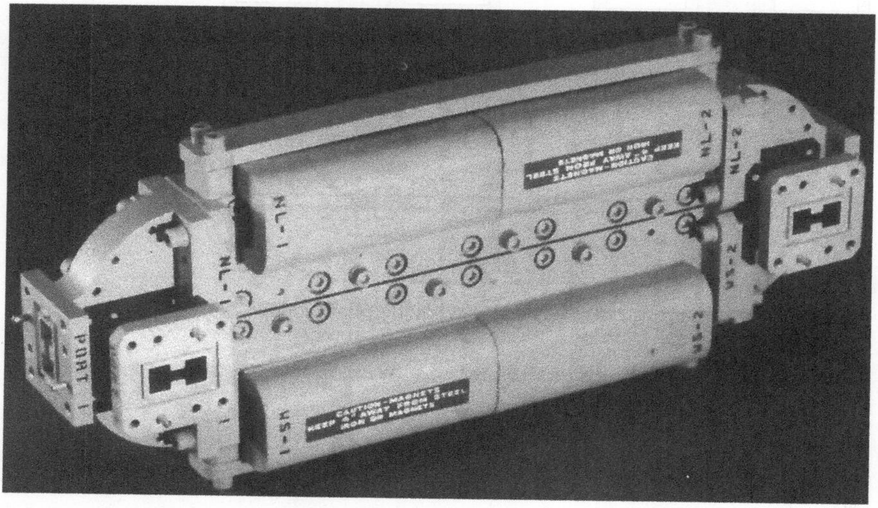

Figure 6.9 Photograph of ridge waveguide differential phase-shift circulator. *Source:* Courtesy Raytheon Co.

6.4 The Stripline Edge Mode Phase-shifter

Planar circuits are, of course, admirably suited for the design of field displacement or edge mode phase-shifters. The parallel plane arrangement has already been studied in some detail in Chapter 3. Figure 6.10 illustrates one stripline configuration.

As is now well understood, operation rests on the fact that a suitably ferrite-filled parallel plate waveguide magnetized perpendicular to the direction of propagation exhibits a TE-type solution of the form:

$$E_y = Ae^{-\alpha x}e^{-j\beta x}$$

$$H_x = \eta E_y$$

$$H_x = 0$$

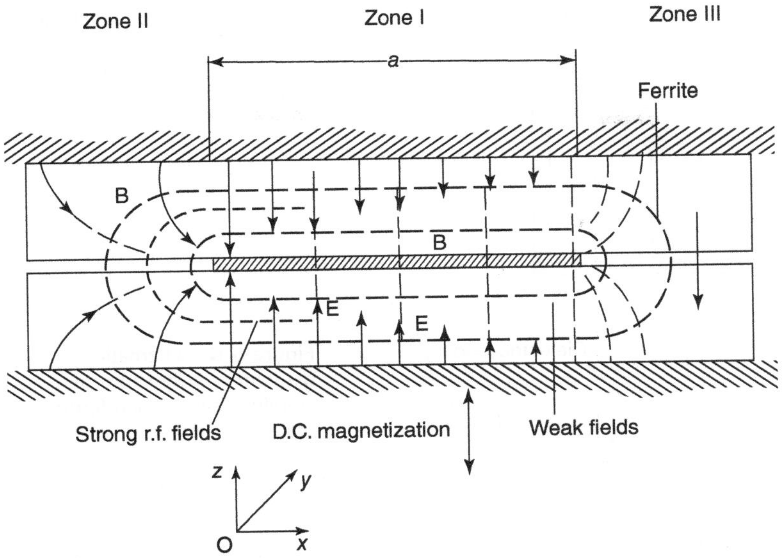

Figure 6.10 Schematic diagram of edge mode stripline phase-shifter.

This solution indicates that the fields decay exponentially across the waveguide but exhibit no attenuation along the direction of propagation. Furthermore, the two edges of this waveguide are decoupled, provided the waveguide is sufficiently wide. In fact, the power is concentrated in the vicinity of one edge in the forward direction of propagation and is displaced to the other one in the reverse direction of propagation. Lining one of the two edges with some resistive material allows edge mode isolators to be fabricated. Lining one of them with sane dielectric material permits nonreciprocal phase-shifters to be constructed.

6.5 Latched Quasi-TEM Phase-shifters

The possibility of latching ferrite phase-shifters is, of course, always of some interest. Suitable coaxial and ridge structures are illustrated in Figures 6.11 and 6.12.

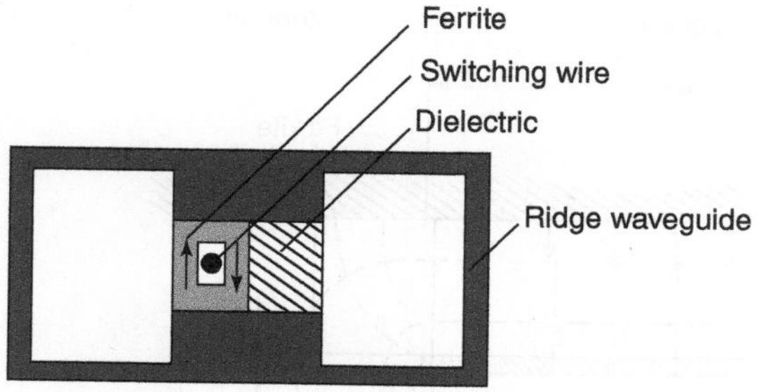

Figure 6.11 Schematic diagram of nonreciprocal ridge waveguide using closed ferrite magnetic circuits.

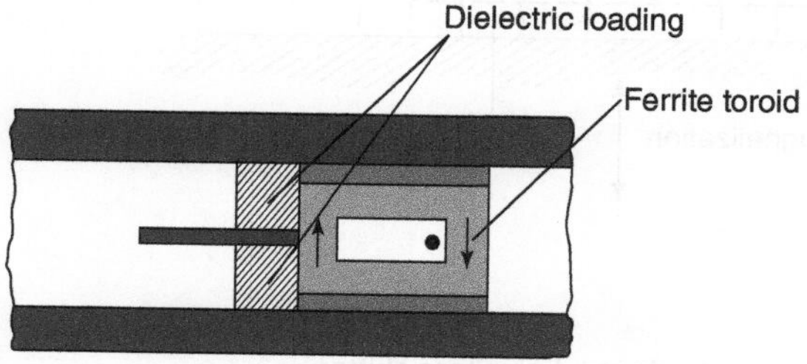

Figure 6.12 Schematic diagram of nonreciprocal stripline using closed ferrite magnetic circuits.

Bibliography

Anderson, W.W. and Hines, M.E. (1961). Wide band resonance isolator. *IRE Trans. Microw. Theory Tech.* **MTT-9**: 63.

Button, K.J. (1958). Theory of non-reciprocal ferrite phase shifters in dielectric loaded coaxial line. *J. Appl. Phys.* **29**: 998.

Clark, J. and Harrison, G.R. (1962). Miniaturized coaxial ferrite devises. *Microw. J.* (June): 108–118.

Cohn, M. (1959). Propagation in a dielectric-loaded parallel plane waveguide. *IRE Trans. Microw. Theory Tech.* **MTT-7**: 202–208.

Duncan, B.J., Swern, L., Tomiyasu, K., and Hannwacker, J. (1957). Design considerations for broadband ferrite coaxial line isolators. *Proc. IRE* **45**: 483–490.

Fleri, D. and Hanley, G. (1959). Non-reciprocity in dielectric loaded TEM mode transmission lines. *IRE Trans. Microw. Theory Tech.* **MTT-7**: 23–27.

Freiberg, L. (1958). Coaxial isolator utilising yttrium iron garnet. *IRE Trans. Microw. Theory Tech.* **MTT-4**: 454.

Grimes, E.S. Jr., Bartholomew, D.D., Scott, D.C., and Sloan, S.C. (1960). Broadband ridge waveguide ferrite devices. *IRE Trans. Microw. Theory Tech.* **MTT-8**: 489.

Hastings, W.I. and Helszajn, J. (1987). WRD200 90° waveguide differential phase shift section. *Proc. Inst. Electr. Eng.* **134**: 223–225.

Helszajn, J. (1982). 3.7–4.2 GHz 90° coaxial ferrite differential phase shifter. *IEE Proc.* **129** (Part H): 199–202.

Hines, M.E. (1971). Reciprocal and nonreciprocal modes of propagation in ferrite stripline and microstrip devices. *IEEE Trans. Microwave Theory Tech.* **MTT-19**: 442–451.

Jones, R.R. (1966). A slow wave digital ferrite strip transmission line phase shifter. *IEEE Trans. Microw. Theory Tech.* **MTT-14** (12): 684–688.

Seidel, H. (1957). Ferrite slabs in transverse electric mode waveguide. *J. Appl. Phys.* **28**: 218–226.

Simon, J.W., Alverson, W.K., and Pippin, J.E. (1966). A reciprocal TEM latching ferrite phase shifter. *IEEE MTT-S International Microwave Symposium Digest*, pp. 241–246.

Soohoo, R.F. (1960). *Theory and application of ferrites*. London: Prentice Hall International Inc.

Sucher, M. and Carlin, H.J. (1957). Coaxial line non-reciprocal phase shifters. *J. Appl. Phys.* **28**: 921.

Suhl, H. and Walker, L.R. (1954). Topics in guided wave propagation through gyromagnetic media. *Bell Syst. Tech. J.* **33**: 1133.

Sullivan, D.J. and Parkes, D.A. (1960). Stepped transformer for partially filled transmission lines. *IRE Trans. Microw. Theory Tech.* **MTT-8**: 212–217.

Treuhaft, M.A. and Silber, L.M. (1958). Use of microwave ferrite toroids to eliminate external magnets and reduce switching power. *Proc. IRE* **46** (8).

Whicker, L.R. and Jones, R.R. (1965). A digital latching ferrite strip transmission line phase shifter. *IEEE Trans. Microw. Theory Tech.* **MTT-13**: 781–784.

7

Finite Element Adjustment of the Rectangular Waveguide-latched Differential Phase-shifter

Joseph Helszajn[1] *and Mark McKay*[2]

[1] *Heriot Watt University, Edinburgh, UK*
[2] *Honeywell, Edinburgh, UK*

7.1 Introduction

The main topic of this chapter is the propagation in rectangular waveguides containing one or two transversely magnetized ferrite slabs or sheets. Figure 7.1a and b illustrate the two basic arrangements. The origin of the non-reciprocal phase-shift in this type of waveguide may be readily understood by recalling that the alternating magnetic fields on either side of the symmetry plane of the waveguide are circularly polarized with opposite senses of rotation. Furthermore, the senses of the polarization are interchanged if the direction of propagation is reversed. A single thin ferrite slab will therefore exhibit a scalar permeability with a positive value in one direction of propagation and a negative value in the other.

Oppositely magnetized slabs on either side of the symmetry plane will behave in a like manner. The operation of the arrangement in Figure 7.2 is identical to that of Figure 7.1b, except that it relies on the existence of planes of circular polarization produced at the interface between two different dielectric media. A latching geometry is obtained here by adding horizontal ferrite members to the geometry in Figure 7.1b in order to produce a closed magnetic circuit. The problem has historically been addressed by satisfying Maxwell equations in each region of the geometry and separately satisfying the boundaries between each. The approach utilized here relies on a Finite Element (FE) formulation of the problem.

Microwave Polarizers, Power Dividers, Phase Shifters, Circulators, and Switches, First Edition. Joseph Helszajn.

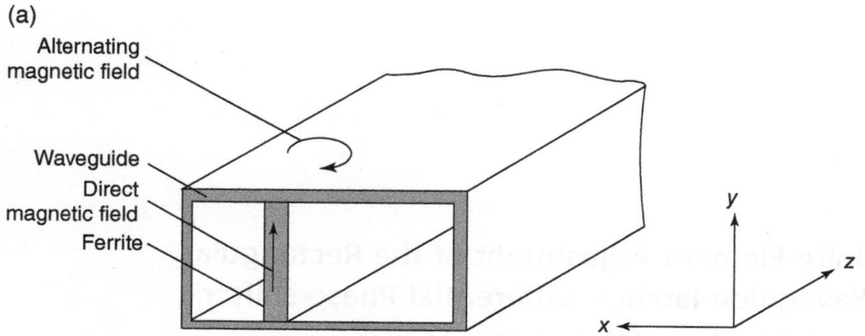

Figure 7.1 Rectangular waveguide phase-shifters: (a) loaded with a ferrite slab and (b) symmetrically loaded with two oppositely magnetized ferrite slabs.

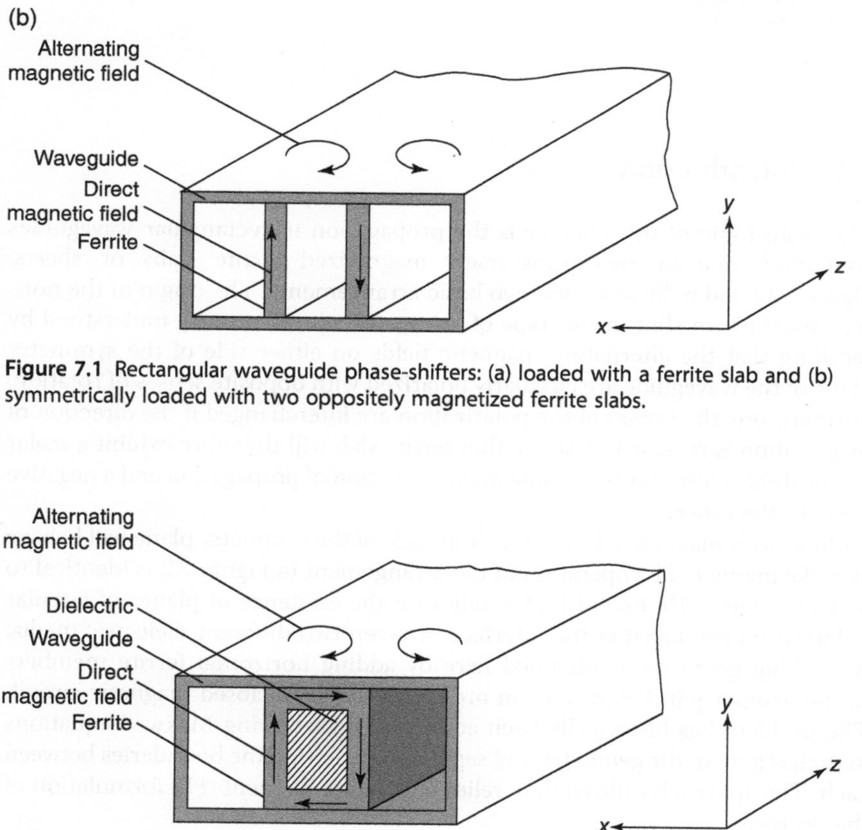

Figure 7.2 Toroidal phase-shifter in rectangular waveguide.

7.2 FE Discretization of Rectangular Waveguide Phase-shifters

The cross-sections of some typical rectangular waveguide ferrite geometries met in practice are illustrated in Figures 7.1 and 7.2. The problems under consideration are inhomogenous ones consisting of isotopic and gyromagnetic regions. A toroidal arrangement met in practice is illustrated in Figure 7.2. The numerical method adopted here is the FE one. Each region is typically divided into thousands of tetrahedral elements. A typical tetrahedral element is depicted in Figure 7.3. A feature of the discretization is that the density of the mesh will be larger in the regions of high field intensity, however, the number of elements on either side of a typical boundary have to be equal. Figure 7.4 illustrates a typical discretization of the geometry in Figure 7.1a by way of example. The calculation of the cutoff numbers of this class of geometry do not involve the introduction of ports, however, the calculation of the phase constants do. The full development of this class of phase-shifter is a two-step procedure. The first involves a calculation of the cutoff number and propagation

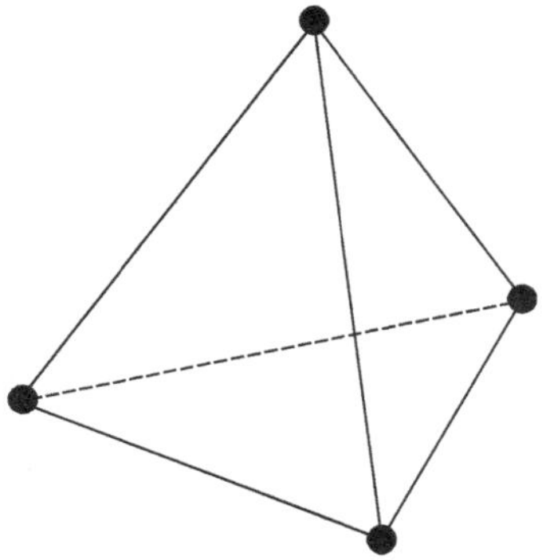

Figure 7.3 Typical tetrahedral finite element.

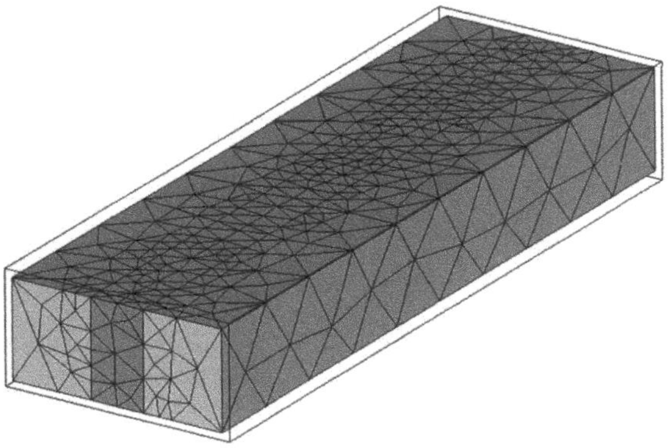

Figure 7.4 Discretization of waveguide loaded with a ferrite slab.

constant of the reciprocal waveguide. The second step involves the introduction of the gyrotropy in the ferrite regions.

7.3 LS Modes Limit Waveguide Bandwidths

Practical latching ferrite phase-shifters rely on the existence of planes of circular polarization at any plane between two different dielectric media. The modes encountered with this geometry are the so-called LS ones. Figure 7.5 is a

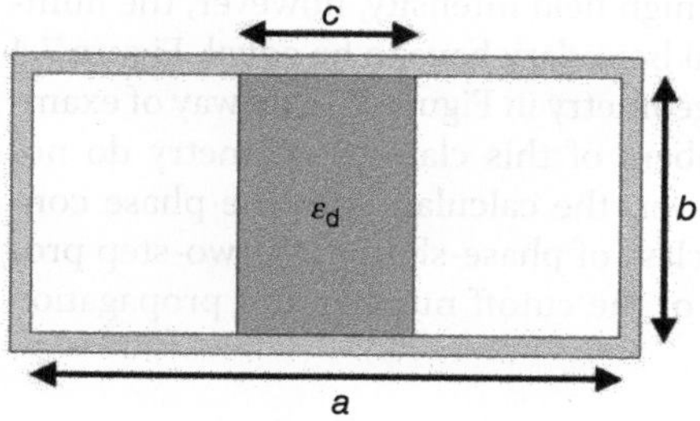

Figure 7.5 Dielectric-loaded rectangular waveguide.

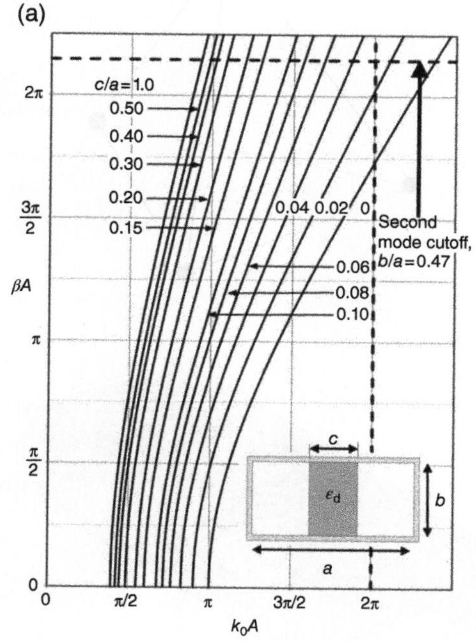

Figure 7.6 LS modes in dielectric-loaded rectangular waveguide: (a) $\varepsilon_d = 6.5$ and (b) $\varepsilon_d = 16.0$.

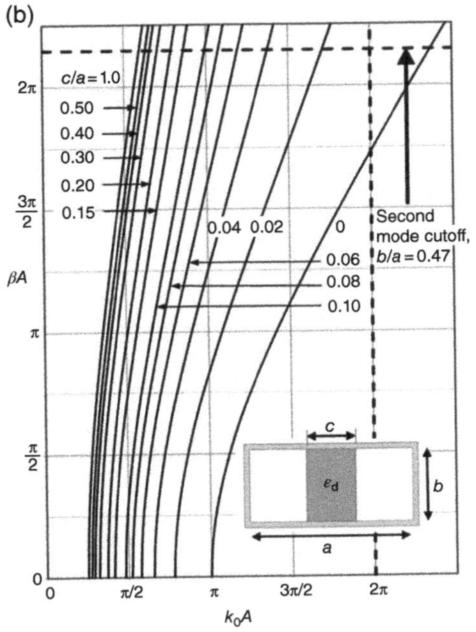

Figure 7.6 (Continued)

schematic diagram of the structure under consideration. The onset of higher-order modes in this type of waveguide is of importance in the design of this class of phase-shifter. Figure 7.6a and b depict some data.

7.4 Cutoff Numbers and Split Phase Constants of a Twin Slab Ferrite Phase-shifter

Some typical calculations on the cutoff numbers and phase constants of the twin slab phase-shifter depicted in Figure 7.1b are summarized in Figures 7.7 and 7.8. This is done for a typical value of gyrotropy equal to ±0.50. The gyrotropy in a saturated material is given in the usual way by the off-diagonal element of the tensor permeability

$$\kappa = \frac{\gamma M_0}{\mu_0 \omega}$$

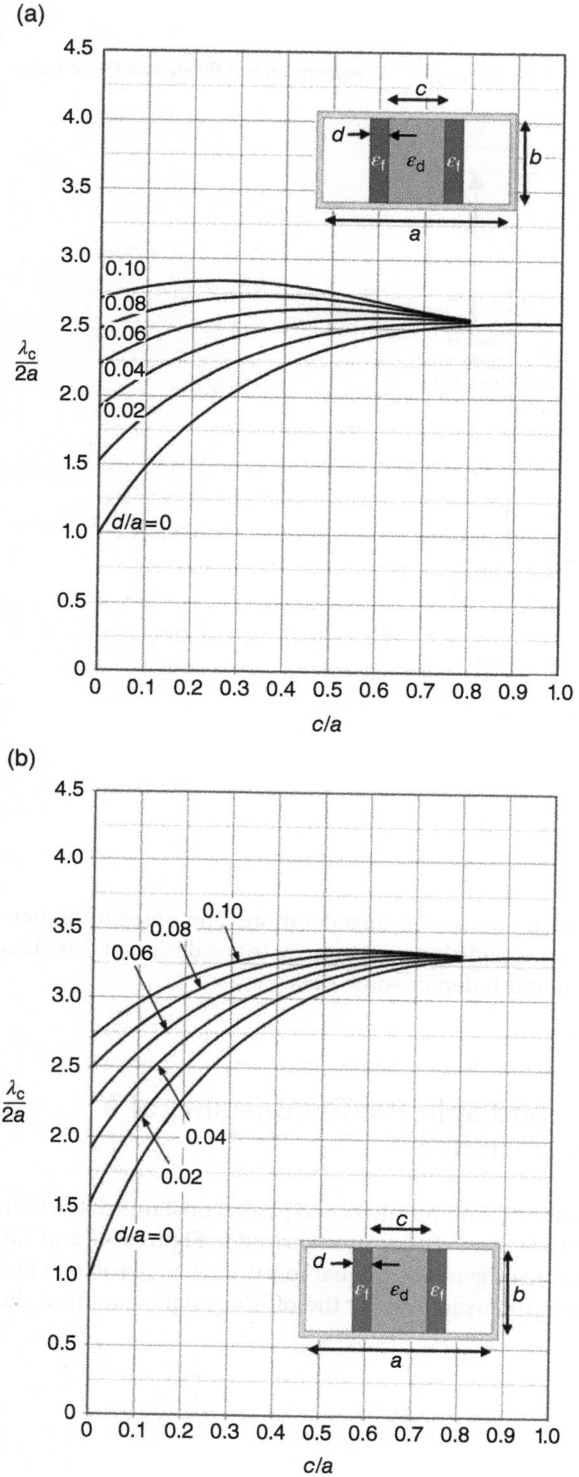

Figure 7.7 Cutoff numbers of twin slab ferrite phase-shifter $\varepsilon_f = 16$: (a) $\varepsilon_d = 6.5$ and (b) $\varepsilon_d = 11$.

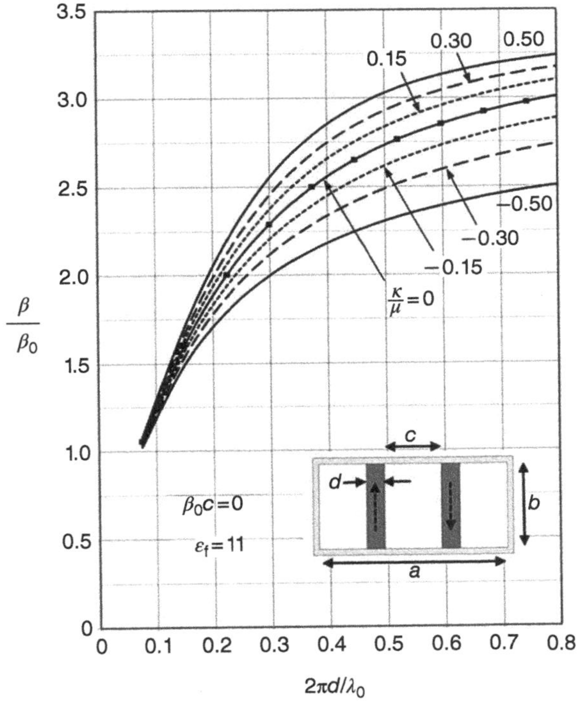

Figure 7.8 Split phase constants of a twin slab ferrite phase-shifter ($\varepsilon_d = 1$).

The diagonal element of the tensor permeability is

$$\mu = 1$$

The tensor permeability is

$$[\mu] = \begin{bmatrix} \mu & -j\kappa & 0 \\ j\kappa & \mu & 0 \\ 0 & 0 & 1 \end{bmatrix}$$

γ is the gyromagnetic ratio (2.21×10^5 rad s^{-1} per A m^{-1}). M_0 is the saturation magnetization of the magnetic insulator (A m^{-1}), ω is the radio frequency (rad s^{-1}), and μ_0 is the free space permeability ($4\pi \times 10^{-7}$ H m^{-1}).

Figure 7.9 indicates the coordinates of the arrangement employed in Figures 7.7 and 7.8. Figures 7.10 depicts the differential phase for two values of dielectric constant. The calculations here and elsewhere in the chapter are undertaken at a frequency of 3 GHz in WR187 waveguide.

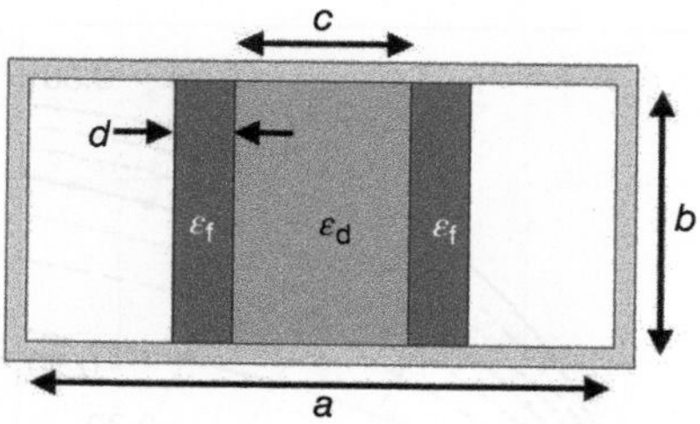

Figure 7.9 Coordinate system of rectangular waveguide loaded with two ferrite slabs.

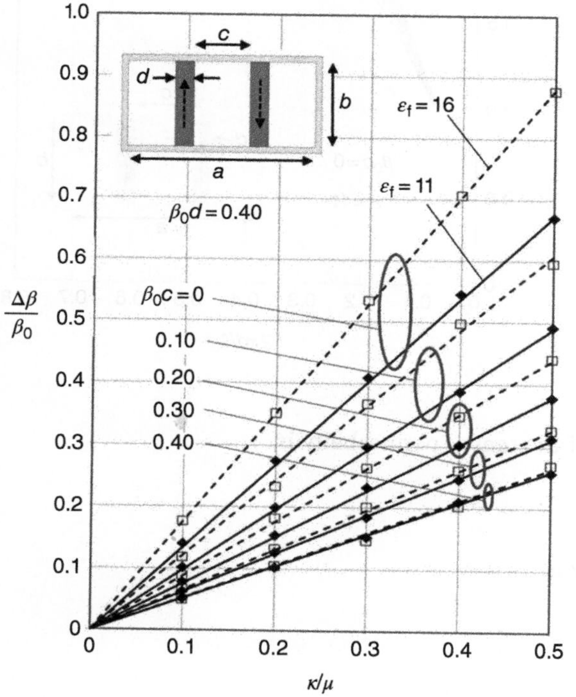

Figure 7.10 Differential phase in twin slab ferrite phase-shifter versus gyrotropy ($\varepsilon_d = 1$).

7.5 The Waveguide Toroidal Phase-shifter

The practical industrial twin slab waveguide phase-shifter is the toroidal one illustrated in Figure 7.2. Figure 7.11 depicts the differential phase-shift obtained by closing the magnetic path of the twin phase-shifter. A similar result with the inner core filled with a dielectric material with a constant of 6.5 is also shown in Figure 7.11.

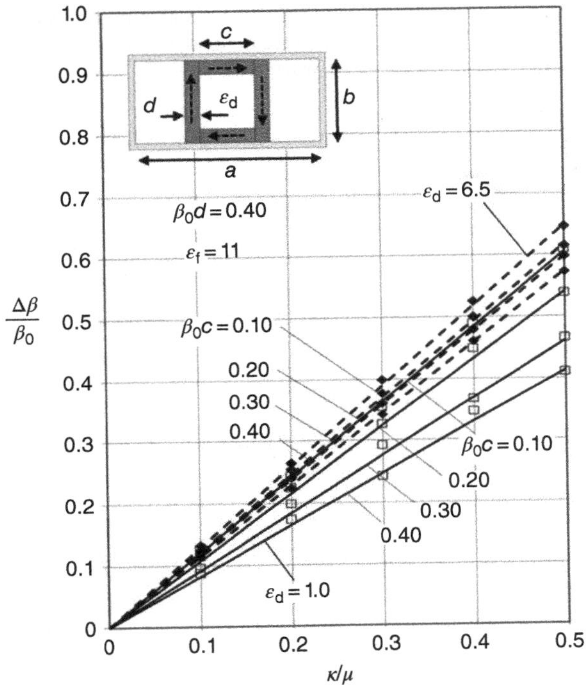

Figure 7.11 Differential phase in toroidal phase-shifter versus gyrotropy.

7.6 Industrial Practice

In practice, the phase sections in Figures 7.1 and 7.2 have to be matched at both the input and output ports to standard rectangular waveguide. Figure 7.12 shows a typical arrangement. Figure 7.13 illustrates one means of avoiding propagation of higher-order modes in this type of phase-shifter. Figure 7.14 depicts a 4-bit structure housed in an undersized waveguide. Figure 7.15 shows a block diagram of a typical electronic driver circuit for a multi-toroid latching waveguide phase-shifter.

7.7 Magnetic Circuits Using Major and Minor Hysteresis Loops

Microwave ferrite phase-shifters rely for their operation on the relationship between a direct magnetic field and the insertion phase of a suitably magnetized ferrite-loaded transmission line. The direct magnetic field can be

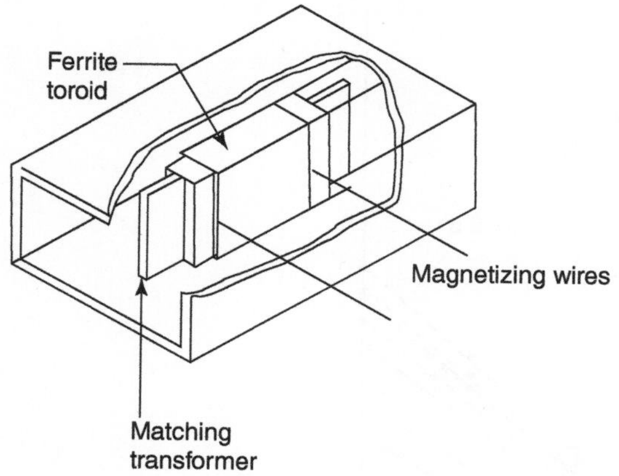

Figure 7.12 Matching between dielectric loaded and regular waveguides.

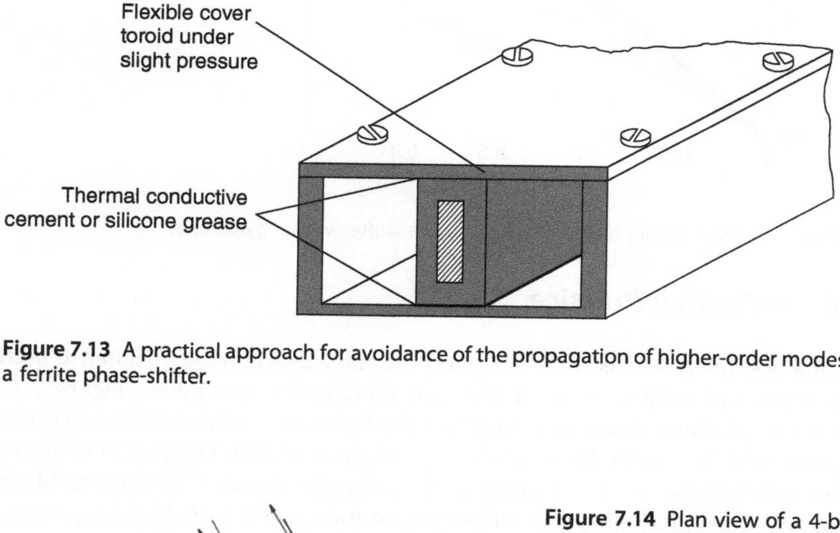

Figure 7.13 A practical approach for avoidance of the propagation of higher-order modes in a ferrite phase-shifter.

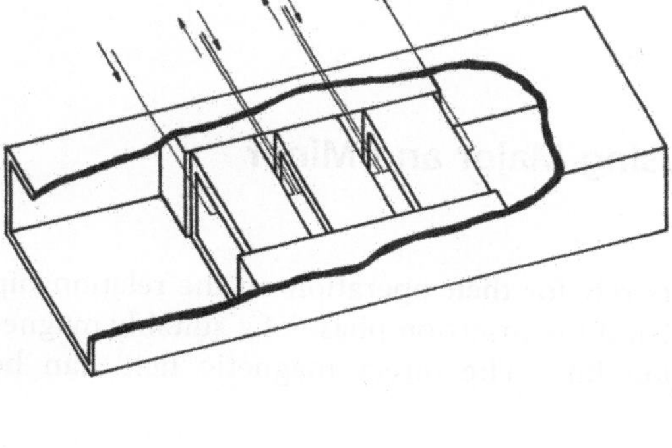

Figure 7.14 Plan view of a 4-bit toroidal phase-shifter in undersized waveguide.

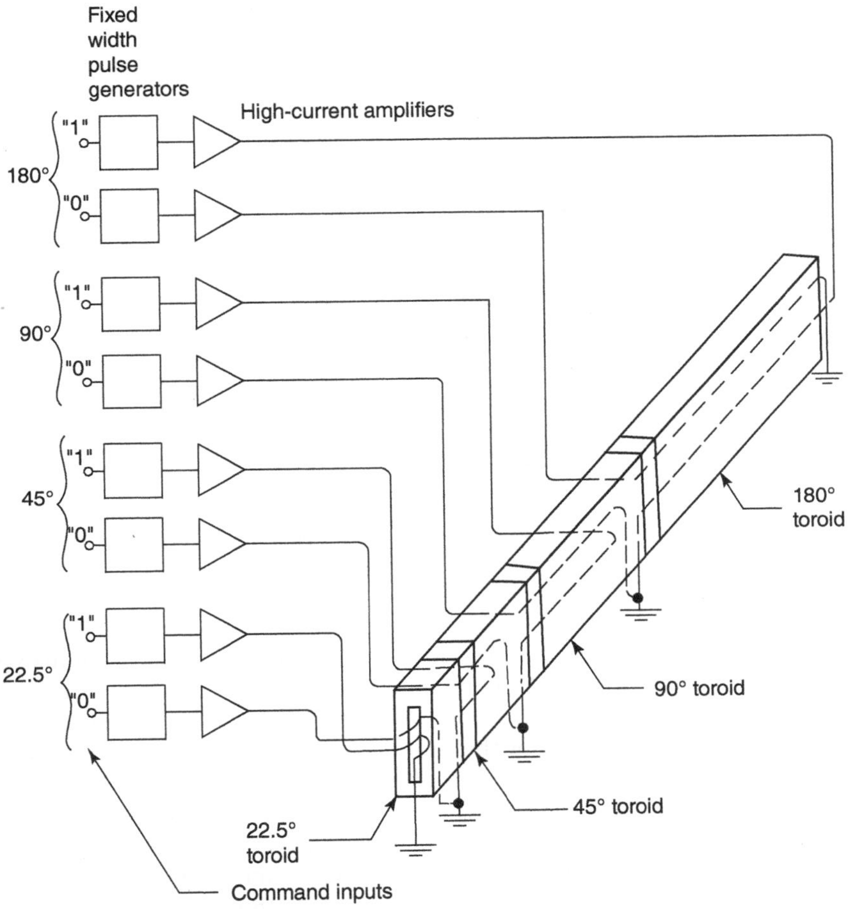

Figure 7.15 Block diagram of a typical electronic driver circuit for a multi-toroid latching waveguide phase-shifter.

established using either an external electromagnet or it can be switched by current pulses through a magnetizing wire between two remanent states of the major or indeed of a minor hysteresis loop of a closed magnetic circuit. The former arrangement requires a holding current to hold the device in a given state. In the latter one, however, no holding current is necessary; the device remains latched in a given state until another switching operation is required. The advantages and disadvantages of each type of circuit are understood.

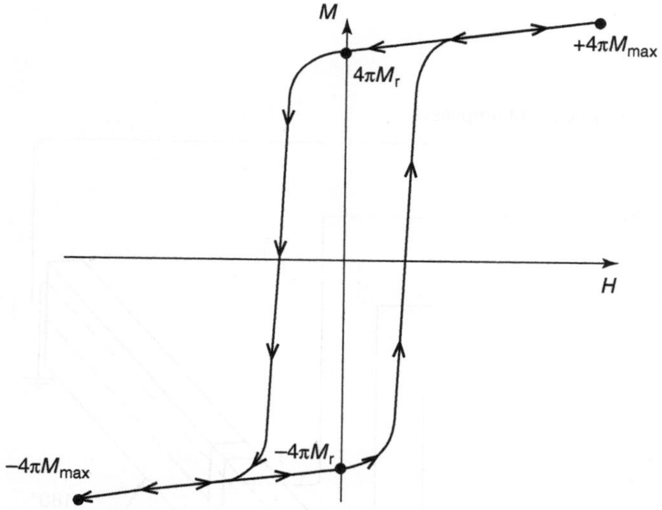

Figure 7.16 Typical hysteresis loop of a latching phase-shifter operating with a major loop switching.

Operation on the major hysteresis loop may be understood by scrutinizing the hysteresis loop in Figure 7.16, provided it is recognized that the size and shape of this loop may vary with the speed of the switching process. In this situation, the magnetization of the toroid is driven between two remanent states ($\pm 4\pi M_0$) equidistant from the origin by the application of a current pulse sufficiently large to produce a field perhaps three or five times that of the coercive force. After this point is reached, the current pulse is removed and the magnetization will move to the remanent value ($\pm 4\pi M_0$) and remain there until another switching operation is desired. In this example, two phase-shift states are available, corresponding to the two possible values of remanent magnetization.

7.8 Construction of Latching Circuits

A large number of phase states can be achieved by connecting a number of bits in cascade, each of which has a different physical length (and consequently phase-shift). This is illustrated in Figure 7.15 in connection with a nonreciprocal 4-bit waveguide phase-shifter. Each electronic driver circuit is relatively simple since it is only required that the toroids be driven back and forth between their major loop remanent states, but the total number of electronic parts may be quite high, especially in units having five or more bits. Furthermore, a large number of latching wires must enter and leave the waveguide and several changes in cross-section may occur through the length of the ferrite/dielectric section. The complexity of the microwave circuit makes it difficult to achieve very low values of insertion loss and restricts the bandwidth of the device; it also leads to manufacturing difficulties. This technique, however, possesses one very significant advantage in some system applications; it can be switched very rapidly (typically less than ½ μs). Operation on a minor hysteresis loop only requires one toroid; intermediate values of phase-shift may be obtained by latching it to an intermediate or minor loop value of remanent magnetization.

7.9 Temperature Compensation Using Composite Circuits

The magnetization of ferrite materials, as already noted, depends on temperature and, indeed, vanishes at the so-called Curie temperature. Figure 7.17 illustrates the *B*–*H* loop characteristics of a YIG material at several temperatures. The use of this material in a microwave latching circuit is obviously incompatible with the realization of temperature-insensitive ferrite devices such as phase-shifters and circulators.

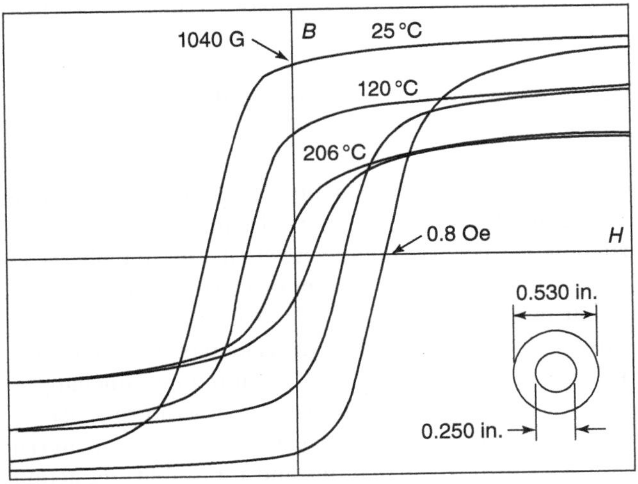

Figure 7.17 Major *B*–*H* loop characteristics of YIG at several temperatures. *Source:* Stern and Ince (1967).

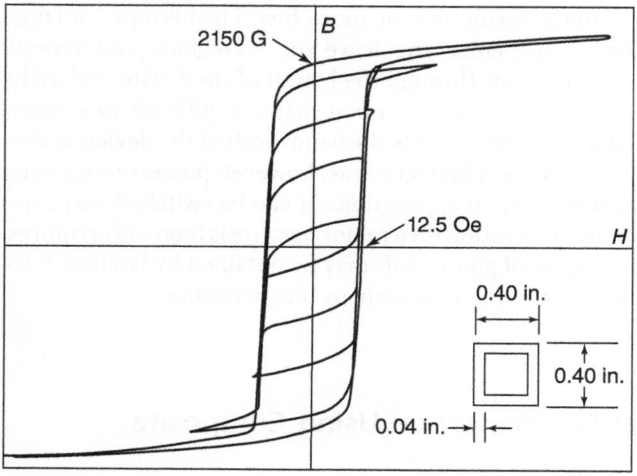

Figure 7.18 Major and minor *B–H* loops of nickel–cobalt ferrite at 25 °C. *Source:* Stern and Ince (1967).

One way to overcome this drawback is to utilize a composite magnetic circuit comprising a temperature-sensitive material within the microwave housing and a temperature-stable non-microwave material with a square hysteresis loop in the external circuit; the microwave material is operated in a minor hysteresis loop and the non-microwave one on its major loop (Figure 7.18). The operating point of the composite magnetic circuit is governed by the magnetic circuit relationships:

$$\mathrm{NI} = H_1 l_1 + H_2 l_2$$

which describes the magnetomotive force (mmf) around the circuit and by

$$\phi = B_1 A_1 = B_2 A_2$$

so that the total number of lines of magnetic induction (ϕ) through a given area (A) is a constant in each region of a closed magnetic circuit.

The mmf (NI) and load conditions of the composite circuit are initially set so that when the mmf is removed the microwave and non-microwave parts of the circuit are operated on their minor and major loops, respectively. As the temperature is raised above room temperature the hysteresis loop of the microwave material shrinks but the flux through the composite circuit remains unchanged until it coincides with the major hysteresis loop of the temperature-sensitive part of the circuit. Above that temperature value, the microwave material can no longer support the flux through the circuit and the stabilizing action of the composite circuit breaks down. The operation of the composite circuit is illustrated in Figure 7.19.

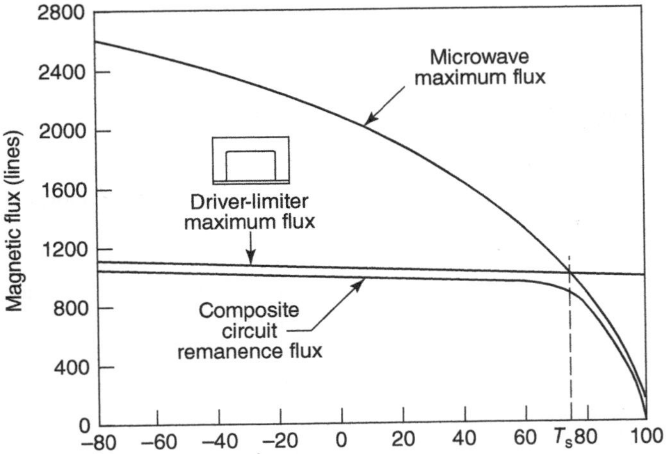

Figure 7.19 Maximum flux composite circuit versus temperature. *Source:* Stern and Ince (1967).

Bibliography

Allen, J.L. (1973). Phase and loss characteristics of high average power ferrite phasers. *IEEE Trans. Microw. Theory Tech.* **MTT-21**: 543–544.

Calvin, A. (1955). High power ferrite load isolators. *IRE. Trans. Microw. Theory Tech.* **MTT-3**: 38–43.

Degenford, J.E., Whicker, L.R., and Wantuch, E. (1967). Millimeter wavelength latching ferrite phase shifters. *NEREM Rec.* **9**: 60–61.

Fox, A.G., Miller, S.E., and Weiss, M.T. (1955). Behaviour and applications of ferrites in the microwave region. *Bell Syst. Tech. J.* **34**: 5.

Gardiol, F.E. (1970). Anisotropic slabs in rectangular waveguide. *IEEE Trans. Microw. Theory Tech.* **MTT-18**: 461–467.

Gardiol, F.E. (1973). Computer analysis of latching phase shifters in rectangular waveguide. *IEEE Trans. Microw. Theory Tech.* **MTT-21**: 57–61.

Ince, W.J. and Stern, E. (1967). Nonreciprocal remanence phase shifters in rectangular waveguide. *IEEE Trans. Microw. Theory Tech.* **MTT-15**: 87–95.

Ince, W.J. and Temme, D.H. (1969). Phasers and time delay elements. In: *Advances in Microwaves*, vol. **3** (ed. L. Young), 38–85. New York: Academic Press.

Ince, W.J., Temme, D.H., and Willwerth, F.G. (1971a). Toroid corner chamfering as a method of improving the figure of merit of latching ferrite phasers. *IEEE Trans. Microw. Theory Tech. (Correspondence)* **MTT-19**: 563–564.

Ince, W.J., DiBartolo, J., Temme, D.H., and Willwerth, F.G. (1971b). A comparison of two nonreciprocal latching phaser configurations. *IEEE Trans. Microw. Theory Tech. (Correspondence)* **MTT-19**: 105–107.

Jordan, A.K. (1962). Some effects of dielectric loading on ferrite phase shifters in rectangular waveguide. *IRE Trans. Microw. Theory Tech.* **MTT-10**: 83–84.

Lax, B. and Button, K. (1962). *Microwave Ferrites and Ferrimagnetics*. New York: McGraw-Hill.

Lax, B., Button, K.J., and Roth, L.M. (1954). Ferrite phase shifters in rectangular waveguide. *J. Appl. Phys.* **25**: 1413–1421.

Mizobuchi, A. and Kurebayashi, H. (1978). New configurations of the nonreciprocal remanence ferrite phase shifter. *IEEE MTT Symposium Digest*, Ottawa, ON, Canada (27–29 June 1978), pp. 97–99.

Rodrigue, G.P., Allen, J.L., Lavedan, L.J., and Taft, D.R. (1967). Operating dynamics and performance limitations of ferrite digital phase shifters. *IEEE Trans. Microw. Theory Tech. (1967 Symposium Issue)* **MTT-15**: 709–713.

Schlomann, E. (1960). On the theory of the ferrite resonance isolator. *IRE Trans. Microw. Theory Tech.* **MTT-8**: 199–206.

Schlomann, E. (1966). Theoretical analysis of twin-slab phase shifters in rectangular waveguide. *IEEE Trans. Microw. Theory Tech.* **MTT-14**: 15–23.

Seidel, B.H. (1957). Ferrite slabs in transverse electric mode waveguide. *J. Appl. Phys.* **28**: 218–226.

Soohoo, R.F. (1961). Theory of dielectric loaded and tapered field ferrite devices. *IRE Trans. Microw. Theory Tech.* **9**: 220–224.

Stern, R.A. and Ince, W.J. (1967). Design of composite magnetic circuits for temperature compensation of microwave ferrite devices. *IEEE Trans. Microw Theory Tech.* **MTT-15**: 295–300.

Vartanian, P.H. (1956). A broadband ferrite microwave isolator. *IRE Trans. Microw. Theory Tech.* **MTT-4**: 8–13.

Vartanian, P.H. and Jaynes, E.T. (1956). Propagation in ferrite-filled transversely magnetized waveguide. *IRE Trans. Microw. Theory Tech.* **MTT-4**: 140–143.

Weisbaum, S. and Boyet, M. (1956). Broadband nonreciprocal phase shifts – analysis of true ferrite slabs in rectangular guide. *J. Appl. Phys.* **27**: 519–524.

Weiss, M.T. (1956). Improved rectangular waveguide resonance isolators. *IRE Trans. Microw. Theory Tech.* **MTT-4**: 240–243.

Whicker, L.R. (1966). Recent advances in digital latching ferrite devices. *IRE International Convention Record*, New York (21–25 March 1966), pp. 49–57.

Whicker, L.R. and Boyd, C.R. Jr. (1971). A new nonreciprocal phaser for use at millimetre wavelengths. *IEEE MTT Symposium Digest*, Washington, DC (16–19 May 1971), pp. 102–103.

8

Edge Mode Phase-shifter

Joseph Helszajn[1] *and Henry Downs*[2]

[1] *Heriot Watt University, Edinburgh, UK*
[2] *Mega Industries, LLC, Gorham, ME, USA*

A classic feature of many magnetized ferrite transmission lines or waveguides is that of a nonreciprocal field or edge mode effect. One simple transmission line that readily exhibits this property is the ferrite-loaded parallel plate waveguide, illustrated in Figure 8.1a, with the direct magnetic field perpendicular to the direction of propagation. The power distribution in this waveguide is displaced towards one edge of the ferrite region in one direction of propagation and towards the other in the opposite direction. An edge mode phase-shifter is readily constructed by loading one or the other of the walls by some dielectric material; a nonreciprocal attenuator or isolator is likewise realized by lining one of the edges by a resistive wall. This class of structure also displays magnetic fields, which are nearly circularly polarized with opposite hands at each edge of the ferrite region. This property, however, is not essential for the description of the phase-shifter and isolator illustrated in Figure 8.1b. Such geometries therefore display both edge mode effects and planes of counterrotating circular polarization.

The simple model of the edge mode device described in this chapter indicates that the attenuation coefficient in the transverse plane is proportional to the product of the width of the ferrite region and the off-diagonal element of the tensor permeability. This suggests that the width of the parallel plate waveguide is fixed by the required decoupling between the two edges (say 10 or 15 dB) and that wide strips are required for design unless ferrite materials with large values of magnetization are employed.

Microwave Polarizers, Power Dividers, Phase Shifters, Circulators, and Switches,
First Edition. Joseph Helszajn.

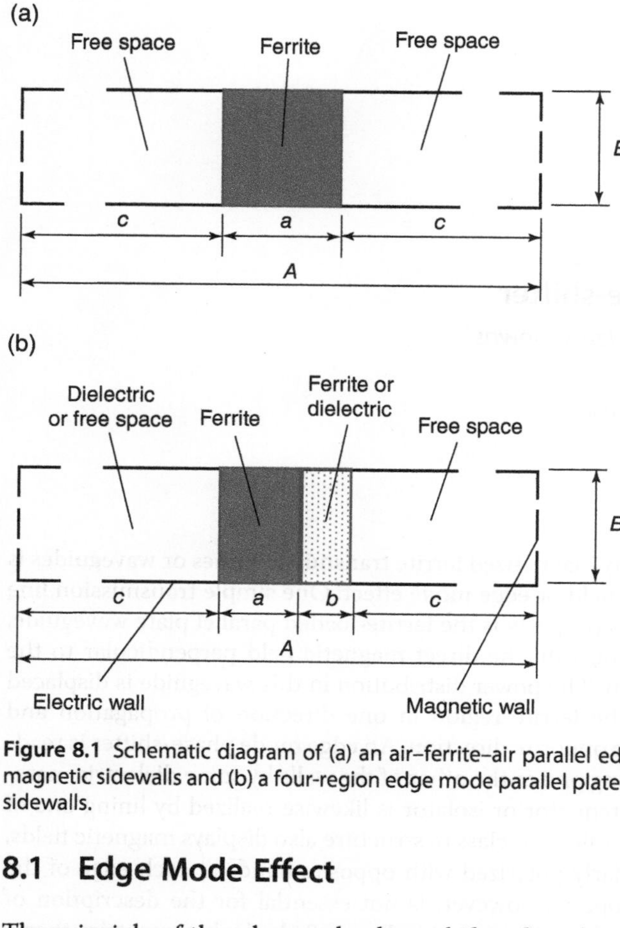

Figure 8.1 Schematic diagram of (a) an air–ferrite–air parallel edge mode prototype with magnetic sidewalls and (b) a four-region edge mode parallel plate waveguide with magnetic sidewalls.

8.1 Edge Mode Effect

The principle of the edge mode phase-shifter, first described by Hines, is well rehearsed and will only be reviewed here for tutorial purposes. The basic phenomena of this device may be demonstrated without difficulty by assuming a quasi-TE solution in the magnetized ferrite region:

$$E_y \neq 0,\;\; H_x \neq 0,\;\; H_z \neq 0 \tag{8.1a}$$

$$E_x = E_z = H_y = 0 \tag{8.1b}$$

and by assuming that the spatial variations of the field patterns are

$$\frac{\partial}{\partial x} = \alpha_x \tag{8.2a}$$

$$\frac{\partial}{\partial y} = 0 \tag{8.2b}$$

$$\frac{\partial}{\partial z} = -j\beta_z \tag{8.2c}$$

Taking the direct magnetic field perpendicular to the direction of propagation also gives the tensor permeability as suppressor:

$$[\mu] = \begin{bmatrix} \mu & 0 & -j\kappa \\ 0 & 1 & 0 \\ j\kappa & 0 & \mu \end{bmatrix} \tag{8.3}$$

Maxwell's first curl equation then gives

$$\begin{bmatrix} \bar{a}_x & \bar{a}_y & \bar{a}_z \\ -\alpha_x & 0 & -j\beta_z \\ 0 & E_y & 0 \end{bmatrix} = -j\omega\mu_0 \begin{bmatrix} \mu & 0 & -j\kappa \\ 0 & 1 & 0 \\ j\kappa & 0 & \mu \end{bmatrix} \begin{bmatrix} H_x \\ 0 \\ H_z \end{bmatrix} \tag{8.4}$$

Solving this equation for H_z and H_x, in terms of E, gives

$$\begin{bmatrix} H_x \\ H_z \end{bmatrix} = \frac{1}{j\omega\mu_0} \begin{bmatrix} H_x & -j\kappa \\ j\kappa & \mu \end{bmatrix} \begin{bmatrix} -j\beta_z & E_y \\ \alpha_x & E_y \end{bmatrix} \tag{8.5}$$

and

$$H_x = \frac{E_y}{\omega\mu_0\mu_{\text{eff}}}\left(-\beta_z + \alpha_x\frac{\kappa}{\mu}\right) \tag{8.6a}$$

$$H_z = \frac{jE_y}{\omega\mu_0\mu_{\text{eff}}}\left(\beta_z\frac{\kappa}{\mu} - \alpha_x\right) \tag{8.6b}$$

where

$$\mu_{\text{eff}} = \frac{\mu^2 - \kappa^2}{\mu}$$

The relationship between the separation constants is given with the aid of the wave equation by

$$\alpha_x^2 - \beta_z^2 + \omega^2\mu_0\mu_{\text{eff}}\varepsilon_f = 0 \tag{8.7}$$

The required result may now be demonstrated by placing a magnetic wall boundary condition at the plane $x = 0$ and assuming a unidirectional wave confined to this edge, i.e.

$$E_y = Ae^{(-\alpha_x x)}e^{(-j\beta_z z)} \tag{8.8}$$

$$H_x = 0 \quad \text{at} \quad x = 0 \tag{8.9}$$

Introducing the latter relationship in Eq. (8.6b) gives

$$\alpha_x = \frac{\kappa}{\mu}\beta_z \tag{8.10}$$

The dispersion equation is satisfied with μ_{eff} either positive or negative with

$$\beta_z = \omega\sqrt{\mu_0\varepsilon_0\varepsilon_{\text{f}}\mu} \tag{8.11}$$

$$\alpha_x = \omega\frac{\kappa}{\mu}\sqrt{\mu_0\varepsilon_0\varepsilon_{\text{f}}\mu} \tag{8.12}$$

The field components in the transverse plane therefore decay exponentially with a coefficient α_x proportional to κ/μ and H_z is zero everywhere:

$$E_y = Ae^{(-\alpha_x x)}e^{(-j\beta_z z)} \tag{8.13}$$

$$H_x = \zeta E_y \tag{8.14}$$

$$H_z = 0 \tag{8.15}$$

where ζ is the wave admittance.

$$\zeta = \sqrt{\frac{\varepsilon_0\varepsilon_{\text{f}}}{\mu_0\mu}} \tag{8.16}$$

The above model supports a TEM solution and displays a low-frequency cut-off number. Furthermore, since μ_{eff} does not enter directly into the description of this solution it may smoothly straddle the two regions where μ_{eff} is either positive or negative.

If the material is just saturated so that $H_0 - N_z M_0/\mu_0 \approx 0$ (as it must be if μ_{eff} is negative), then

$$\kappa \approx \frac{\omega_{\text{m}}}{\omega} \tag{8.17}$$

then

$$\mu = 1 \tag{8.18}$$

Equations (8.11) and (8.12) may also be written as

$$\beta_z = \omega\sqrt{\mu_0\varepsilon_0\varepsilon_{\text{f}}} \tag{8.19}$$

$$\alpha_x = \omega_{\text{m}}\sqrt{\mu_0\varepsilon_0\varepsilon_{\text{r}}} \tag{8.20}$$

The edge mode effect is therefore completely frequency independent and the decoupling between the two edges is merely dependent upon the relationship between ω_{m} and the width (a) of the ferrite section; the only frequency limitation in this class of device is thus the onset of higher-order modes.

A complete solution of the edge mode device also requires, of course, a definition of characteristic impedance. Impedances based on power–voltage and power–current relations are given by Hines.

8.2 Edge Mode Characteristic Equation

The principle of the edge mode effect in a simple ferrite region with magnetic sidewalls has been derived in closed form in Section 8.4. The model adopted in the discussion here is due to Bolle and Talisa (1979) and consists of imposing magnetic wall boundary conditions at two terminal planes sufficiently far removed from the main ferrite region to ensure convergence of the phase constants of the device. Its characteristic equation may be derived by forming the overall transmission matrix of the waveguide in the transverse plane. This method is somewhat more straightforward than that used historically, which relies on matching the fields at each boundary. For the four-region geometry illustrated in Figure 8.1b, the result is

$$\begin{bmatrix} A & B \\ C & D \end{bmatrix} = \begin{bmatrix} A_1 & B_1 \\ C_1 & D_1 \end{bmatrix} \begin{bmatrix} A_2 & B_2 \\ C_2 & D_2 \end{bmatrix} \begin{bmatrix} A_3 & B_3 \\ C_3 & D_3 \end{bmatrix} \begin{bmatrix} A_4 & B_4 \\ C_4 & D_4 \end{bmatrix} \tag{8.21}$$

The two outside free-space regions support decaying waves and the ***ABCD*** parameters of these regions are described by hyperbolic functions. The inner ferrite and dielectric regions support unattenuated waves and the corresponding parameters involve trigonometric variables. The required characteristic equation of the phase-shifter is then formed by imposing magnetic wall boundary conditions at the sidewalls of the waveguide. The result is

$$C = 0 \tag{8.22}$$

The split roots of this equation correspond to the phase constants of the device.

The field patterns and the power densities in the two directions of propagation are then calculated in the usual way from a knowledge of the corresponding phase constants.

8.3 Fields and Power in Edge Mode Devices

The absolute power down any waveguide is given by forming the Poynting vector in terms of the transverse components of the fields and the cross-section of the structure defined by its overall width (A) and ground plane spacing (B) by

$$P_t = \frac{1}{2}\mathrm{Re}\left[\int_0^A \int_0^B \left(E_t \times H_t^*\right)\right] d_x d_y \tag{8.23}$$

Both the fields and the power flow down the waveguide are evaluated in reduced units in order not to restrict the results to any particular frequency; the onset of spinwave instability in the ferrite region, which may occur at large power levels, may therefore be estimated from these data. P_t is evaluated numerically here by subdividing the overall cross-section into 500 subregions.

The power, P_t', in reduced units is formed from a knowledge of P_t by substituting k_0A and k_0B for A and B in the preceding integral:

$$P_t' = k_0^2 P_t \tag{8.24}$$

where

$$k_0 = \frac{2\pi}{\lambda_0} \tag{8.25}$$

P_t' is now evaluated numerically in terms of E_y in arbitrary units (say 1) at one of the magnetic sidewalls and its absolute value is subsequently set by equating P_t' to unity, i.e.

$$\frac{P_t'}{1} = \frac{1}{E_y^2} \tag{8.26}$$

In forming the power flow in reduced units it is also assumed that $k_0B = 1$, so that the results are readily scaled to any situation for which k_0B has some different value.

Figure 8.2 depicts the field and power distributions in the forward and backward directions of propagation for an air–ferrite–air configuration with magnetic sidewalls. It clearly illustrates both an edge mode effect and the quasi-circularly polarized nature of the alternating magnetic field at the planes of the ferrite and/or dielectric regions as well as outside them.

Some perspective of the scales used to display the power and field plots in these illustrations may be formed by noting that k_0A for ordinary rectangular waveguides lies between 3.8 and 5.7. The wide dimensions (A) of the edge mode waveguides studied here (in reduced units) have therefore been sized somewhat below the recommended frequency interval of standard waveguide, in order to cater for the ferrite loading:

$$k_0A = k_0(a + b + c + \cdots) = 2.50\,\text{rad} \tag{8.27a}$$

where a, b, c, etc., are the linear dimensions of the different sections of the edge mode waveguide.

The narrow dimension of the waveguide (B) is fixed throughout this work, as already noted, as

$$k_0B = 1 \tag{8.27b}$$

The aspect ratio of the waveguide

$$\frac{A}{B} = 2.50 \tag{8.27c}$$

is therefore of the order of standard waveguides.

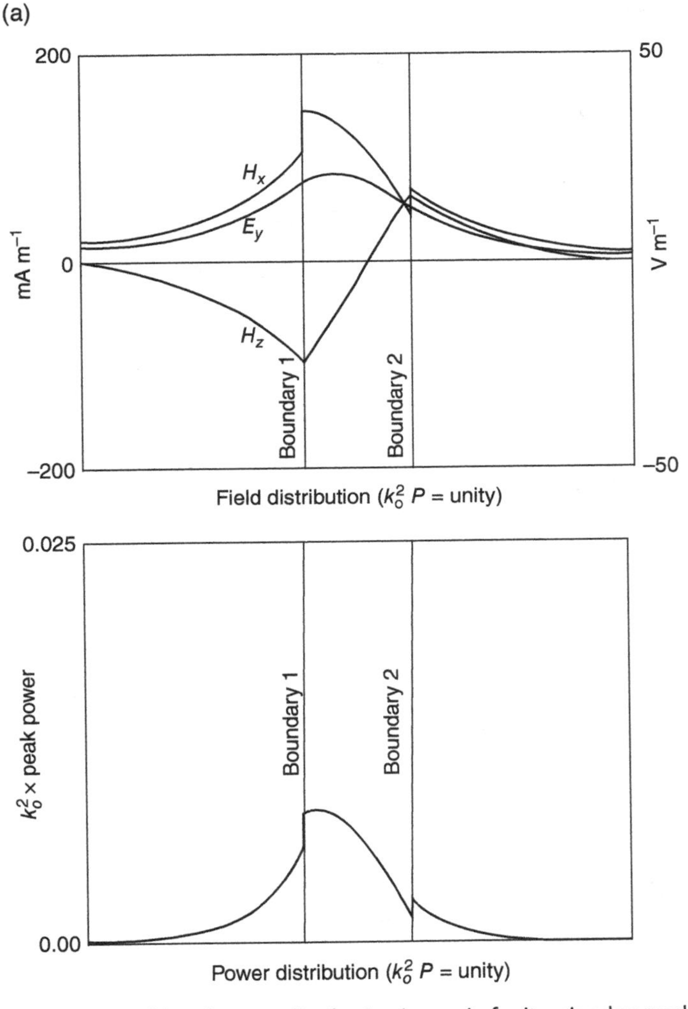

Figure 8.2 Field and power distribution in an air–ferrite–air edge mode prototype with magnetic sidewalls for propagation in (a) the forward direction and (b) the reverse direction with $k_0a = 0.50$, $c/a = 2$, $\varepsilon_f = 15$, and $\kappa = -0.40$. *Source:* Helszajn and Downs (1987).

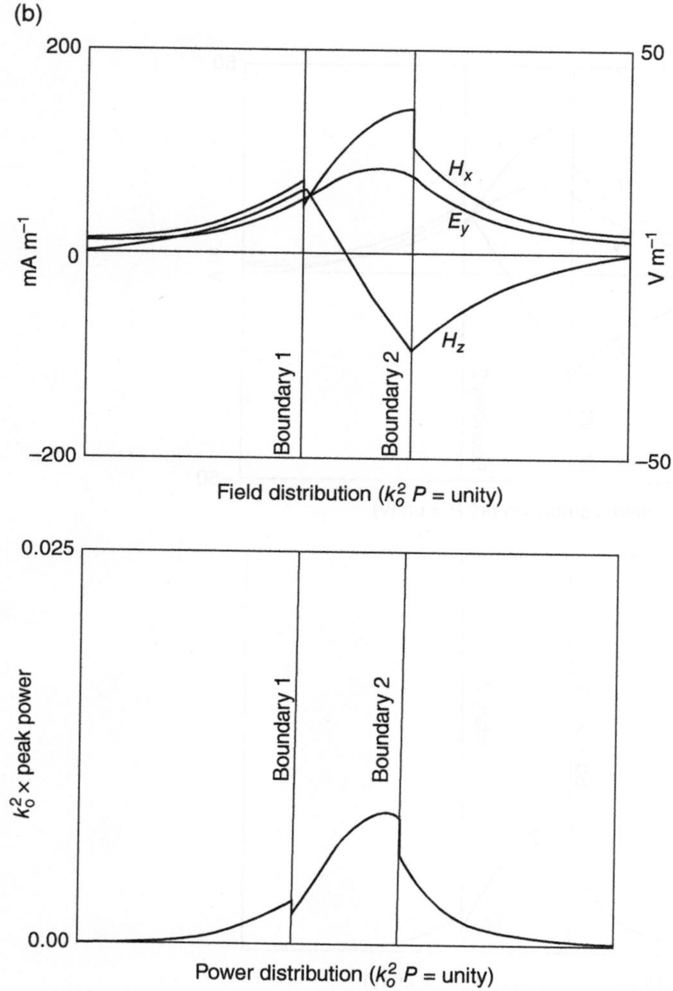

Figure 8.2 (Continued)

8.4 Circular Polarization and the Edge Mode Effect

Scrutiny of the magnetic fields of the geometry in Figure 8.2 indicates that it is nearly circularly polarized at the two planes between the ferrite–air regions. Inspection of the nature of the other solutions indicates similar features. Interestingly enough, the power flow is in each instance displaced towards the edge that exhibits the scalar permeability ($\mu + \kappa$) and away from the edge that displays the scalar permeability $(\mu - \kappa)$; $(\mu \pm \kappa)$ are the eigenvalues of the tensor permeability exhibited by counterrotating eigenvectors. The connection between the quality of the circular polarization and the edge mode effect is qualitatively understood. The polarization at the edges of the ferrite region and everywhere outside it in a demagnetized air–ferrite–air structure is a standard problem, provided the fields are assumed to decay outside the ferrite region and that the sidewalls are at plus and minus infinity. The result is described by

$$\frac{H_x}{H_z} = j\frac{\beta_z}{\alpha_{xa}} \tag{8.28}$$

where β_z is the propagation coefficient along the z direction in both the ferrite and free-space regions and α_{xa} is the x-directed attenuation constant in the outer air regions. Thus,

$$\beta_z^2 = \alpha_{xa}^2 - k_0^2, \quad \beta_{xa} = 0 \tag{8.29}$$

$$\beta_z^2 = k_0^2 \mu_e \varepsilon_f - \beta_{xf}^2, \quad \alpha_{xf} = 0 \tag{8.30}$$

Figure 8.3 indicates the polarization at the ferrite–air edge and everywhere outside it for a typical situation.

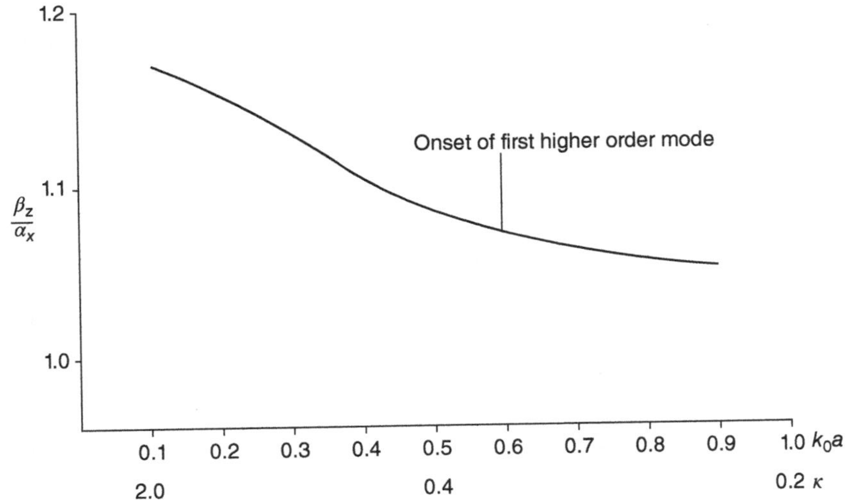

Figure 8.3 Quality of circular polarization versus frequency and k_0 for an air–ferrite–air edge mode prototype with $c/a = 2$.

8.5 Edge Mode Phase-shifter

The edge mode phase-shifter is constructed by loading one of the edges by a slow wave structure, usually a dielectric lamination with some different relative dielectric constant from that of the ferrite material. The simple theory of the edge mode device indicates that the ferrite region should be relatively wide and that its bandwidth can be very large. Figure 8.4 depicts a typical result for one value of ε_d, but it is not suggested that this solution is best in any sense of the word. Also superimposed on this illustration is the onset of the first higher-order quasi-TE mode in this type of waveguide. The maximum bound on the linear dimensions of this device is of course determined by the onset of the higher-order mode; the minimum bound by the realizable microwave specification. Octave band frequency intervals may, for instance, be specified from illustrations such as that in Figure 8.4 with κ between 0.10 and 0.20 (say), 0.20 and 0.40, 0.40 and 0.80, 0.30 and 0.60, and so forth, and with k_0a between similar intervals for each choice of a/b, ε_f, ε_d. The optimum solution is of course not clear-cut and is outside the remit of this work. Figure 8.5 indicates one result which straddles the two regions where μ_e is positive or negative ($2.70 \geq \kappa \geq 0.30$) but with a saturated material. This illustration suggests that there is a range of structures with ε_d between 5 and 15 that is suitable for the design of a 90° ± 4.5° phase bit over one octave band. It is noted, in passing, that well-formed planes of circular polarization exist on this geometry at both band edges of its frequency response. Figure 8.6 gives the field and power

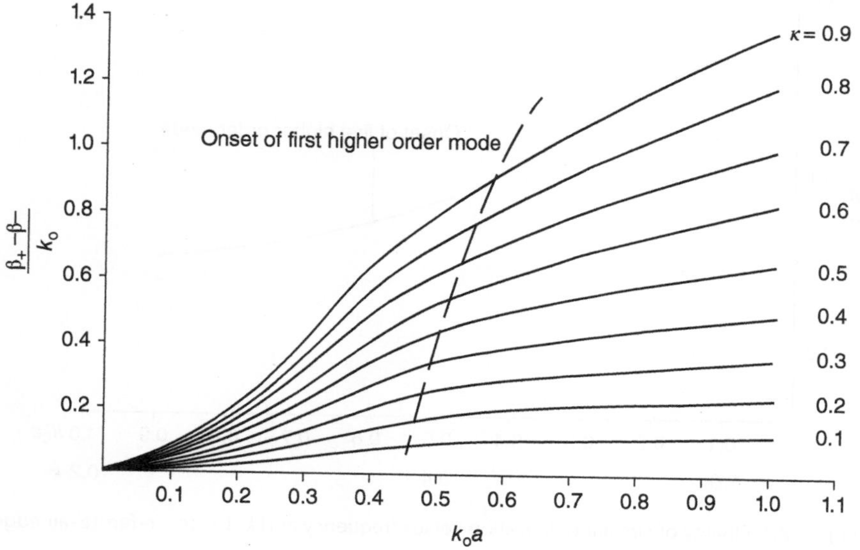

Figure 8.4 Normalized differential phase-shift for an air–ferrite–dielectric–air parallel plate edge mode waveguide with magnetic sidewalls with $\varepsilon_f = 15$, $\varepsilon_d = 25$, $a/b = 4$, $c/(a+b) = 2$, and parametric values of κ.

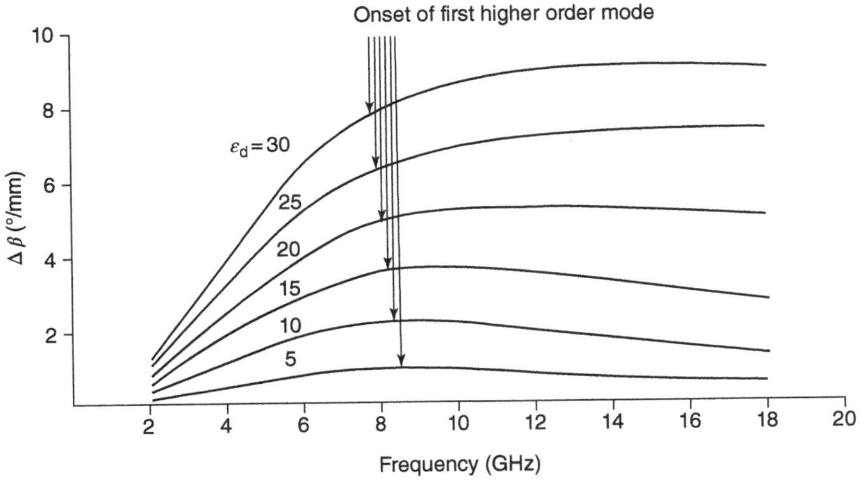

Figure 8.5 Frequency response of air–ferrite–dielectric–air edge mode differential phase-shifter with magnetic sidewalls with $\varepsilon_f = 15$, $\varepsilon_d = 25$, $a/b = 4$, $c/(a+b) = 2$, and parametric values of ε_d.

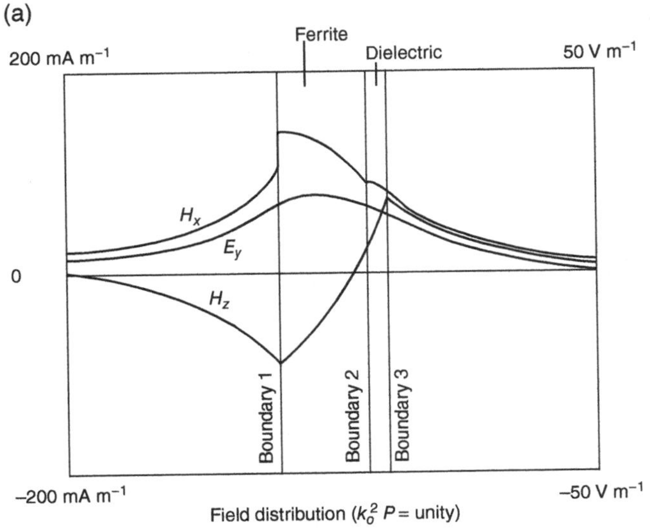

Figure 8.6 Field and power distributions in an air–ferrite–dielectric–air parallel plate edge mode waveguide with magnetic sidewalls for propagation in (a) the forward direction and (b) the reverse direction with $k_0a = 0.40$, $k_0b = 0.10$, $c/(a+b) = 2$, $\varepsilon_f = 15$, $\varepsilon_d = 25$, and $\kappa = -0.40$.

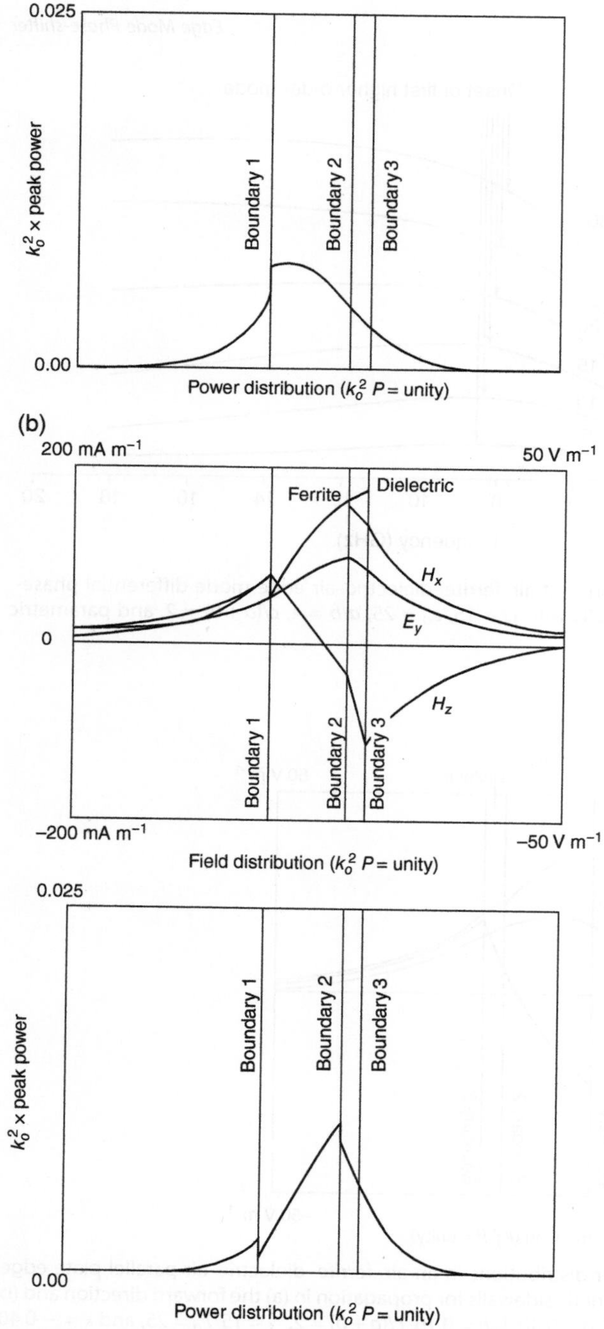

Figure 8.6 (Continued)

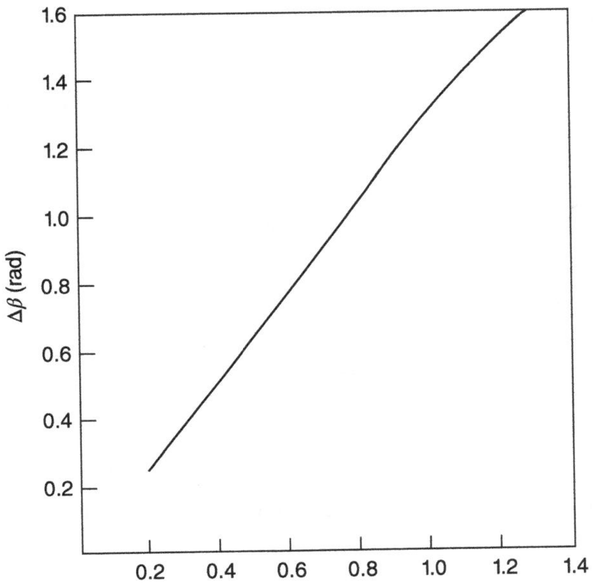

Figure 8.7 Differential phase-shift for an air–ferrite–dielectric–air parallel plate edge mode waveguide with magnetic sidewalls with $\varepsilon = 15$, $\varepsilon_d = 15$, $k_0a = 2.5$, $a/b = 4$, and $c/(a + b) = 2$.

distributions for the two directions of propagation and Figure 8.7 the differential phase shift in this type of waveguide.

The *ABCD* description of the problem treated here does not cater for higher-order modes; the data in this and the other illustrations therefore do not apply under that situation unless some suitable mode suppressor is utilized.

8.6 Edge Mode Isolators, Phase-shifters, and Circulators

The edge feature in a wide stripline or microstrip line has, in practice, been used in the construction of wideband isolators, phase-shifters, and three- and four-port junction circulators. The problem, in each case, is the transition between a standard stripline or microstrip circuit with a narrow strip width and that of the edge circuit with a large strip width. Figure 8.8 illustrates some typical microstrip configurations.

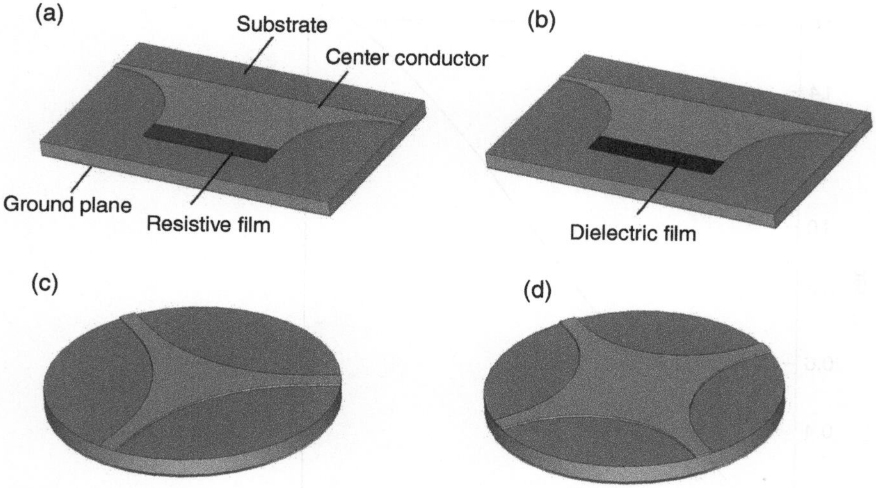

Figure 8.8 Edge mode (a) isolator, (b) phase-shifter, (c) three-port circulator, and (d) four-port circulator.

Bibliography

Bolle, D.M. (1976a). The edge guide mode on ferrite loaded stripline. *IEEE MTT-S International Microwave Symposium*, Cherry Hill, NJ (June 1976), Digest of Papers, pp. 257–259.

Bolle, D.M. (1976b). The peripheral or edge modes on the inhomogeneously and homogeneously ferrite loaded stripline. *Proceedings of 6th European Microwave Conference*, Rome, Italy (September 1976), pp. 560–564.

Bolle, D.M. and Talisa, S.H. (1979). The edge guide mode nonreciprocal phase shifter. *IEEE Trans. Microw. Theory Tech.* **MTT-27**: 878.

Chiron, B., Forterre, G., and Rannou, C. (1971). Nouveaux dispositifs non réciproques à très grande largeur de bande utilisant des ondes de surface électromagnétiques. *Onde Électr.* **51** (9): 816–818.

Clark, W.P., Hering, K.N., and Charlton, D.A. (1966). TE mode solutions for partially ferrite filled rectangular waveguide using ABCD matrices. *IEEE Int. Conv. Rec. (USA)* **14**: 39–48.

Cohn, M. (1966). Propagation in a dielectric loaded parallel plane waveguide. *IRE Trans. Microw. Theory Tech.* **MTT-7**: 202–208.

Courtois, L., Chiron, B., and Forterre, G. (1974). Improvement in broadband ferrite isolators. In: *Proceedings of 1974 AIP Conference on Magnetism and Magnetic Materials*, 501–502. New York: American Institute of Physics.

Courtois, L., Bernard, N., Chiron, B., and Forterre, G. (1976). A new edge mode isolator in the very high frequency range. *IEEE Trans. Microw. Theory Tech.* **MTT-24** (3): 129–135.

De Santis, P. (1976). Edge guided waves five years later. *IEEE MTT-S International Microwave Symposium*, Cherry Hill, NJ (June 1976), Digest of Papers, pp. 248–250.

De Santis, P. (1978). *A Unified Treatment of Edge Guided Waves*. NRL Report 8158. Washington, DC: Naval Research Laboratories.

Dydyk, M. (1977a). Edge guide: one path to wideband isolator design, part 1. *Microwaves* (February): 54–58.

Dydyk, M. (1977b). Edge guide: one path to wideband isolator design, part 2. *Microwaves* (January): 50–56.

Helszajn, J. and Downs, H. (1987). Field displacement, circular polarization, scalar permeabilities and differential phase shift in edge mode ferrite devices. *Microwellen Mag.* **14** (3): 269–278.

Helszajn, J., Murray, R.W., Davidson, E.G.S., and Suttie, R.A. (1977). Microwave subsidiary resonance ferrite limiters. *IEEE Trans. Microw. Theory Tech.* **MTT-25**: 190–195.

Hines, M.E. (1971). Reciprocal and nonreciprocal modes of propagation in ferrite stripline and microstrip devices. *IEEE Trans. Microw. Theory Tech.* **MTT-19**: 442–451.

Talisa, S.H. and Bolle, D.M. (1971). On the modelling of the edge guided mode stripline isolators. *IEEE Trans. Microw. Theory Tech.* **MTT-27** (6): 584–591.

9

The Two-port On/Off *H*-plane Waveguide Turnstile Gyromagnetic Switch

Joseph Helszajn[1], Mark McKay[2], Alicia Casanueva[3], and Angel Mediavilla Sánchez[3]

[1] *Heriot Watt University, Edinburgh, UK*

[2] *Honeywell, Edinburgh, UK*

[3] *Communication Engineering Department, University of Cantabria, Santander, Spain*

9.1 Introduction

The two-port on/off switch described in this chapter is a tee junction consisting of an *E*-plane Faraday rotation section at the junction of two *H*-plane rectangular waveguides. The junction is often referred to as an *H*-plane geometry in the literature, but the convention adopted here is that introduced by Dicke.

The arrangement under consideration is illustrated in Figure 9.1. The structure has a passband when the side port is 90° long. It has a stopband when it is 180° long. The former is here obtained, provided the demagnetized Faraday rotator is 90° long. The latter is achieved by rotating the polarization on the rotator by 90° in the positive direction of propagation and a further 90° rotation in the reverse direction for a total of 180°

The two-port switch supports, in keeping with all microwave circuits, a multiplicity of higher-order solutions. The chapter includes one example which displays a passband in the demagnetized state and a stopband in the magnetized state.

9.2 Two-port *H*-plane Turnstile On/Off Switch

The *H*-plane tee junction is also known as a series tee junction. A subset of the Dicke junction is the two-port geometry obtained by closing the side port. Its schematic diagram is depicted in Figure 9.2.

Microwave Polarizers, Power Dividers, Phase Shifters, Circulators, and Switches, First Edition. Joseph Helszajn.

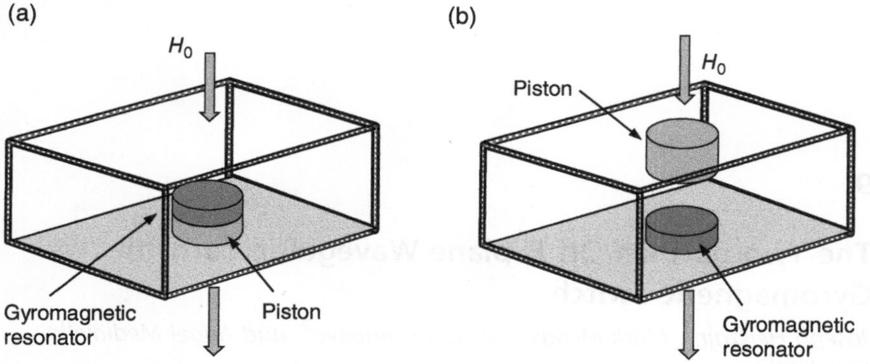

Figure 9.1 Two-port *H*-plane waveguide reflection switches using reentrant (a) and inverted reentrant (b) single quarter-wave-long gyromagnetic resonators.

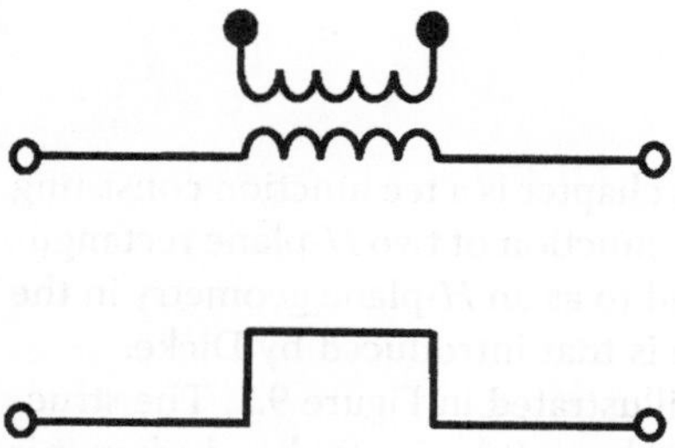

Figure 9.2 *H*-plane series tee junction.

The operation of the on or demagnetized and off or magnetized switch is shown in Figure 9.3a and b. The first topology produces a series lumped element resonator between the ports of the main waveguide. The second produces a shunt circuit there. The effect of rotating the polarization of the rotator by 90° in the forward direction of propagation and a further 90° in the reverse one for a total of 180° by magnetizing the junction is to add a 90° unit element (UE) in cascade with that of the demagnetized rotator. The even mode of the circuits, not shown, does not propagate along the gyromagnetic waveguide. As the odd mode coupling is between the transverse magnetic fields in the main waveguide and rotator section it is necessary to introduce an additional 90° UE at the secondary terminals of the transformer to account for these being in quadrature with the longitudinal magnetic field for the even mode.

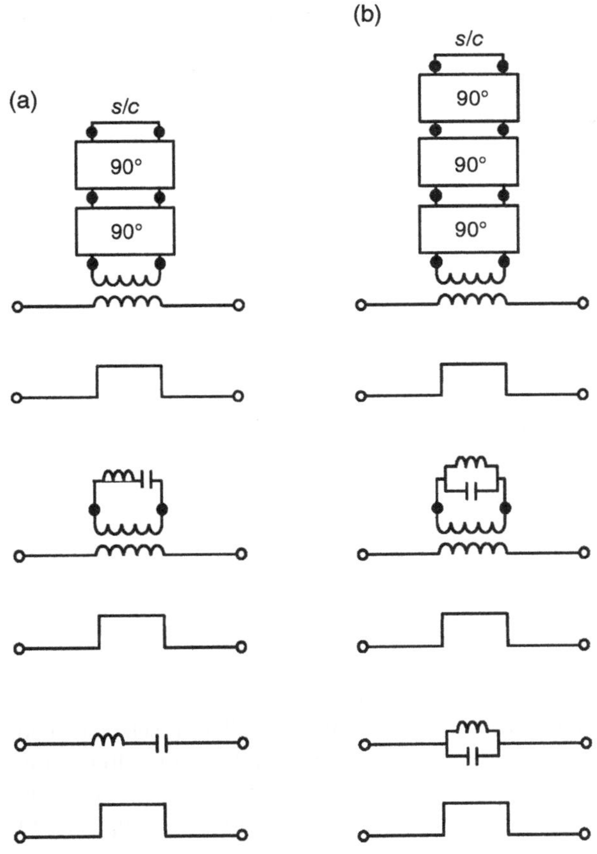

Figure 9.3 Lumped element models of *H*-plane switches: (a) demagnetized and (b) magnetized resonators.

9.3 Even and Odd Eigenvectors of *E*-plane Waveguide Tee Junction

The adjustment of the switch is facilitated by decomposing a single generator setting at one port into in-phase and out-of-phase (even and odd) eigenvectors settings at both ports. The even or in-phase eigenvector establishes a finite electric field across the symmetry plane of the rectangular waveguides and a null in the alternating magnetic field across the open flat face of the circular waveguide; it does not produce any propagation along the side waveguide. Its geometry is a two-layer planar circular resonator consisting of a dielectric region with a

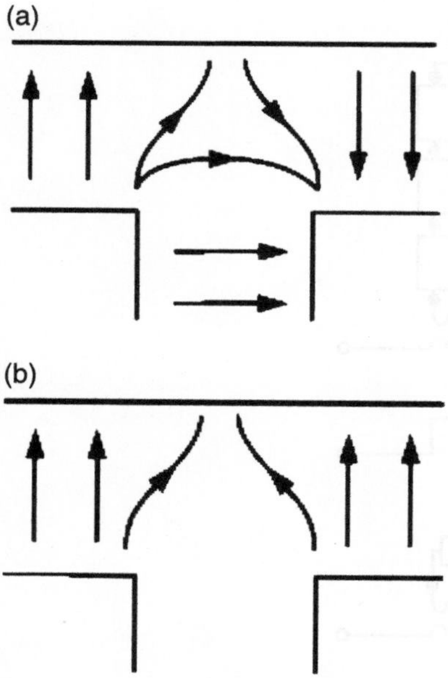

Figure 9.4 *E*-plane waveguide tee junction: (a) odd eigenvector and (b) even eigenvector.

dielectric constant ε_f and a gap region with a dielectric constant ε_d. The odd or out-of-phase eigenvector, on the other hand, produces a null in the electric field at the symmetry plane of the rectangular waveguides and a finite alternating magnetic field across the circular one, which may be decomposed into counter-rotating fields along the side waveguide. The two situations are depicted in Figure 9.4.

9.4 Eigenvalue Adjustment of Turnstile Plane Switch

The scattering parameters of the two-port arrangement are specified in terms of the even and odd reflections eigenvalues of the geometry:

$$S_{11} = \frac{\rho_{\text{even}} + \rho_{\text{odd}}}{2} \tag{9.1a}$$

$$S_{21} = \frac{\rho_{\text{even}} - \rho_{\text{odd}}}{2} \tag{9.1b}$$

where

$$\rho_{\text{even}} = 1.\exp -j2(\theta_{\text{even}}) \tag{9.2a}$$

$$\rho_{odd} = 1.\exp - j2\left(\theta_{odd} + \frac{\pi}{2}\right) \tag{9.2b}$$

θ_{even} and θ_{odd} are the electrical lengths of the UEs. θ_{even} or θ_{odd} are zero on the axis of the junction and 90° at the edge of the resonator. On the axis ρ_{even} or ρ_{odd} is −1 for a *s/c* UE and ρ_{even} or ρ_{odd} is +1 for an *o/c* UE.

The reflection eigenvalues coincide with even and odd voltage settings at the ports of the network. The eigen-networks at the input terminals obtained in this way are indicated in Figure 9.5.

The odd mode propagates along the Faraday rotation section whereas the even one does not. The former has an additional angle $\theta°$ in the positive direction of propagation, a similar angle in the reverse direction, and a 180° phase reversal at the short-circuit plate. The even mode does not propagate along the Faraday rotation section.

The passband condition coincides with

$$\rho_{even} = -1, \quad \theta_{even} = \frac{\pi}{2} \tag{9.3a}$$

$$\rho_{odd} = +1, \quad \theta_{odd} = \frac{\pi}{2} \tag{9.3b}$$

This gives

$$S_{11} = 0 \tag{9.4a}$$

$$S_{21} = -1 \tag{9.4b}$$

The stopband condition is obtained for this network, provided

$$\rho_{even} = -1, \quad \theta_{even} = \frac{\pi}{2} \tag{9.5a}$$

$$\rho_{odd} = -1, \quad \theta_{odd} = \frac{\pi}{2} \tag{9.5b}$$

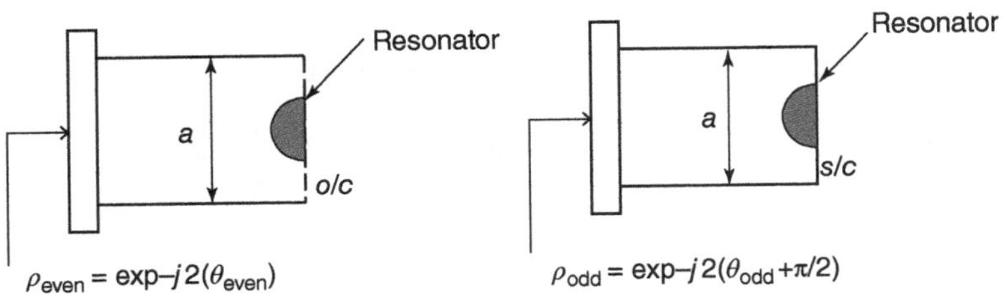

Figure 9.5 Plan view of even and odd eigen-networks of an *H*-plane two-port reflection switch.

This gives

$$S_{11} = -1 \tag{9.6a}$$

$$S_{21} = 0 \tag{9.6b}$$

The eigenvalues under considerations are illustrated in Chapter 11.

9.5 Eigen-networks

The nature of the even and odd reflection coefficients of the demagnetized and magnetized H-plane switch may be deduced by bisecting its equivalent circuit by electric and magnetic walls. The eigen-networks obtained thus are indicated in Figures 9.6 and 9.7. The reflection coefficients are in the first instance are −1 and +1 at the edge of the resonator. The latter conditions are compatible with a passband frequency response. The reflection coefficients of the magnetized switch are both −1. The latter quantities are compatible with a stopband frequency response.

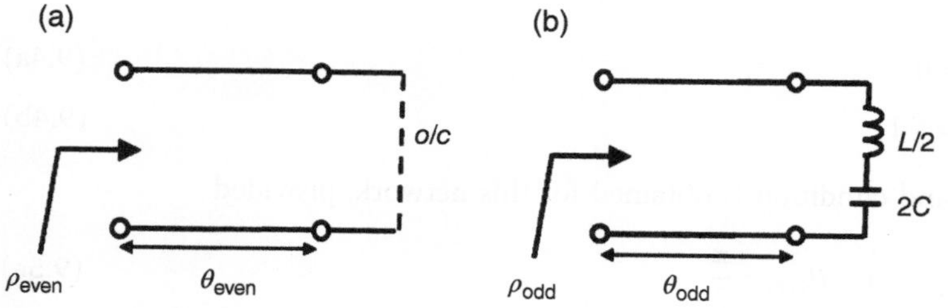

Figure 9.6 (a) Even mode demagnetized eigen-network. (b) Odd mode demagnetized eigen-network.

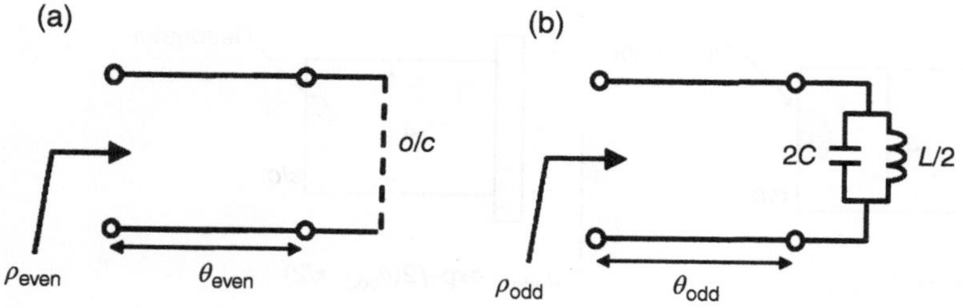

Figure 9.7 (a) Even mode magnetized eigen-network. (b) Odd mode magnetized eigen-network.

9.6 Numerical Adjustments of Passbands

The eigenvalue adjustment of the switch starts by making use of the connection between the reflection eigenvalues and the scattering parameters of the junction at the reference plane of the junction:

$$\rho_{\text{even}} = S_{11} + S_{21} \tag{9.7a}$$

$$\rho_{\text{odd}} = S_{11} - S_{21} \tag{9.7b}$$

The relationship between k_0R and q_{eff} is not unique. It continues by constructing polynomial solution $P(k_0R)$ and $Q(k_0R)$ connecting q_{even} and q_{odd} to k_0R at which $\rho_{\text{even}} = -1$ and $\rho_{\text{odd}} = +1$.

$$q_{\text{even}} = P(k_0R), \quad \rho_{\text{even}} = -1, \quad \frac{R}{L} = \text{constant}, \quad k_0 = \text{constant} \tag{9.8a}$$

$$q_{\text{odd}} = Q(k_0R), \quad \rho_{\text{odd}} = +1, \quad \frac{R}{L} = \text{constant}, \quad k_0 = \text{constant} \tag{9.8b}$$

The required characteristic equation at the reference plane of the resonator for the unknown product k_0R is

$$P(k_0R) - Q(k_0R) = 0 \tag{9.9}$$

This condition is satisfied, provided

$$q_{\text{eff}} = q_{\text{even}} = q_{\text{odd}} \tag{9.10}$$

where

$$q_{\text{even}} = \frac{L}{L + S_{\text{even}}} \tag{9.11a}$$

$$q_{\text{odd}} = \frac{L}{L + S_{\text{odd}}} \tag{9.11b}$$

$$q_{\text{eff}} = \frac{L}{L + S} \tag{9.11c}$$

k_0R is obtained by solving the characteristic equation in Eq. (9.9) and q_{eff} is obtained thereafter by having recourse to Eq. (9.10). The reference plane is in the process restricted to the terminals of the resonator. Repetitive calibrations of the reference plane of the junction are minimized by varying q_{eff} for parametric values of k_0R rather than the converse choice. A typical calculation involves partitioning the k_0R interval into m segments and the q_{eff} one into n segments. A typical solution is obtained with $m = 6$ and $n = 4$ implying 24 problem drawings and six calibration steps.

The adjustment procedure for either process starts by fixing the rectangular waveguide, the frequency of the switch under consideration, and the dielectric constant of the resonator:

$$\text{WR75}, \quad a = 2b$$

$$f_0 = 13.25\,\text{GHz}$$

$$\frac{f_0}{f_c} = 1.68$$

$$\varepsilon_f = 15.0$$

A passband is here taken by way of an example. A stopband is obtained by reversing the sign of ρ_{odd}.

$$\rho_{even} = -1$$

$$\rho_{odd} = +1$$

k_0 and R/L are here the independent variables and k_0R and the gap factor q_{eff} as the dependent ones.

$$k_0 = 0.277\,\text{rad}\,\text{mm}^{-1},$$

$$\frac{R}{L} = 2.0$$

The solution here is q_{eff}= 0.5364, k_0R= 0.807

Figure 9.8a shows the comparison between calculated and experimental return and insertion loss (Helszajn et al. 2010). Figure 9.8b depicts the stopband frequency response of the two-port on–off gyromagnetic switch (Helszajn et al. 2010).

9.7 An Off/On *H*-plane Switch

An off/on *H*-plane switch is also readily realized. The off state is here obtained by replacing the 90° demagnetized rotator section by a 180° one. The on state is achieved by rotating the polarization by 90° by magnetizing the junction. The latter operation adds an additional 90° section to the arrangement thereby producing a passband between the ports. Figure 9.9 illustrates the developments of both the stop and passband obtained in this way.

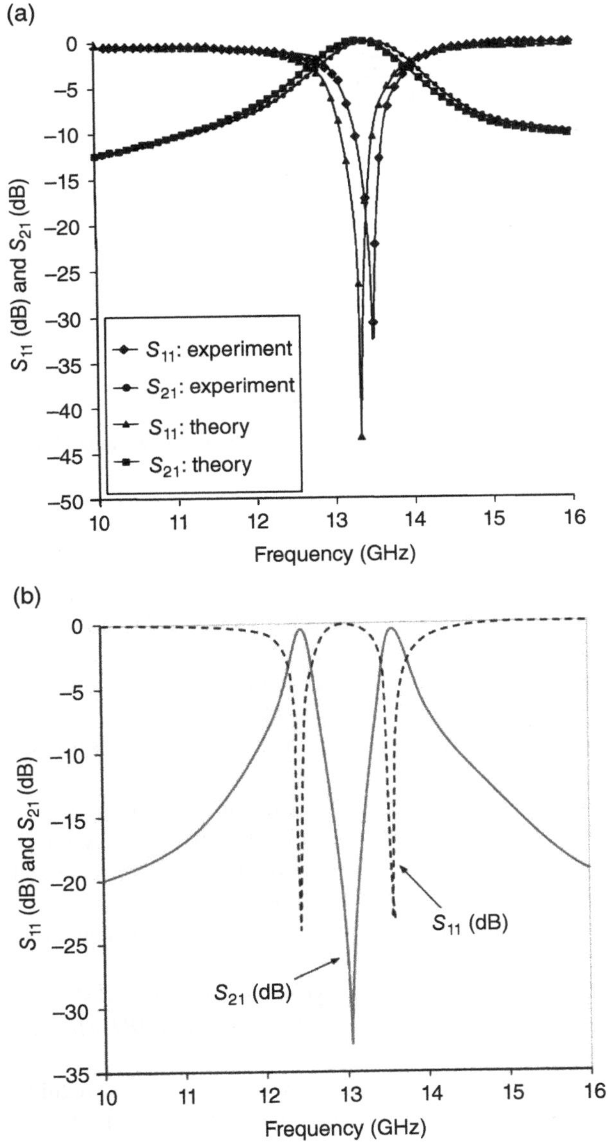

Figure 9.8 (a) Comparison between calculated and experimental passband return loss and insertion loss ($R/L = 2.0$, $k_0R = 0.807$, $q_{eff} = 0.5364$) (Helszajn et al. 2010). (b) Experimental stopband frequency response of two-port on–off gyromagnetic switch ($B_0/\mu_0 M_0 = 0.34$) (Helszajn et al. 2010).

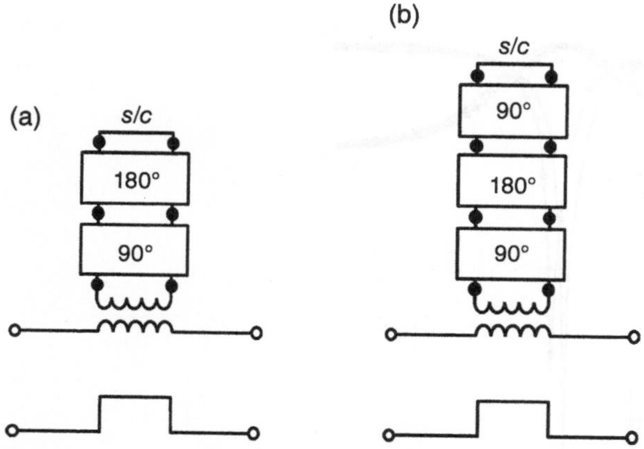

Figure 9.9 Off and On circuits of *H*-plane switches using 180° long (a) demagnetized and (b) magnetized resonators.

Bibliography

Akaiwa, Y. (1974). Operation modes of a waveguide Y-circulator. *IEEE Trans. Microw. Theory Tech.* **MTT-22**: 954–959.

Clavin, A. (1963). Reciprocal and nonreciprocal switches utilizing ferrite junction circulators. *IEEE Trans. Microw. Theory Tech.* **MTT-11**: 217–218.

Freiberg, L. (1961). Pulse operated circulator switch. *IRE Trans. Microw. Theory Tech.* **9** (3): 266–266.

Goodman, P.C. (1965). A latching ferrite junction circulator for phased array switching applications. *IEEE, GMTT Symposium*, Clearwater, FL (5–7 May 1965).

Green, J.J. and Sandy, F. (1974). Microwave characterization of partially magnetized ferrites. *IEEE Trans. Microw. Theory Tech.* **MTT-22**: 541–645.

Helszajn, J. (1994). Experimental evaluation of junction circulators: a review. *Proc. IEE Microw. Antennas Propag.* **141** (5): 351–358.

Helszajn, J. and Sharp, J. (2012). Cut-off space of a gyromagnetic planar disk resonator with a triplet of stubs with up and down magnetization. *IET Microw. Antennas Propag.* **6** (5): 569–576.

Helszajn, J., Casanueva, A., Mediavilla, A. et al. (2010). A 2-port WR75 waveguide turnstile gyromagnetic switch. *IEEE Trans. Microw. Theory Tech.* **MTT-58**: 1485–1492.

Katoh, I., Konishi, H., and Sakamoto, K. (1980). A 12 GHz broadband latching circulator. *Conference Proceedings, European Microwaves*, Poland, pp. 360–364.

Passaro, W.C. and McManus, J.W. (1966). A 35 GHz latching switch. Presented at the IEEE International Microwave Symposium, Palo Alto, CA (19 May 1966).

10

Off/On and On/Off Two-port *E*-plane Waveguide Switches Using Turnstile Resonators

Joseph Helszajn[1], *Mark McKay*[2], *and John Sharp*[3],

[1] *Heriot Watt University, Edinburgh, UK*
[2] *Honeywell, Edinburgh, UK*
[3] *Napier University, Edinburgh, UK*

10.1 Introduction

The operation of the *E*-plane waveguide tee junction is the topic of this chapter. It consists of a quarter-wave long 90° Faraday rotator section on one narrow waveguide wall separated from the other by a gap. Its schematic diagram is illustrated in Figure 10.1a. The junction is often referred to as an *E*-plane geometry in the literature, but the convention adopted here is that introduced by Dicke. The dual *E*-plane geometry obtained by placing the resonator on the wide wall of the waveguide is also shown for completeness sake in Figure 10.1b (Yoshida 1959). The demagnetized tee junction under consideration experimentally supports stop and pass band filter characteristics when demagnetized and either off/on or on/off switching characteristics when magnetized. The large-gap demagnetized geometry is a stop band filter, the small one is a pass band. The frequency transformation between the two has been demonstrated by Omori (1968). The filters are converted into off/on and on/off switches by replacing the dielectric resonators by 90° Faraday rotators. The stop band filter is here mapped into an off/on switch, the pass band filter into an on/off one. An understanding of the operation of this class of circuit requires an appreciation of the eigenvalue problem. It supports one-port even and odd eigen-networks and reflection coefficients obtained by bisecting the network by electric and magnetic walls. Stop bands are obtained between the rectangular waveguide ports whenever the even and odd reflection coefficients are in phase. Pass bands exist whenever the reflection coefficients are out of phase. The even eigen network is a two-layer quasi-planar resonator characterized by a gap-dependent dielectric

Microwave Polarizers, Power Dividers, Phase Shifters, Circulators, and Switches, First Edition. Joseph Helszajn.

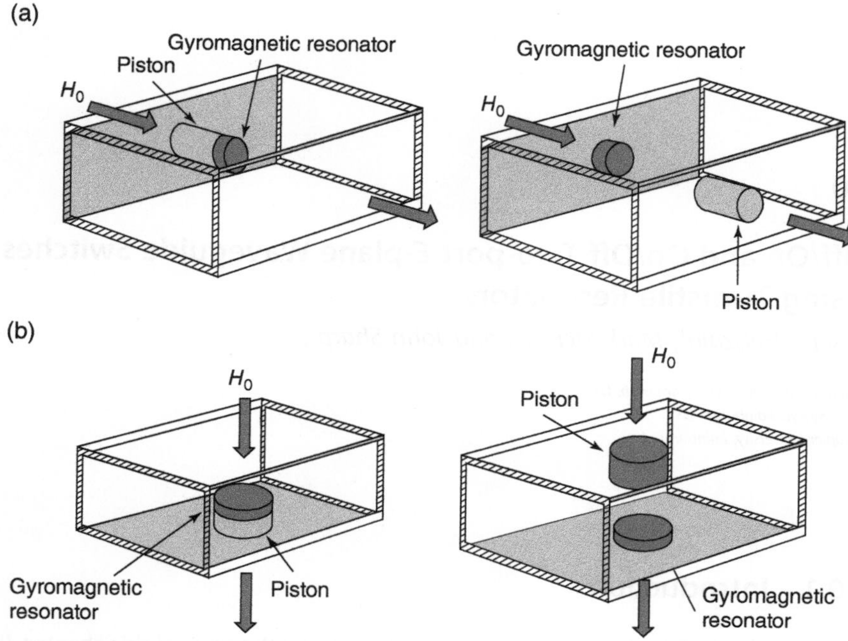

Figure 10.1 Schematic diagrams of reentrant and inverted reentrant ferrite switches using (a) *E*-plane tee junctions and (b) *H*-plane tee junctions.

constant and it supports propagation along the dielectric or gyromagnetic resonator. The odd eigen-network does not propagate along the Faraday rotator section and is unaffected by the gap. The dynamic range of the gap-dependent dielectric constant is bracketed between that of free space and that of the ferrite insulator. The off/on and on/off states may be exchanged by shifting either the even or odd reflection coefficients by 180°. The gap of the junction provides a means of reversing the even one; replacing the 90° rotator by a 180° section steps the even reflection coefficient by a similar amount. The chapter includes one numerical calculation on the frequency responses of an off/on switch.

10.2 The Shunt *E*-plane Tee Junction

The *E*-plane tee junction dealt with here is also known as a shunt circuit. Its equivalent circuit is shown in Figure 10.2. A short-circuit 90° unit element (UE) will produce an *o/c* at the secondary and primary terminals of the ideal transformer. However, the even mode coupling is from the longitudinal magnetic field in the main waveguide to the transverse magnetic field in the rotator section and it is necessary to introduce an additional 90° UE at the secondary

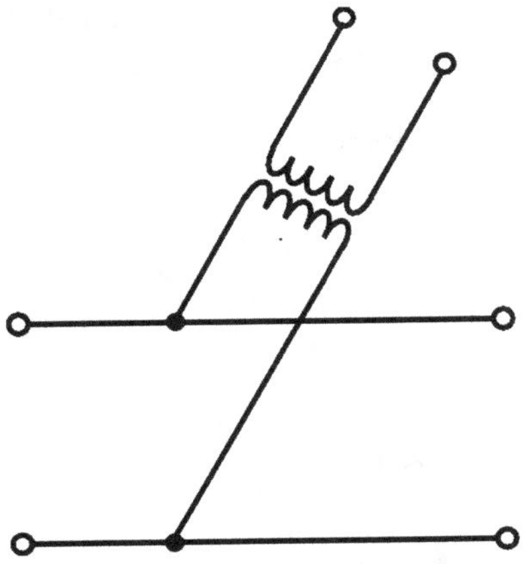

Figure 10.2 Shunt *E*-plane tee junction.

terminals to account for the fields being in quadrature. The net effect is to produce a s/c at the secondary and primary terminals of the transformer and a stop band between the ports of the demagnetized junction.

The effect of rotating the polarization of the Faraday rotator by 90° by magnetizing the junction is to introduce an additional 90° UE in cascade with that of the rotator. The load at the secondary and primary terminals of the transformer

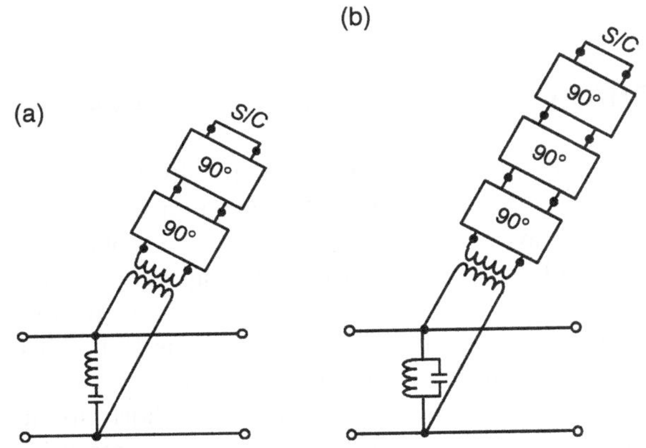

Figure 10.3 (a) *E*-plane stop band (demagnetized). (b) *E*-plane pass band (magnetized).

is now an open-circuit. The net result is a pass band between the input and output terminals of the waveguide. The two steps are shown in Figure 10.3a and b.

The demagnetized stop band may be converted to a pass band by replacing the 90° Faraday rotation section in the side waveguide by a 180° one. The stop band thereafter is obtained by rotating the polarization of the Faraday rotation section by 90°. This establishes an additional 90° section in cascade with that of the demagnetized one for a total of 360° resulting in a stop band between the ports. The two steps are indicated in Figure 10.4a and b.

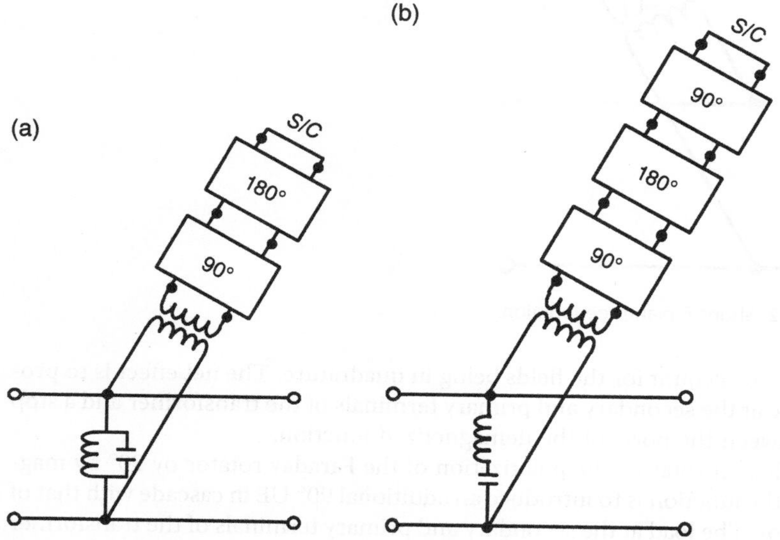

Figure 10.4 (a) *E*-plane pass band (demagnetized). (b) *E*-plane stop band (magnetized).

10.3 Operation of Off/On and On/Off *E*-plane Switches

The off/on switch is a large gap or non-evanescent geometry. Its off state is obtained by optimizing the details of a quarter-wave long demagnetized Faraday section on one narrow wall of the waveguide together with the details of a two-layer quasi-planar resonator. The latter fixes the size of the gap between the open face of the resonator and the opposite narrow waveguide wall or piston. The reflection coefficient of the rotator section at its input plane is 180°. This angle is made up of a 180° angle in the forward direction of propagation, a further 180° angle in the reverse direction, and a 180° angle at the short-circuit plate. The demagnetized Faraday rotation section has, therefore, no essential effect on the odd reflection coefficient at the symmetry plane of the junction. The odd reflection coefficient is here +1 at the terminals of the resonator. It is +1 in the case of the even reflection coefficient. The on state is separately realized by inverting the polarization of the even eigenvector by magnetizing the 90° rotator. This effect may be understood by noting that the rotator produces a 90° rotation of the polarization in the forward direction of propagation and a further 90° in the reverse direction for a total of 180°. The odd eigenvector does not couple in either the demagnetized or magnetized resonator. The reflection coefficients are here −1 and +1 for the even and odd modes respectively.

10.4 Even and Odd Eigenvector of *H*-plane Waveguide Tee Junction

The nature of the even and odd field patterns of the junction are reviewed and clarified in this section. This is done in order to avoid confusion between the two excitations. A tee junction is referred to as an *E*-plane geometry when its side waveguide is on the narrow wall. It is described as an *H*-plane one when it is on the wide wall. The junction here is an *E*-plane geometry. The even eigenvector propagates down the side waveguide of the switch whereas the odd one does not. An understanding of this feature may be obtained by scrutinizing the fields supported by a regular *H*-plane tee junction in Figure 10.5a and b. These illustrate that the odd eigen-network does not produce a field along the side waveguide of the junction, whereas the even eigen-network does.

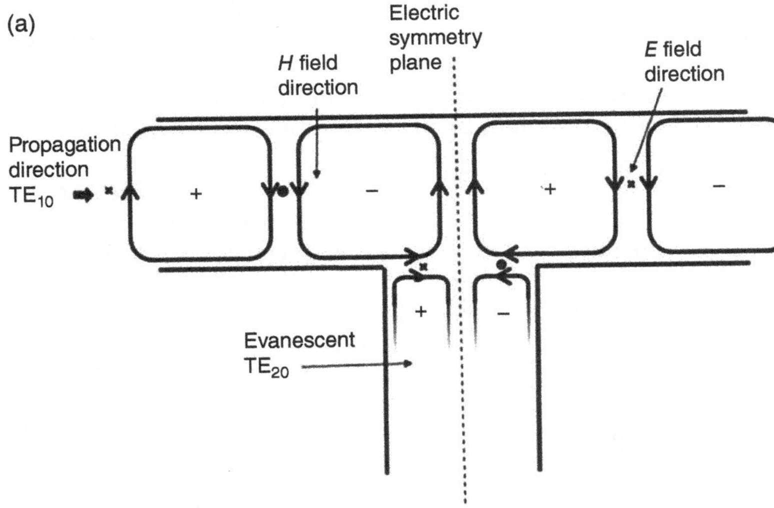

Figure 10.5 *H*-plane waveguide tee junction: (a) odd eigenvector and (b) even eigenvector.

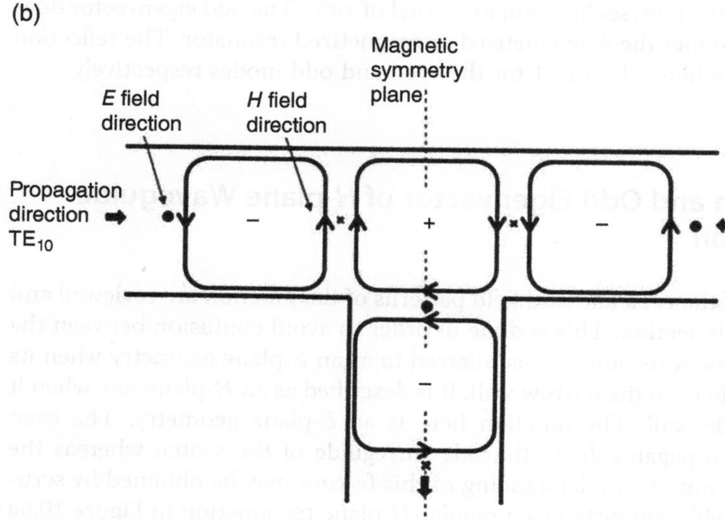

Figure 10.5 (Continued)

10.5 Phenomenological Description of Two-port Off/On and On/Off Switches

The geometry under consideration is an *H*-plane tee junction consisting of two rectangular waveguides and a closed circular dielectric waveguide mounted on the flat face of a circular piston. A classic property of this sort of junction is that it is characterized by stop and pass bands between the main waveguide ports depending upon the position of the back plate of the circular waveguide.

The operation of a large-gap two-port switch is depicted in Figure 10.6. Its reciprocal equivalent circuit is a series resonator in shunt with the waveguide. The development of the magnetized state starts by decomposing the resonator into its normal mode form.

It proceeds by removing the degeneracy between the frequencies of the circuit by replacing the dielectric resonator by a magnetized magnetic insulator. The equivalent circuit obtained in this way is a shunt resonator in shunt with the waveguide. The transition between the off/on states is summarized in Figure 10.6. The dual problem of the small-gap arrangement is indicated in Figure 10.7.

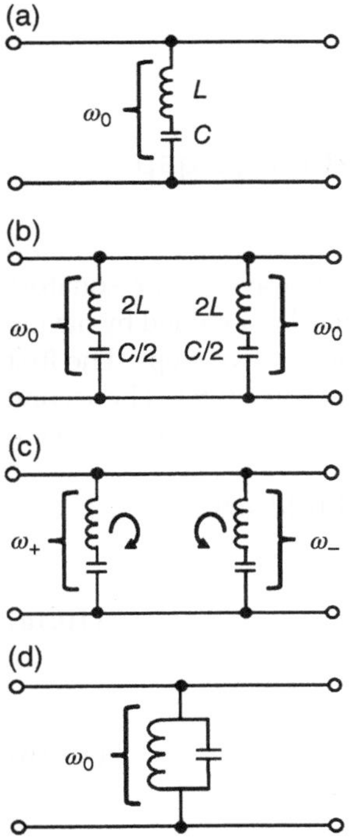

Figure 10.6 (a–d) Operation of the two-port off/on *E*-plane switch.

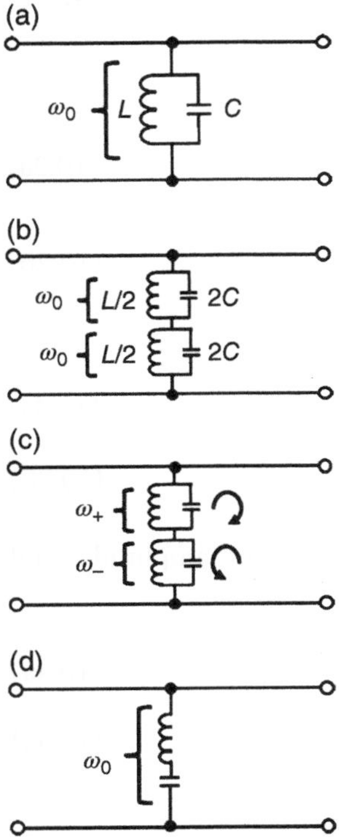

Figure 10.7 (a–d) Operation of the two-port on/off *E*-plane switch.

10.6 Eigenvalue Diagrams of Small- and Large-gap *E*-plane Waveguide Tee Junction

The operation of *E*-plane ferrite switches using quarter-wave long resonators mounted in the *H*-plane of a rectangular waveguide may be obtained by having recourse to its eigenvalue problem. Its adjustment involves two steps. The first fixes the reciprocal stop or pass band condition of the arrangement. The second replaces the dielectric resonator by a gyromagnetic one. This section deals with the first of the two.

The eigenvectors are here defined by even and odd field settings.

$$\bar{U}_{\text{even}} = \frac{1}{\sqrt{2}}\begin{bmatrix}1\\1\end{bmatrix} \tag{10.1a}$$

$$\bar{U}_{\text{odd}} = \frac{1}{\sqrt{2}}\begin{bmatrix}1\\-1\end{bmatrix} \tag{10.1b}$$

and

$$\bar{U}.\bar{U}^{\text{T}} = 1 \tag{10.1c}$$

The scattering parameters are

$$S_{11} = \frac{\rho_{\text{even}} + \rho_{\text{odd}}}{2} \tag{10.2a}$$

$$S_{21} = S_{12} = \frac{\rho_{\text{even}} - \rho_{\text{odd}}}{2} \tag{10.2b}$$

or

$$\rho_{\text{even}} = S_{11} + S_{21} \tag{10.3a}$$

$$\rho_{\text{odd}} = S_{11} - S_{21} \tag{10.3b}$$

The even and odd one-port reflection coefficients have unit amplitudes and angles ϕ_{even} and ϕ_{odd}, respectively.

$$\rho_{\text{even}} = 1.\exp(-j\phi_{\text{even}}) \tag{10.4a}$$

$$\rho_{\text{odd}} = 1.\exp(-j\phi_{\text{odd}}) \tag{10.4a}$$

In the case of a short-circuit eigen-network. ϕ is

$$\phi = 2\left(\theta + \frac{\pi}{2}\right) \tag{10.5a}$$

In the case of an open-circuit one, it is

$$\phi = 2\theta \tag{10.5b}$$

θ is the electrical length of the eigen-networks.

10.7 Eigenvalue Diagrams of *E*-plane Waveguide Tee Junction

The eigenvalue diagrams of the on/off and off/on switches are depicted in Figure 10.8. The one-port even and odd eigen-networks obtained here are illustrated in Figure 10.9.

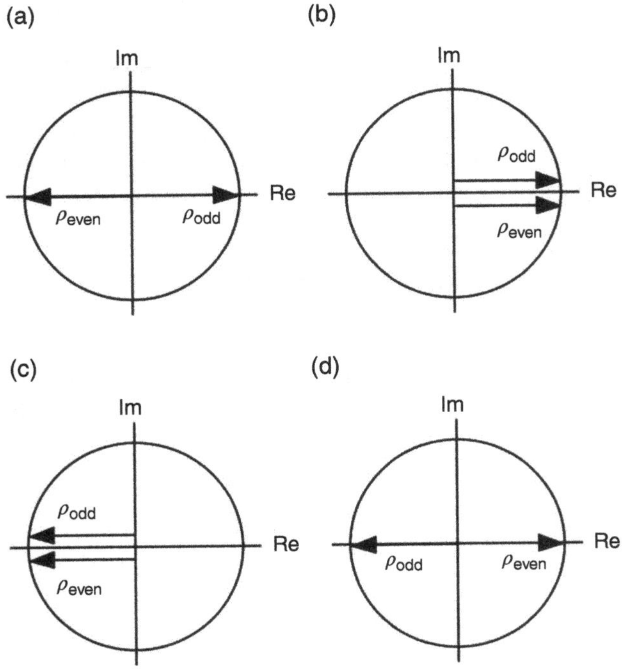

Figure 10.8 (a–d) Even and odd eigenvalue diagrams of two-port *E*-plane switches at a reflection plane 90° away from the symmetry plane.

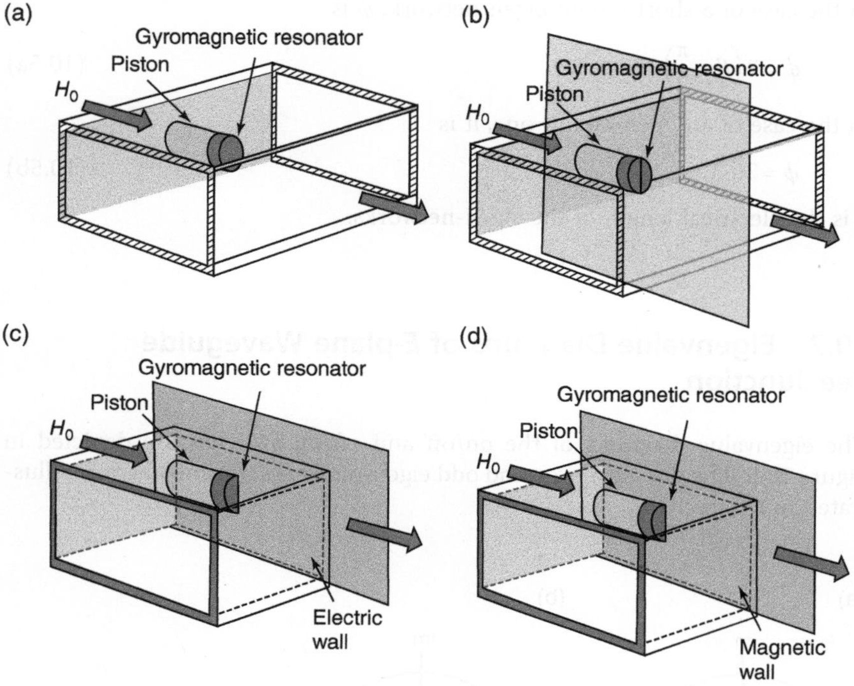

Figure 10.9 (a) Schematic diagram of reentrant switch using *E*-plane tee junction. (b) Symmetry plane of reentrant switch using *E*-plane tee junction. (c) Odd mode eigen-network. (d) Even mode eigen-network.

10.8 Eigen-networks of *E*-plane Tee Junction

The lumped element demagnetized and magnetized eigen-networks of the *E*-plane tee junction are indicated in Figures 10.10 and 10.11. The even and odd reflection angles on the axis of the junction with $\omega_0^2 LC = 1$ are

$$\rho_{\text{even}} = -1$$

$$\rho_{\text{odd}} = -1$$

and

$$\rho_{\text{even}} = +1$$

$$\rho_{\text{odd}} = +1$$

at the terminals of the network.

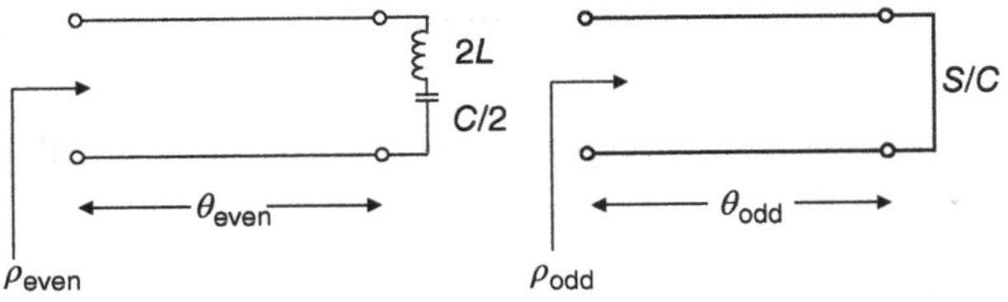

Figure 10.10 Eigen-networks of demagnetized *E*-plane junction.

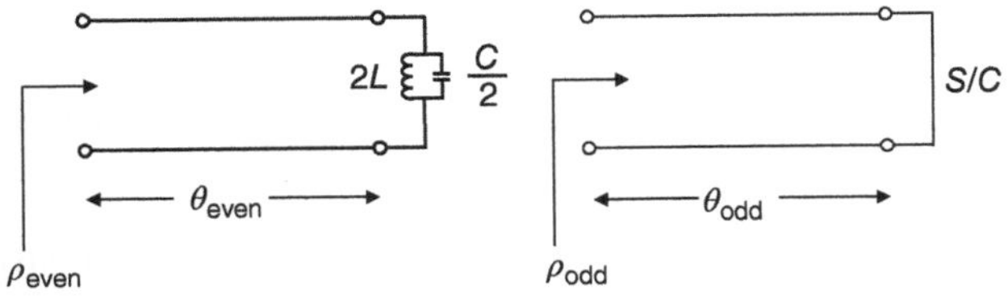

Figure 10.11 Eigen-networks of magnetized *E*-plane junction.

The corresponding magnetized reflection coefficients are

$$\rho_{\text{even}} = +1$$

$$\rho_{\text{odd}} = -1$$

and

$$\rho_{\text{even}} = -1$$

$$\rho_{\text{odd}} = +1$$

The odd reflection coefficient is, in this instance, identical to that met in the description of the demagnetized junction. That of the even one, however, has its sign reversed. This feature is responsible for resetting the stop band of the off/on switch to a pass band here.

10.9 Eigenvalue Algorithm

The conditions considered here produce two gap-resonator ratios for each trial value of k_0R from which the gap S_{eff} and q_{eff} may be extracted. The eigenvalues of the large-gap stop band at the reference plane of the resonator are

$$\rho_{\text{even}} = +1 \tag{10.6a}$$

$$\rho_{\text{odd}} = +1 \tag{10.6b}$$

The eigenvalues at the small-gap pass band conditions at the same possible pairs of terminals are indicated below,

$$\rho_{\text{even}} = -1 \tag{10.7a}$$

$$\rho_{\text{odd}} = +1 \tag{10.7b}$$

together with k_0 and f_0/f_c constants

$$k_0 R = (k_0 R)_{\text{even}} = (k_0 R)_{\text{odd}}, \quad \frac{R}{L} = \text{constant} \tag{10.8a}$$

$$\frac{S}{L} = \frac{S_{\text{even}}}{L} = \frac{S_{\text{odd}}}{L}, \quad \frac{R}{L} = \text{constant} \tag{10.8b}$$

S_{even} and S_{odd} are even and odd gaps and L is the length of the resonator.

10.10 Pass and Stop Bands in Demagnetized *E*-plane Waveguide Tee Junction

A typical stop or pass band filter may be converted to a pass or stop band by either retuning its even or odd eigen-networks. The transition between the stop and pass bands indicated by Omori is an example of the former transformation. The odd reflection coefficient of the junction is here fixed by the Faraday rotation section and is unaffected by the gap between the flat face of the resonator and the opposite narrow wall of the waveguide. The even reflection coefficient is separately characterized by a gap-dependent effective dielectric constant produced by its quasi-planar two-layer circuit. Its dynamic range is compatible with a 180° phase reversal of the reflection coefficient thereby exchanging the demagnetized stop and pass bands of the junction. The gap effective dielectric constant resides between that of the ferrite insulator and free space. The dominant solution, in the case of the *E*-plane circuit, is a stop band in keeping with a series equivalent circuit in shunt with the main waveguide. Figure 10.12a and b illustrate some simulations.

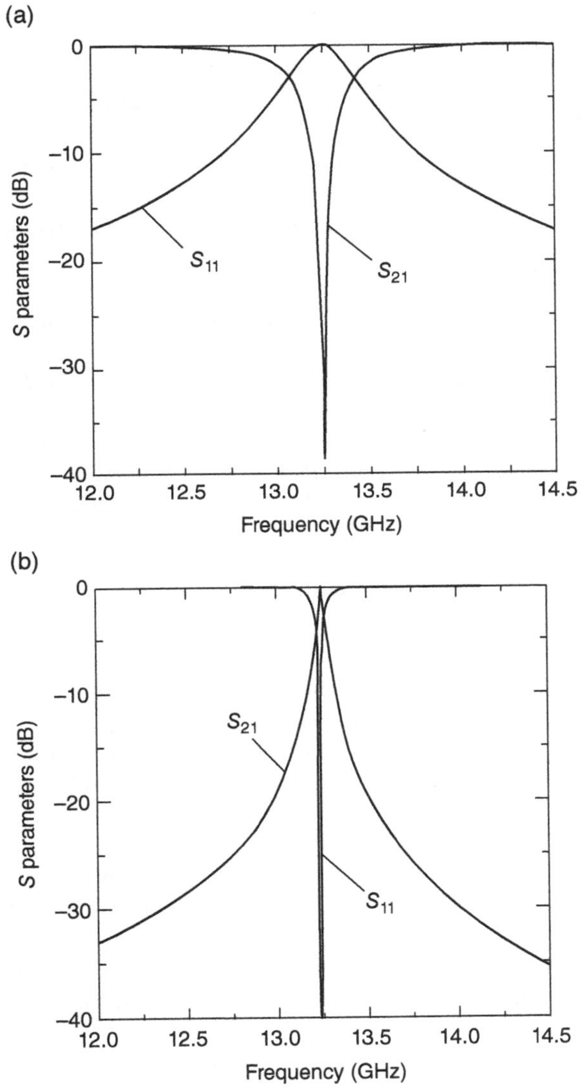

Figure 10.12 (a) Stop band frequency response of demagnetized large-gap *E*-plane junction (Helszajn et al. unpublished). (b) Pass band frequency response of demagnetized small gap *H*-plane junction (Helszajn et al. unpublished).

Bibliography

Allanson, J.T., Cooper, R., and Cowling, T.G. (1946). The theory and experimental behaviour of right-angled junctions in rectangular-section wave guides. *IEE Proc.* 177–187. doi: 10.1049/ji-3-2.1946.0028.

Altman, J.L. (1964). *Microwave Circuits*, Van Nostrand series in Electronics and Communications. Princeton, NJ: Van Nostrand.

Auld, B.A. (1959). The synthesis of symmetrical waveguide circulators. *IRE Trans. Microw. Theory Tech.* **MTT-7**: 238–246.

Branner, G.R., Kumar, B.P., and Thomas, D.G. Jr. (1995). Design of microstrip T junction power divider circuits for enhanced performance. *IEEE Conference*, Rio de Janeiro, Brazil (13–16 August 1995).

Buchta, G. (1966). Miniaturized broadband E-tee circulator at X-band. *Proc. IEEE* **54**: 1607–1608.

Casanueva, A., Leon, A., Mediavilla, A., and Helszajn, J. (2013). Characteristic planes of microstrip and unilateral finline tee-junctions. *Progress in Electromagnetics Research Symposium Proceedings*, Stockholm, Sweden (12–15 August 2013), pp. 173–179.

Davis, L.E. and Longley, S.R. (1963). E-plane three-port X-band waveguide circulators. *IEEE Trans. Microw. Theory Tech.* **MTT-11**: 443–445.

DeCamp, E.E. Jr. and True, R.M. (1971). 1-MW four-port E-plane junction circulator. *IEEE Trans. Microw. Theory Tech.* **MTT-19**: 100–103.

Franco, A.G. and Oliner, A.A. (1962). Symmetric strip transmission line tee junction. *IRE Trans. Microw. Theory Tech.* **MTT-10** (2): 118–124.

He, F.F., Wu, K., Hong, W. et al. (2008). A planar magic-T structure using substrate integrated circuits concept and its mixer application. *IEEE Microw. Wirel. Compon. Lett.* **18** (6): 386–388.

Helszajn, J. (1974). Two-port dipolar switch. *Electr. Lett.* **10**: 46–47.

Helszajn, J. (2015). The electrically symmetric solution of the 3-port H-plane waveguide tee junction at the Dicke ports. *IET Proc. Microw. Antennas Propag.* **9** (6): 561–568.

Helszajn, J. and Cheng, S. (1990). Aspect ratio of open resonators in the design of evanescent mode E-plane circulators. *IEE Proc. Microw. Antennas Propag.* **137**: 55–60.

Helszajn, J., Caplin, M., Frenna, J., and Tsounis, B. (2014). Characteristic planes and scattering matrices of E and H-plane waveguide tee junctions. *IEEE Microw. Wirel. Compon. Lett.* **24** (4): 209–211.

Helszajn, J., Sharp, J., Carignan, L.P., and McKay, M. Two-port waveguide switches using re-entrant and inverted re-entrant H-plane tee junctions (unpublished).

Mansour, R.R. and Dude, J. (1992). Analysis of microstrip T-junction and its application to the design of transfer switches. *IEEE Conference*, Albuquerque, NM (1–5 June 1992).

McGrown, J.W. and Wright, W.H. Jr. (1967). A high power, Y-junction E-plane circulator. *G-MTT International Microwave Symposium Digest*, Boston, MA (8–11 May 1967), pp 85–87.

Montgomery, C.G., Dicke, R.H., and Purcell, E.M. (1948). *Principles of Microwave Circuits*, MIT Radiation Laboratory Series, vol. **VIII**, 432. New York: McGraw-Hill Book Co.

Omori, S. (1968). An improved E-plane waveguide circulator. *G-MTT International Microwave Symposium Digest*, Detroit, MI (20–22 May 1968), pp 228–236.

Rong, Y., Yao, H., Zaki, K.A., and Dolan, T.G. (1999). Millimeter-wave Ka-band H-plane diplexers and multiplexers. *IEEE Trans. Microw. Theory Tech.* **MTT-47** (12): 2325–2330.

Saenz, E., Cantora, A., Ederra, I. et al. (2007). A metamaterial T-junction power divider. *IEEE Microw. Wirel. Compon. Lett.* **17** (3): 172–174.

Yao, H.-W., Abdelmonem, A.E., Liang, J.F. et al. (1993). Wide-band waveguide T-junctions for diplexer applications. *IEEE Trans. Microw. Theory Tech.* **MTT-41** (12): 2166–2173.

Yoshida, S. (1959). E-type T circulator. *Proc. IRE* **47**: 208.

11

Operation of Two-port On/Off and Off/On Planar Switches Using the Mutual Energy–Finite Element Method*

Joseph Helszajn[1] *and David J. Lynch*[2]

[1] *Heriot Watt University, Edinburgh, UK*
[2] *Filtronic Wireless Ltd, Salisbury, MD, USA*

11.1 Introduction

One technique that may be utilized to evaluate the immittance matrices of an arbitrary n-port microwave network is the Green's method. Another is the mutual energy approach. The two differ in that the former involves reducing a nonhomogeneous boundary value problem into a homogeneous one while the latter consists of solving a number of homogeneous and nonhomogeneous one-port problem regions. The eigenvectors appearing in the evaluation of the cutoff spaces of isotropic and gyromagnetic planar circuits are therefore invalid in this instance and must be recalculated from first principles. Since a typical off-diagonal entry in an impedance or an admittance matrix of any n-port may be obtained by considering a typical pair of ports, the solution of such a circuit is sufficient for the description of an n-port problem. Scrutiny of such a two-port indicates that a typical off-diagonal element may be deduced from the knowledge of a calculation on a one-port nonhomogeneous circuit. The evaluation of a typical diagonal element merely requires, of course, a calculation of a one-port homogeneous circuit. A complete characterization of this sort of circuit requires, therefore, the solutions of one-port networks only. If the electric fields due to each distinct one-port excitation are obtained using the finite element method, then the immittance matrices may be expressed in terms of the same variables. The related scattering parameters of the circuit may then be

---

* Reprinted with permission (Lynch, D. and Helszajn, J. (1997). Frequency response of N-port planar gyromagnetic circuits using the mutual energy – finite element method. *IEE Proc. Microw. Antennas Propag.* **144**: 221–228).

Microwave Polarizers, Power Dividers, Phase Shifters, Circulators, and Switches, First Edition. Joseph Helszajn.

found by using the usual conformal mapping between the two descriptions. This chapter employs the mutual energy technique in conjunction with the finite element method to tackle this type of problem. The chapter includes the design of On–Off and Off–On two-port stripline switches using a commercial solver. The geometry utilized here has the ferrite puck in one half-space replaced by a metal plug.

11.2 Impedance and Admittance Matrices from Mutual Energy Consideration

The development of the open- or short-circuit parameters of a two-port planar circuit begins with the conservation of power immittance definitions. Taking the open-circuit parameters for the two-port circuit illustrated in Figure 11.1 by way of an example gives

$$I_j^j Z_{ij} I_i^i = \iint_{\Omega_i} \left(\bar{E}_i^j \times \bar{H}_i^{i*}\right) \cdot \bar{k}\, d\Omega \tag{11.1}$$

$$I_i^i Z_{ii} I_i^i = \iint_{\Omega_i} \left(\bar{E}_i^i \times \bar{H}_i^{i*}\right) \cdot \bar{k}\, d\Omega \tag{11.2}$$

provided

$$\bar{E}_i^j = V_i^j \bar{f}_i(x,y) \tag{11.3}$$

$$\bar{H}_i^j = I_i^j \bar{g}_i(x,y) \tag{11.4}$$

and

$$\iint_{\Omega_i} \left[\bar{f}_i(x,y) \times \bar{g}_i(x,y)\right] \cdot \bar{k}\, d\Omega = 1 \tag{11.5}$$

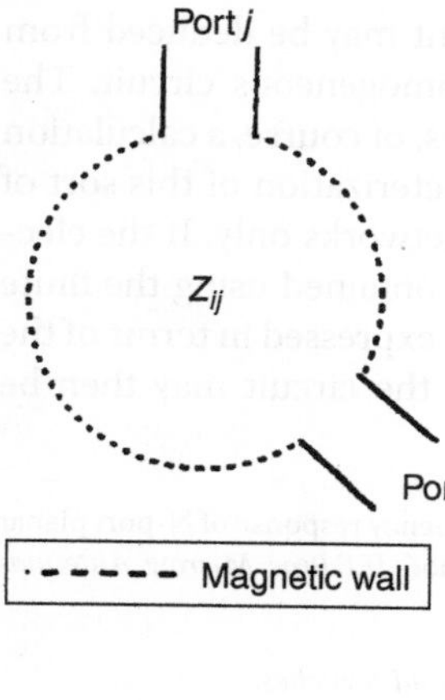

Figure 11.1 Mutual impedance (Z_{ij}) in two-port planar circuit with magnetic sidewalls.

For a stripline circuit:

$$\bar{f}_i(x,y) = \frac{1}{H} \tag{11.6}$$

$$\bar{g}_i(x,y) = \frac{1}{2W} \tag{11.7}$$

and

$$Z_i = \frac{V_i}{I_i} = \eta_0\left(\frac{H}{2W}\right) \tag{11.8}$$

$\bar{E}_i^j$ is the electric field produced at port i due to a current I_j^j at port j and $\bar{H}_i^i$ is the magnetic field produced at port j due to a current I_i^i at port i. The planar circuits defined by these boundary conditions are separately illustrated in Figure 11.2a and b.

If $\bar{E}_i^j$ and $\bar{H}_i^i$ are separately evaluated with $I_j^j = 1A$ and $I_i^i = 1A$, then

$$Z_{ij} = \iint_{\Omega_i}\left(\bar{E}_i^j \times \bar{H}_i^{i*}\right)\cdot\bar{k}\,\mathrm{d}\Omega \quad I_j^j = 1, I_i^i = 1 \tag{11.9}$$

$$Z_{ii} = \iint_{\Omega_i}\left(\bar{E}_i^i \times \bar{H}_i^{i*}\right)\cdot\bar{k}\,\mathrm{d}\Omega \quad I_i^i = 1 \tag{11.10}$$

The dual quantities for the two-port circuit in Figure 11.3 are

$$Y_{ij} = \iint_{\Omega_i}\left(\bar{E}_i^i \times \bar{H}_i^{j*}\right)\cdot\bar{k}\,\mathrm{d}\Omega \quad V_j^j = 1, V_i^i = 1 \tag{11.11}$$

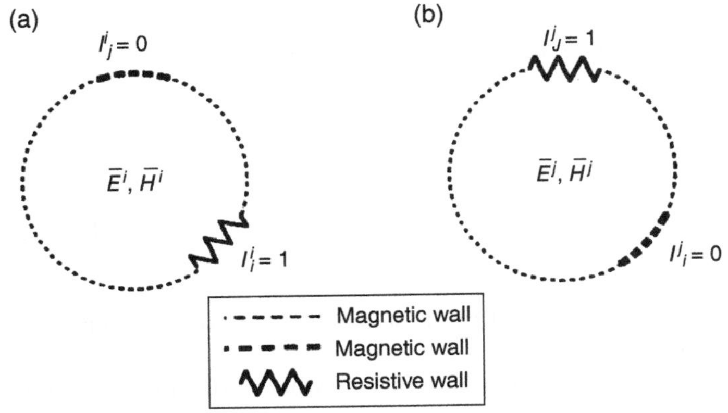

Figure 11.2 Problem region defined by (a) $I_i^i = 1$ and (b) $I_j^j = 1$.

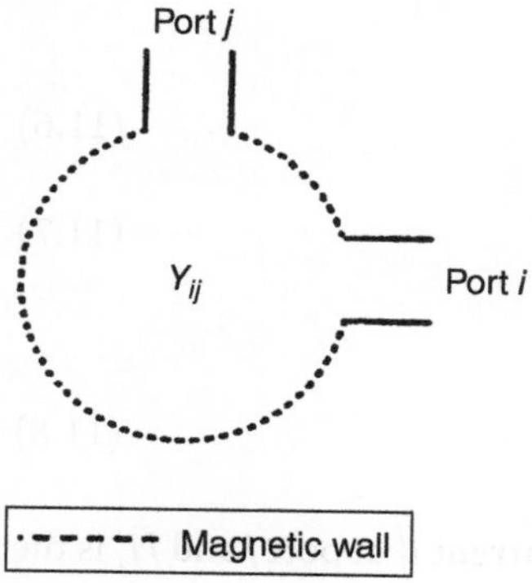

Figure 11.3 Mutual admittance (Y_{ij}) in two-port planar circuit with magnetic sidewalls.

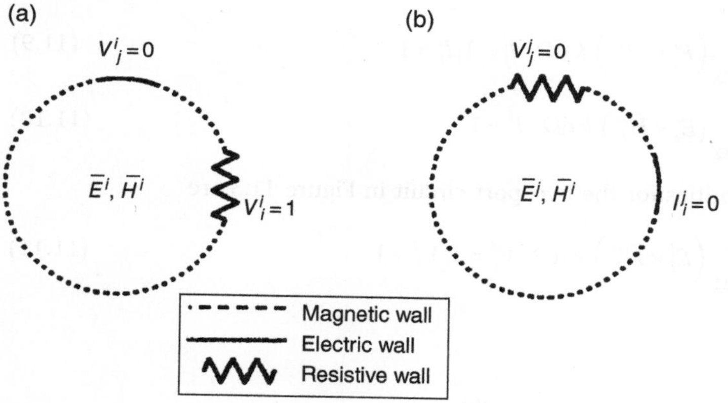

Figure 11.4 Problem region defined by (a) $V_i^i = 1$ and (b) $V_j^j = 1$.

$$Y_{ii} = \iint_{\Omega_i} \left(\bar{E}_i^i \times \bar{H}_i^{i*}\right) \cdot \bar{k}\, d\Omega \quad V_i^i = 1 \tag{11.12}$$

The planar problem regions defined by these sort of two-port problems are illustrated in Figure 11.4.

The topology of the n-port problem differs somewhat in this instance from that of the two-port one. It is depicted in Figure 11.5.

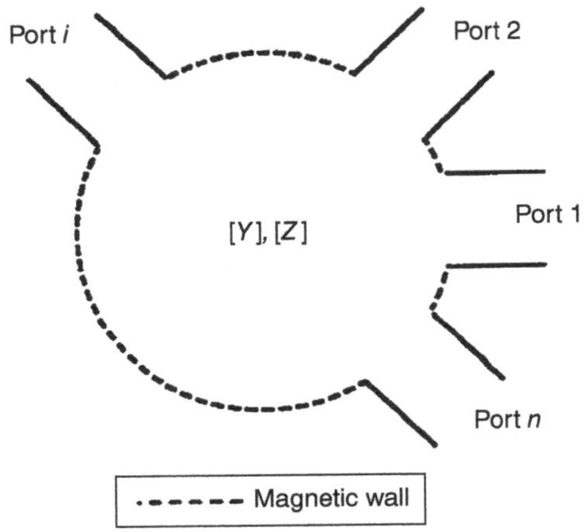

Figure 11.5 n-Port problem region defined by $V_i^j = 1$.

11.3 Impedance and Admittance Matrices for Reciprocal Planar Circuits

The derivation of a typical entry of the impedance matrix of an n-port planar circuit begins by having recourse to the following vector relationship:

$$\nabla\cdot\left(\bar{E}^j \times \bar{H}^{i*}\right) = \left[\bar{H}^{i*}\cdot\left(\nabla\times\bar{E}^j\right) - \bar{E}^j\cdot\left(\nabla\times\bar{H}^{i*}\right)\right] \tag{11.13}$$

The superscript i refers to the situation when port i is fed with an unit current and all the other ports are open-circuited. Similarly, the superscript j corresponds to the situation with a unit current at port j and all the others open-circuited. Integrating both sides of the preceding equation over the volume of the problem region and applying Gauss's theorem to the left-hand side gives

$$\iint_S \left(\bar{E}^j \times \bar{H}^{i*}\right)\cdot\bar{n}\mathrm{d}s = \iiint_V \left[\bar{H}^{i*}\cdot\left(\nabla\times\bar{E}^j\right) - \bar{E}^j\cdot\left(\nabla\times\bar{H}^{i*}\right)\right]\mathrm{d}v \tag{11.14}$$

where

$$s = s' + \sum_{i}^{n} \Omega_i \quad i = 1,2 \tag{11.15}$$

s is the surface that completely encloses the whole problem region, s' is that of the sidewalls not occupied by the ports, and Ω_i are the terminal or reference planes at the ports of the problem region. In order to reveal a single term in the port summation, all ports except port i or j are terminated in a magnetic or electric wall.

The surface area of integration of the integrand on the left-hand side of Eq. (11.14) can be reduced to that over the ports by having recourse to the following vector identities:

$$(\bar{A} \times \bar{B}) \cdot \bar{n} = (\bar{n} \times \bar{A}) \cdot \bar{B}$$

$$(\bar{A} \times \bar{B}) \cdot \bar{n} = (\bar{B} \times \bar{n}) \cdot \bar{A}$$

and noting that

$$\bar{n} \times \bar{E} = 0 \text{ on electric walls}$$

$$\bar{n} \times \bar{H} = 0 \text{ on magnetic walls}$$

Making use of the preceding conditions allows the surface integral over the surfaces in Eq. (11.14) to be replaced by one over the cross-section Ω_i:

$$\iint_{\Omega_i} \left(\bar{E}^j \times \bar{H}^{i*}\right) \cdot \bar{n} \mathrm{d}\Omega = \iiint_V \left[\bar{H}^{i*} \cdot \left(\nabla \times \bar{E}^j\right) - \bar{E}^j \cdot \left(\nabla \times \bar{H}^{i*}\right)\right] \mathrm{d}v \quad (11.16)$$

The left-hand side of the latter equation is now recognized as the mutual impedance between ports i and j of the circuit in Figure 11.1. This gives

$$Z_{ij} = -\iiint_V \left[\bar{H}^{i*} \cdot \left(\nabla \times \bar{E}^j\right) - \bar{E}^j \cdot \left(\nabla \times \bar{H}^{i*}\right)\right] \mathrm{d}v \quad I_i^i = 1, I_j^j = 1 \quad (11.17)$$

The impedance entry corresponding to a one-port circuit is similarly defined by

$$Z_{ii} = -\iiint_V \left[\bar{H}^{i*} \cdot \left(\nabla \times \bar{E}^i\right) - \bar{E}^i \cdot \left(\nabla \times \bar{H}^{i*}\right)\right] \mathrm{d}v \quad I_i^i = 1 \quad (11.18)$$

The evaluation of the impedance matrix for a reciprocal planar circuit continues by recalling Maxwell's curl equations for an isotropic medium.

$$\nabla \times \bar{H} = j\omega\varepsilon_0\varepsilon_f\bar{E} \quad (11.19)$$

$$\nabla \times \bar{E} = -j\omega\mu_0\bar{H} \quad (11.20)$$

Substitution of these identities into Eq. (11.18) allows it to be written in terms of the electric field only.

$$Z_{ij} = \frac{j}{k_0\eta_0}\iiint_V \left[\left(\nabla \times \bar{E}^{i*}\right) \cdot \left(\nabla \times \bar{E}^j\right) - k_0^2\varepsilon_f\bar{E}^{i*} \cdot \bar{E}^j\right] \mathrm{d}v \quad (11.21)$$

where

$$k_0^2 = \omega^2\varepsilon_0\mu_0$$

$$\eta_0 = \sqrt{\frac{\mu_0}{\varepsilon_0}}$$

For the planar circuit case considered here the fields do not vary along the axis of the resonator and only the z-component of the electric field exists. Introducing this condition into the preceding equation and noting the identity

$$(\nabla \times \bar{A})\cdot(\nabla \times \bar{B}) = \frac{\partial \bar{A}}{\partial x}\frac{\partial \bar{B}}{\partial x} + \frac{\partial \bar{A}}{\partial y}\frac{\partial \bar{B}}{\partial y} = (\nabla_t \bar{A})\cdot(\nabla_t \bar{B})$$

indicates that Z_{ij} may be written as

$$Z_{ij} = \frac{jH}{k_0 \eta_0} \iint_S \left[(\nabla_t E_z^{i*})\cdot(\nabla_t E_z^j) - k_0^2 \varepsilon_f E_z^{ij} \cdot E_z^j \right] \tag{11.22}$$

where s is the surface area of the circuit and H is the thickness of the substrate. Typical port excitations for the planar circuit are illustrated in Figures 11.2 and 11.4.

The expression for Z_{ij} may be further reduced by expanding the first integrand by having recourse to Green's theorem and a suitable vector identity. This gives

$$Z_{ij} = \frac{jH}{k_0 \eta_0} \left[-\iint_S (E_z^{j*} \cdot \nabla_t^2 E_z^i)\,\mathrm{d}s + \int_\xi \left(E_z^{j*} \frac{\partial E_z^i}{\partial n} \right) \mathrm{d}t - k_0^2 \varepsilon_f \iint_S (E_z^{j*} \cdot E_z^i)\,\mathrm{d}s \right] \tag{11.23}$$

The contour integral quantity in the preceding expression is identically zero since

$$\frac{\partial E_z^i}{\partial n} = 0 \quad \text{on the magnetic wall} \tag{11.24a}$$

$$E_z^i = 0 \quad \text{at all ports except port } i \tag{11.24b}$$

$$E_z^j = 0 \quad \text{at all ports except port } j \tag{11.24c}$$

The required expression for Z_{ij} reduces to

$$Z_{ij} = \frac{-jH}{k_0 \eta_0} \left[\iint_S E_z^{i*} \cdot (\nabla_t^2 + k_0^2 \varepsilon_f) E_z^j \,\mathrm{d}s \right] \tag{11.25}$$

Scrutiny of this equation indicates that it has a similar form to that met in connection with the variational expression for an isotropic resonator except that it does not correspond to either a quadratic or a Hermitian form. This feature is in keeping with the fact that Z_{ij} need not correspond to a positive real function. It is recalled that the electric fields are solutions of the problem regions with the ports open rather than short-circuited.

A typical self-impedance term is given without ado by

$$Z_{ii} = \frac{-jH}{k_0 \eta_0} \left[\iint_S E_z^{i*} \cdot (\nabla_t^2 + k_0^2 \varepsilon_f) E_z^i \,\mathrm{d}s \right] \tag{11.26}$$

The entries of the admittance matrix are deduced in a similar fashion by expanding the identity:

$$\nabla\cdot\left(\bar{E}^i \times \bar{H}^{j*}\right) \tag{11.27}$$

where $\bar{E}^i$ and $\bar{H}^j$ are the fields produced with $V_j^j = 1$ or $V_i^i = 1$ and all other ports short-circuited. The required result is

$$Y_{ij} = \frac{-jH}{k_0\eta_0}\left[\iint_S E_z^{i*}\cdot\left(\nabla_t^2 + k_0^2\varepsilon_f\right)E_z^j \mathrm{ds}\right] \quad V_i^i = 1, V_j^j = 1 \tag{11.28}$$

Likewise, the self-admittance of the circuit is

$$Y_{ii} = \frac{-jH}{k_0\eta_0}\left[\iint_S E_z^{i*}\cdot\left(\nabla_t^2 + k_0^2\varepsilon_f\right)E_z^i \mathrm{ds}\right] \quad V_i^i = 1 \tag{11.29}$$

11.4 Immittance Matrices of *n*-Port Planar Circuits Using Finite Elements

The admittance or impedance matrix of a circuit may be calculated once the electric fields due to the two one-port circuits defined by ports i and j are obtained. One method of doing so is again the finite element one.

The fields E_z^j and E_z^i are each solutions of a homogeneous boundary value problem. When the finite element method is used, these are given in the usual way by

$$E_z^j = \sum_{k=1}^{n} u_k^j \alpha_k \tag{11.30a}$$

$$E_z^i = \sum_{k=1}^{n} u_k^i \alpha_k \tag{11.30b}$$

where α_k are the Lagrange interpolation polynomials and u_k^j and u_k^i are the complex coefficients that correspond to the discrete values of electric field at the finite element nodes. For an n-port planar circuit problem the unknown coefficients are determined from the solution of the following set of linear simultaneous equations:

$$[A_{\mathrm{ff}}]\bar{U}_f = -\left[A_{\mathrm{fp}}\right]\bar{U}_{\mathrm{p}} \tag{11.31}$$

where

$$[A_{\mathrm{ff}}] = [D_{\mathrm{ff}}] - k_{\mathrm{e}}^2[B_{\mathrm{ff}}] \tag{11.32a}$$

$$[A_{\mathrm{fp}}] = [D_{\mathrm{fp}}] - k_{\mathrm{e}}^2[B_{\mathrm{fp}}] \tag{11.32b}$$

$\bar{U}_{\mathrm{p}}$ is a column matrix containing the prescribed values of the electric field at the coupling ports; $\bar{U}_{\mathrm{f}}$ is a column matrix containing the unknown values of electric field; k_{e}^2 is the wavenumber and $[B_{\mathrm{ff}}]$, $[D_{\mathrm{fp}}]$ etc. are square or rectangular matrices obtained from partitioning the $[B]$ and $[D]$ matrices.

$\bar{U}^i$ and $\bar{U}^j$ correspond to an input at port i or j, with electric or magnetic walls at the remaining ports.

$$\bar{U}^i = \begin{bmatrix} U_{\mathrm{f}} \\ U_{\mathrm{p}} \end{bmatrix}^i, \quad \bar{U}^j = \begin{bmatrix} \mathrm{U}_{\mathrm{f}} \\ U_{\mathrm{p}} \end{bmatrix}^j \tag{11.33}$$

Substituting the expanded electric field into the expression for the impedance matrix of an n-port isotropic circuit, after some algebraic manipulation yields

$$Z_{ij} = \frac{-jH}{k_0\eta_0}\left[\left(\bar{U}^{i*}\right)^{\mathrm{T}} \cdot \left\{[D] - k_0^2\varepsilon_{\mathrm{f}}[B]\right\}\bar{U}^j\right] \tag{11.34}$$

and

$$Z_{ii} = \frac{-jH}{k_0\eta_0}\left[\left(\bar{U}^{i*}\right)^{\mathrm{T}} \cdot \left\{[D] - k_0^2\varepsilon_{\mathrm{f}}[B]\right\}\bar{U}^i\right] \tag{11.35}$$

The $[D]$ and $[B]$ matrices coincide with the discretization of the energy functional. Since these matrices have already been computed when the fields are evaluated, the admittance and impedance matrices may be obtained by simple matrix operations on the nodal field vector solutions.

The admittance parameters of an n-port planar isotropic circuits are given by duality by

$$Y_{ij} = \frac{jH}{k_0\eta_0}\left[\left(\bar{U}^{j*}\right)^{\mathrm{T}} \cdot \left\{[D] - k_0^2\varepsilon_{\mathrm{f}}[B]\right\}\bar{U}^i\right] \tag{11.36}$$

The diagonal entries of the immittance matrix are obtained by replacing i and j in the corresponding off-diagonal elements.

11.5 Frequency Response of Two-port Planar Circuits Using the Mutual Energy–Finite Element Method

The mutual energy method may be used in conjunction with the finite element method to evaluate the frequency response of any n-port planar circuit. One simple two-port circuit is the stripline band elimination filter depicted in Figure 11.6. It consists of a circular resonator with two orthogonal ports. The frequency response of this circuit may be established by constructing its

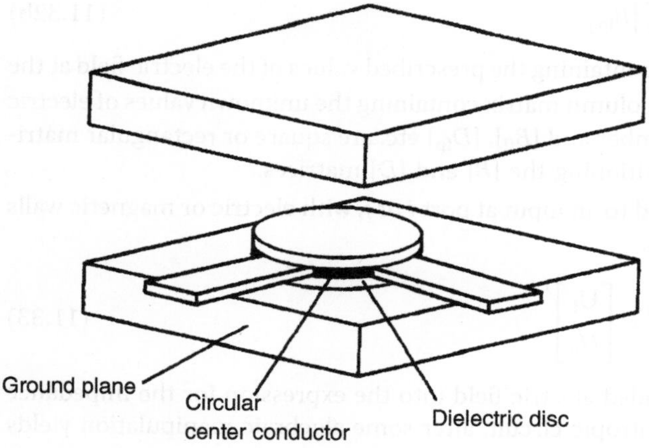

Figure 11.6 Topology of two-port isotropic stripline band elimination filter.

admittance matrix and having recourse to the bilinear mapping between it and its scattering one. In this instance, for an input at port 1 the electric field at the finite element nodes along the periphery subtended by port 2 are identically zero. The dual case holds for an input at port 2. These two problem regions are illustrated in Figure 11.7. The entries of the admittance matrix may be simply evaluated once the 2 one-port circuits have been solved for the electric field vectors $\bar{U}^1$ and $\bar{U}^2$. The scattering parameters are readily computed from the

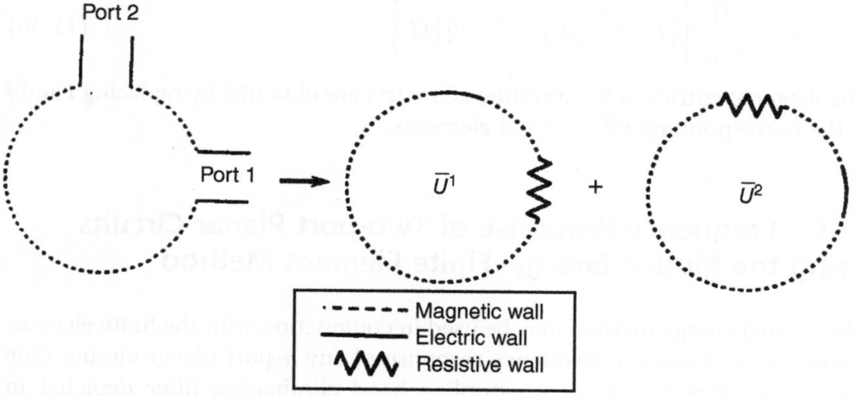

Figure 11.7 Decomposition of band elimination filter into 2 one-port circuits for evaluation of admittance matrix.

normalized admittance ones by having recourse to the bilinear transformation between the scattering and admittance matrices.

$$[S] = \frac{[I] - [Y]}{[I] + [Y]} \tag{11.37}$$

Figure 11.8 illustrates the transmission characteristics of the circuit. Figure 11.9 depicts the finite element mesh used. It consists of 34 second-order triangles connecting 107 interpolation nodes. Some results based on other classic numerical techniques are superimposed on the illustration for the purpose of comparison. The transmission zero evident from this diagram coincides with the frequency of the first pair of degenerate counterrotating modes in this type of resonator.

One further interesting circuit, which is readily assessable to a solution, is the bandpass circuit illustrated in Figure 11.10. It is similar to the band elimination circuit except that its coupling ports are collinear instead of at right angles to each other. Its two problem regions are illustrated in Figure 11.11. The finite element mesh used for the solution of this circuit is illustrated in Figure 11.12. The scattering parameters are separately depicted in Figures 11.13

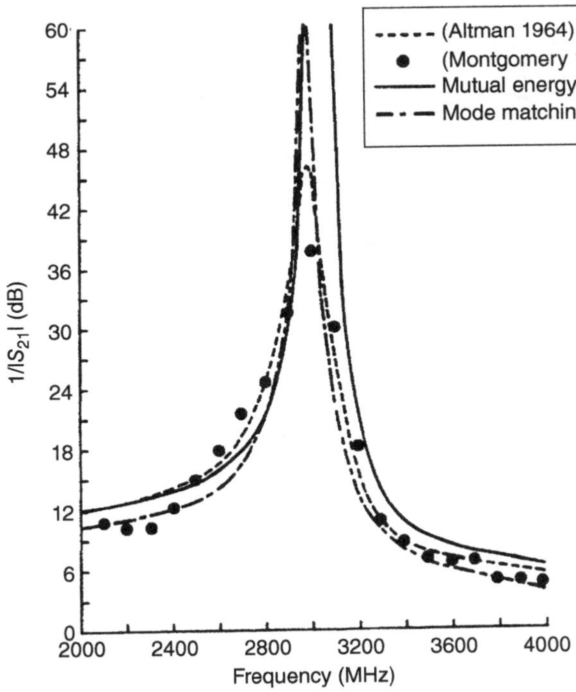

Figure 11.8 Transmission coefficient of stripline band elimination filter (R = 19.06 mm, H = 3.2 mm, W = 7.78 mm, ε_f = 2.4).

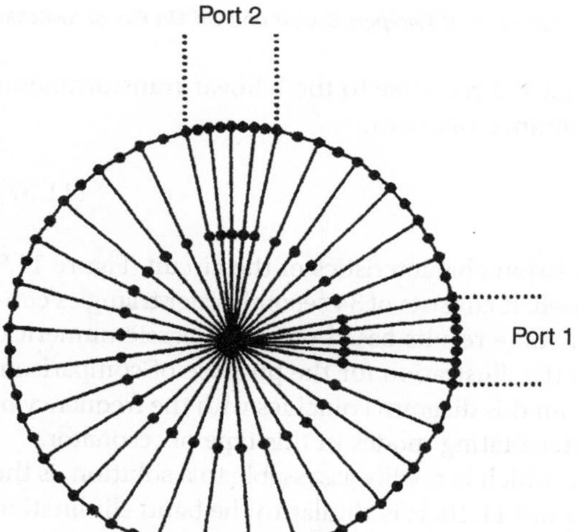

Figure 11.9 Finite element mesh used in solution of stripline band elimination filter.

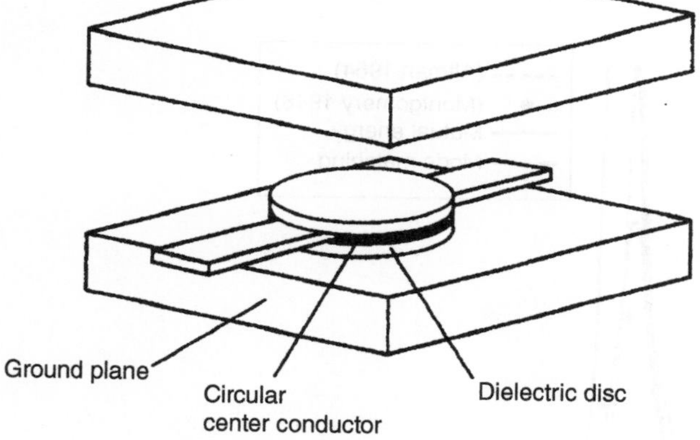

Figure 11.10 Topology of two-port gyromagnetic dipolar switch.

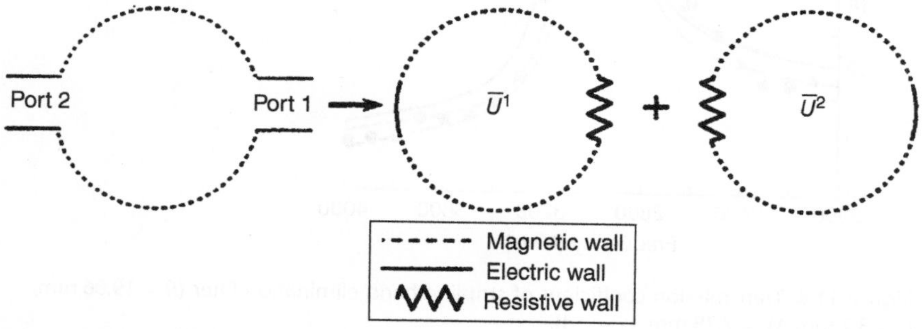

Figure 11.11 Decomposition of band elimination filter into 2 one-port circuits for evaluation of admittance matrix.

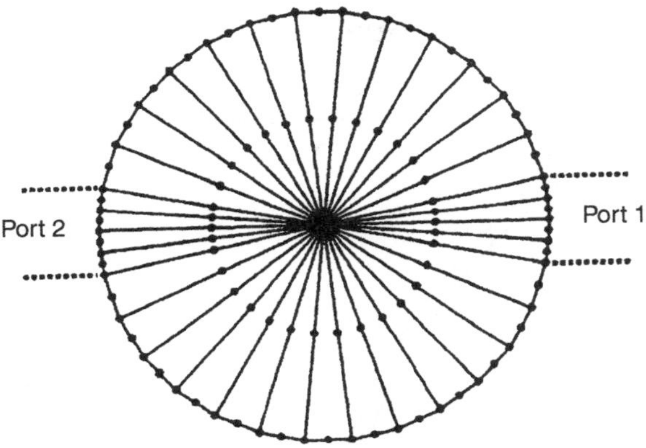

Figure 11.12 Finite element mesh used in solution of stripline dipolar switch.

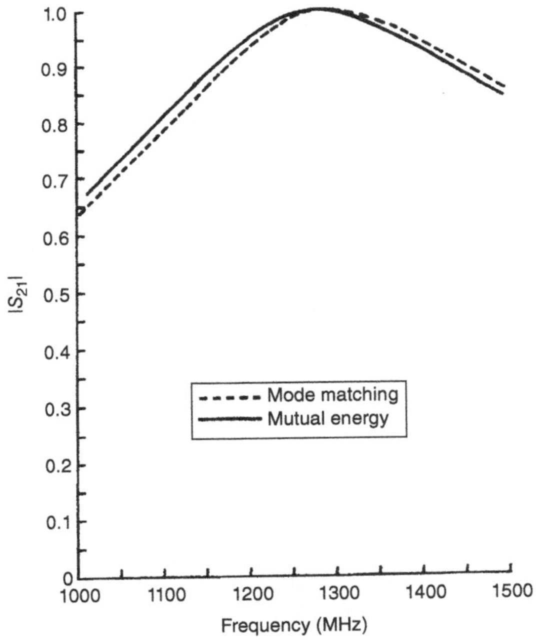

Figure 11.13 Reflection coefficient for two-port dipolar switch for $\kappa/\mu = 1$ using mutual energy and mode-matching formulations.

and 11.14. The equivalent result from the mode-matching method is also shown for comparison. Scrutiny of these results indicates that this structure displays a passband at the first pair of dominant degenerate modes of the decoupled problem region.

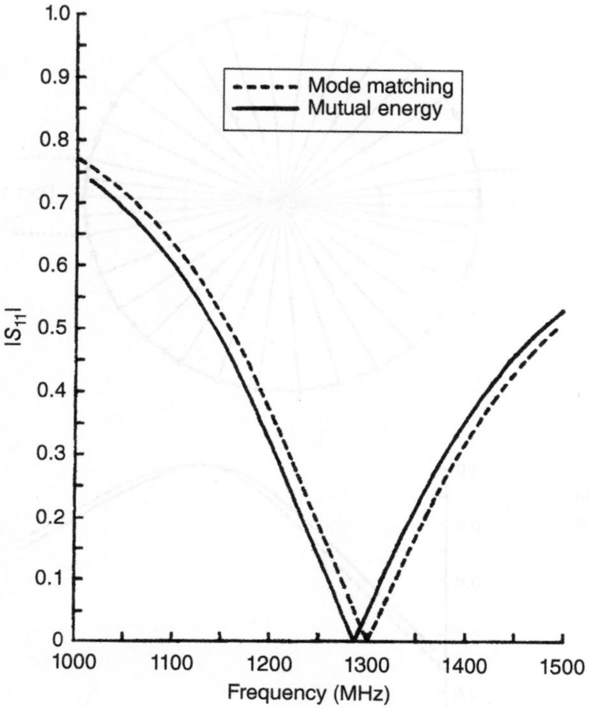

Figure 11.14 Transmission coefficient for two-port dipolar switch for $\kappa/\mu = 1$ using mutual energy and mode-matching formulations.

11.6 Stripline Switch Using Puck/Plug Half-spaces

A two-port On/Off stripline junction may also be realized by replacing one of the two ferrite pucks by a metal plug. Figure 11.15 shows the geometry in question. The junction is described by its radius R, the thickness of the half-spaces H, the width W of the 50Ω striplines and the coupling angle ψ given by

$$\sin\psi = \frac{W}{2R} \tag{11.38}$$

Figure 11.16a displays the reflection and transmission parameters of the demagnetized switch. Figure 11.16b and c indicate the magnitudes and phase angles of the even and odd eigenvalues. Figure 11.16d is the eigenvalue diagram of the circuit. The details employed to obtain these data are $R = 3.18$ mm, $H = 1.5$ mm, $R/H = 2.12$, and $\psi = 0.20$ rad.

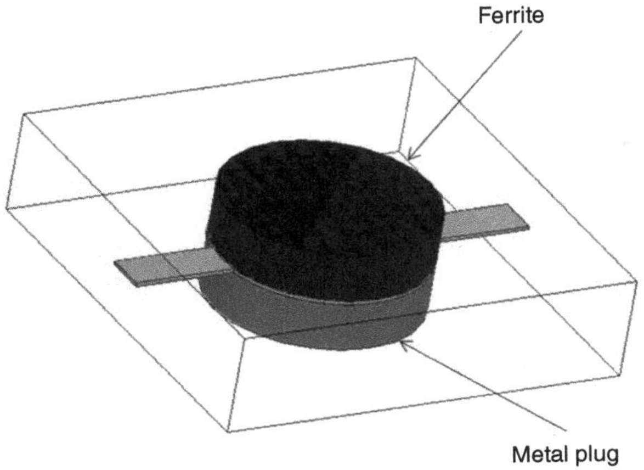

Figure 11.15 Schematic diagram of two-port stripline on/off switch using puck and plug half-spaces.

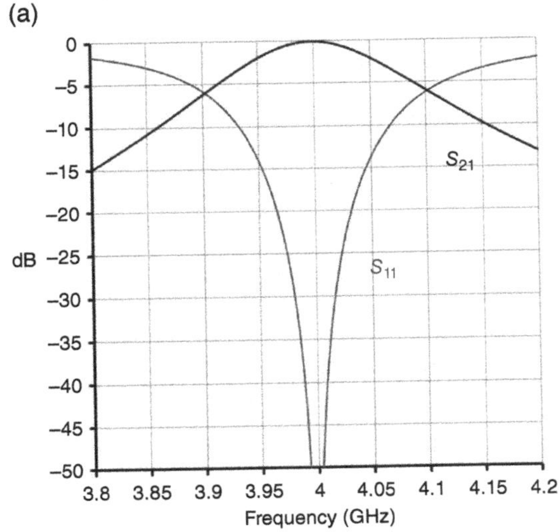

Figure 11.16 (a) Reflection and transmission parameters of the demagnetized switch, (b) magnitude of even and odd eigenvalues, (c) phase of even and odd eigenvalues, and (d) eigenvalue diagram.

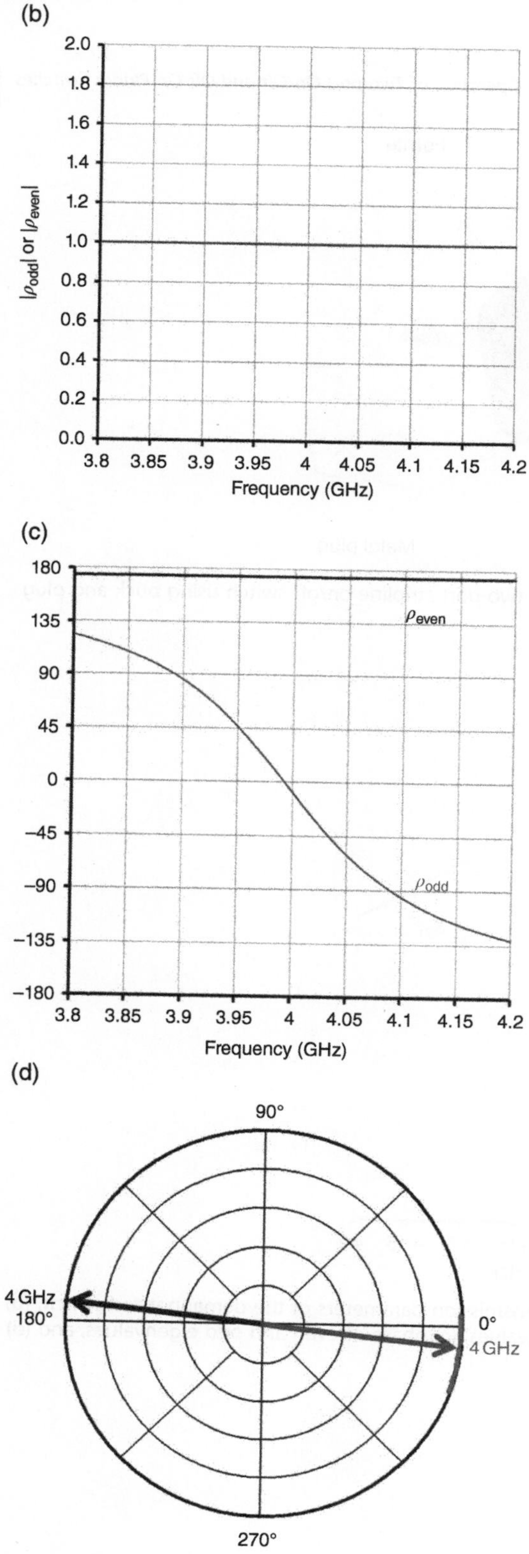

Figure 11.16 (Continued)

Bibliography

Altman, J.L. (1964). *Microwave Circuits*. New York: Van Nostrand.

Collin, R.E. (1966). *Foundations for Microwave Engineering*. New York: McGraw Hill.

Fay, C.E. and Comstock, R.L. (1965). Operation of the ferrite junction circulator. *IEEE Trans. Microw. Theory Tech.* **MTT-13**: 15–27.

Garcia, P. and Webb, J.P. (1990). Optimization of planar devices by the finite element method. *IEEE Trans. Microw. Theory Tech.* **MTT-38**: 48–53.

Hammond, P. (1981). *Energy Methods in Electromagnetism*. Oxford: Clarendon Press.

Kurokawa, K. (1969). *An Introduction to the Theory of Microwave Circuits*. London: Academic Press.

Lynch, D. and Helszajn, J. (1997). Frequency response of N-port planar gyromagnetic circuits using the mutual energy-finite element method. *IEE Proc. Microw. Antennas Propag.* **144**: 221–228.

Montgomery, C.G., Dickie, R.H., and Purcell, E.M. (1948). *Principles of Microwave Circuits*. New York: McGraw Hill.

Ramo, S., Whinnery, J.R., and Van Duzer, T. (1965). *Fields and Waves in Communication Electronics*. New York: Wiley.

Webb, J.P. (1990). Absorbing boundary conditions for the finite element analysis of planar devices. *IEEE Trans. Microw. Theory Tech.* **MTT-38**: 1328–1332.

Webb, J.P. and Parihar, S. (1986). Finite element analysis of H-plane rectangular waveguide problems. *IEE Proc. (Pt. H)* **133** (2): 91–94.

12

Standing Wave Solutions and Cutoff Numbers of Planar WYE and Equilateral Triangle Resonators

Joseph Helszajn

Heriot Watt University, Edinburgh, UK

12.1 Introduction

Important planar resonators with top and bottom electric walls and a magnetic sidewall that have the symmetries of the three-port junction circulator are the wye and equilateral triangle geometries. The chapter summarizes the dominant mode cutoff number and standing wave pattern of each arrangement. This is done for both a dielectric and a gyromagnetic substrate. The degenerate and split cutoff numbers of planar resonators enter in the description of the half-wave long cavities with top and bottom magnetic walls in Chapter 14.

The wye resonator is, in its simplest form, the junction of three quarter-wave long transmission lines at 120°. Its geometry may be dealt with by having recourse to a Finite Element (FE) solver. The quasi-wye resonator formed by a central circular region symmetrically loaded by three open-circuited transmission lines is separately dealt with. One means of dealing with this structure is to have recourse to various numerical techniques. The method utilized here is based on the decomposition of the problem geometry into a symmetrical three-port region bounded by three open-circuited lines. The resonant frequencies of the symmetrical and counterrotating families of modes of the overall circuit are then obtained by satisfying the boundary conditions between the impedance eigenvalues of the circular region and the impedance of a typical open-circuited stub. Some calculations based on a FE solver are included for completeness. It provides one means of constructing standing wave solutions of the various modes of the geometry. A closed-form description of a quasi-wye resonator consisting of an inner gyromagnetic region to which are

Microwave Polarizers, Power Dividers, Phase Shifters, Circulators, and Switches,
First Edition. Joseph Helszajn.

connected three stubs is separately dealt with. It provides one means by which the split frequencies of the resonator under the application of the gyrotropy may be deduced. Although it is difficult to visualize the rotation of the equipotential lines in a magnetized wye resonator, it is nevertheless possible to do so at one of two possible triplets of ports. This may be done by taking suitable linear combinations of those of the demagnetized geometry. The chapter separately deals with the equilateral triangle resonator.

12.2 Cutoff Space of WYE Resonator

The equipotential lines and cutoff numbers of the first three modes of a planar wye resonator have been computed by having recourse to a FE program. The schematic diagram of the resonator under consideration is depicted in Figure 12.1.

The coupling angle (ψ) of a typical stub is in this geometry defined by the width (W) of the stub and the inner radius (r) of the circular plate. The geometry is subdivided into 12 triangular elements and a third-order polynomial approximation is made to the electromagnetic (EM) fields in each triangle. The degree of the polynomial is fixed by the volume of the labor involved in setting up the matrix problem. The number of triangles chosen is determined by the fact that the amount of computer time taken to solve the problem is not linearly dependent upon the number of triangles. A typical segmentation of a wye resonator is illustrated in Figure 12.2. The equipotential lines of the fundamental transverse magnetic (TM) mode in this resonator are indicated in Figure 12.3. Its cutoff number has also been computed and is specified by

$$kR = 1.643 \tag{12.1}$$

where

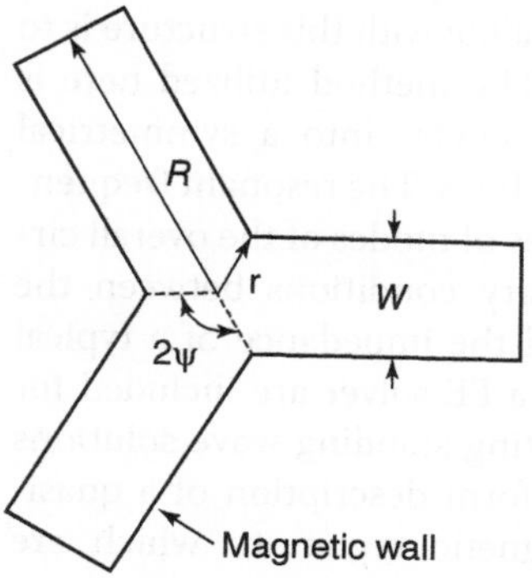

Figure 12.1 Schematic diagram of wye resonator.

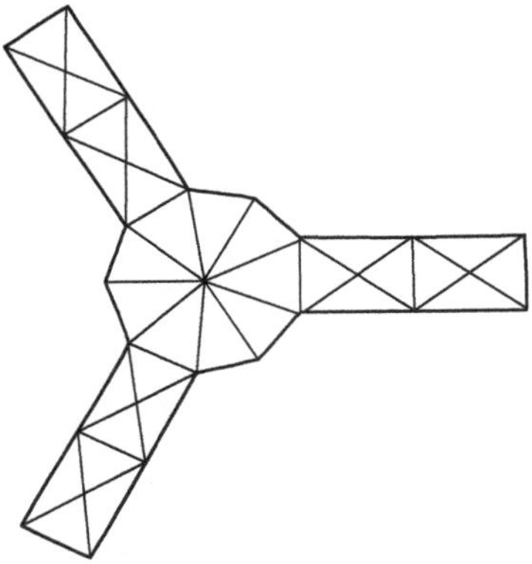

Figure 12.2 Segmentation of wye resonator into finite elements.

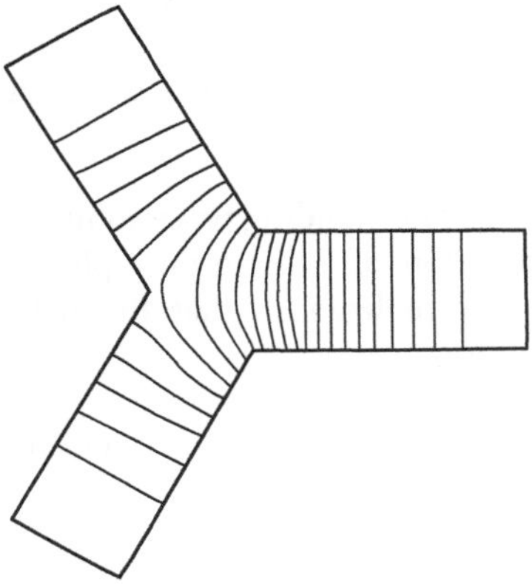

Figure 12.3 Equipotential lines of dominant mode in wye resonator.

$$k = \omega\sqrt{\mu_0\mu_r\varepsilon_0\varepsilon_r} \tag{12.2}$$

ε_r is the relative dielectric constant and μ_r is the relative permeability of the ferrite material. λ_0 is the free-space wavelength in meters. This cutoff number applies for $W/R = 0.40$.

Figure 12.4a and b depict the first symmetric and the first higher-order TM standing wave patterns in this type of geometry. The cutoff numbers are $kR = 3.33$ and $kR = 4.91$, respectively.

The equipotential lines of the symmetric mode in Figure 12.4a have a maximum value at both the center and at the end of the stub, and a zero

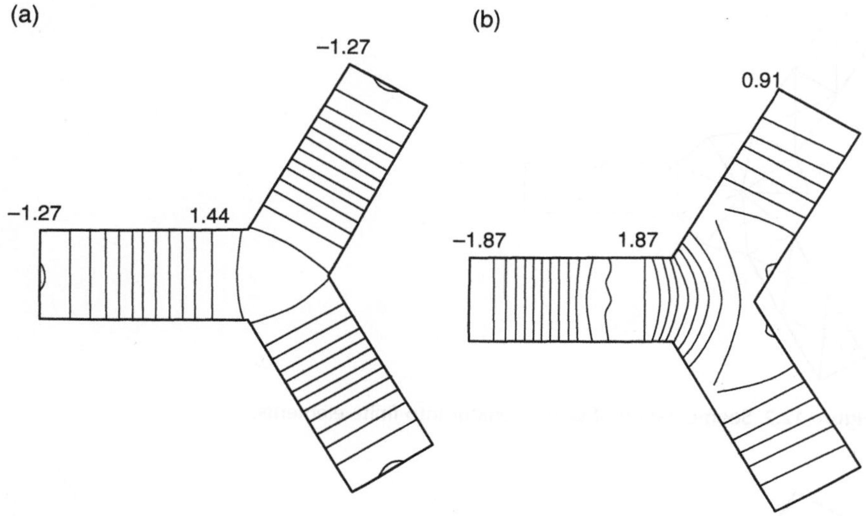

Figure 12.4 (a) First symmetric mode in wye resonator and (b) first higher-order dominant mode in wye resonator.

approximately midway along the stub at $kR = 1.67$. The standing wave solution in Figure 12.4b is also suitable for the construction of a three-port planar circulator. The fields at the nodes have been normalized so that the field distribution of a typical mode satisfies the condition:

$$\iint \phi_n^2 \mathrm{d}s = 1 \tag{12.3}$$

The resonant modes produced by the program are orthogonal.

12.3 Standing Wave Circulation Solution of WYE Resonator

The equipotential lines of the dominant mode in the isotropic resonator have the symmetry encountered in the construction of planar circulators on magnetized substrates. It may therefore be utilized in the construction of a three-port planar junction circulator. Circulation solutions in magnetized wye resonators are constructed by taking a linear combination of two standing wave patterns of the demagnetized wye resonator with one pattern rotated through 120°. The construction is depicted in Figure 12.5 in the case of the dominant mode. It indicates that an ideal circulation condition can be realized by coupling to the wye resonator at one of two triplets of ports as illustrated in Figure 12.6. The arrangement in Figure 12.6b produces a widely used commercial

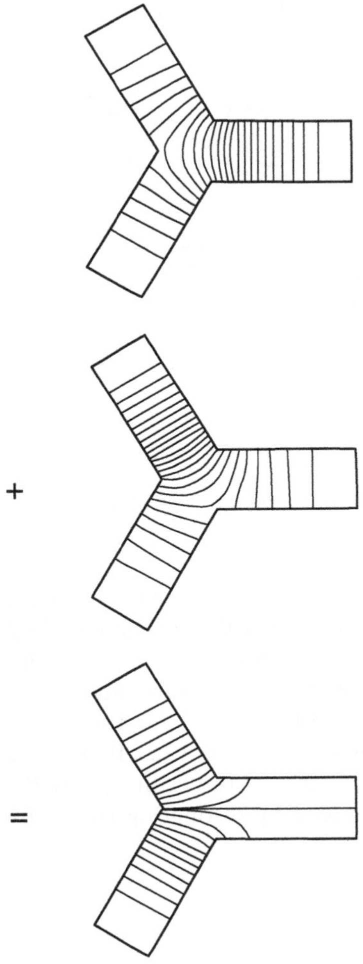

Figure 12.5 Equipotential lines of dominant mode in wye circulator.

quarter-wave-coupled three-port junction circulator whose outside radius is of the order of a quarter-wave at the operating frequency of the device.

12.4 Resonant Frequencies of Quasi-wye Magnetized Resonators

One possible variation of the wye resonator is the disk–stub gyromagnetic arrangement consisting of a circular plate to which are connected three short unit elements (UEs). The frequency may be deduced by visualizing

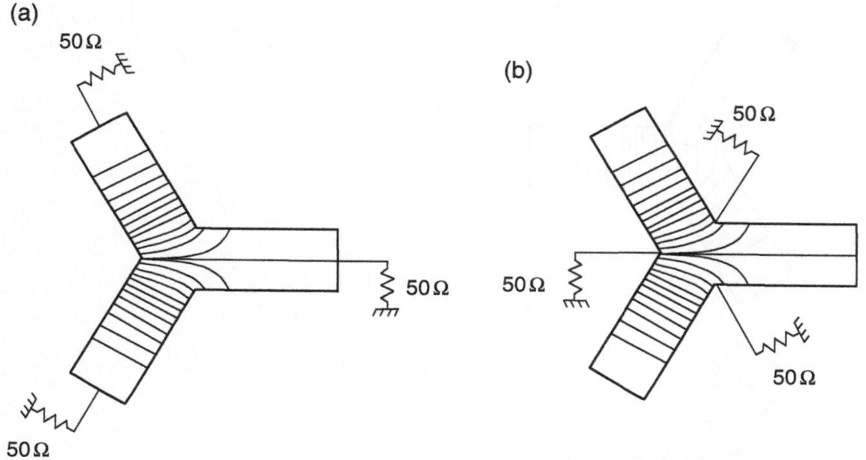

Figure 12.6 Circulation terminals of wye resonator for (a) "aa" terminals and (b) "bb" terminals.

it as a circular region loaded by three UEs or by a six-port arrangement with three of its ports closed by magnetic walls and the other three terminated by suitable stubs. The equivalence between the two models suggests that the first six impedance poles of the problem region are strictly speaking necessary in order to reproduce the boundary conditions of the resonator. The topology under consideration is indicated in Figure 12.7. It is fixed by a coupling angle or shape angle ψ and the ratio of the radii R_i and R_0. Its degenerate or split resonance may be deduced by resonating the stubs

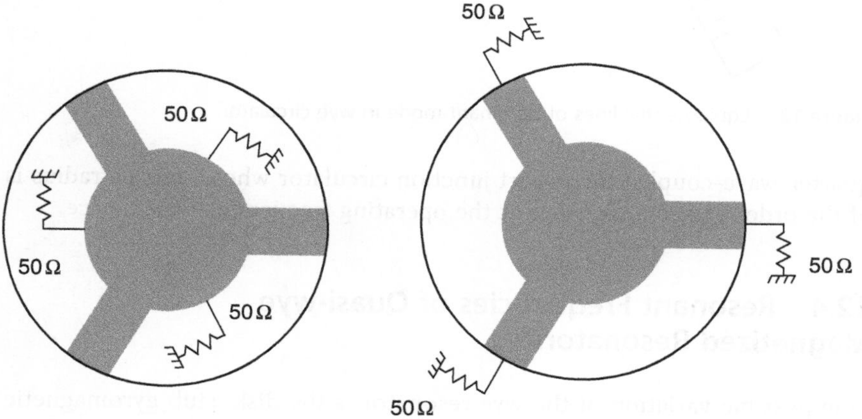

Figure 12.7 Schematic diagram of quasi-wye resonator using a circular disc loaded with UEs.

with the eigen-networks of the circular gyromagnetic region. The description of a typical impedance pole of an m-port symmetrical region is a standard result in the literature.

A loosely coupled junction may be visualized as a symmetrical six-port network with alternate ports open-circuited or as a three-port network.

If the eigenvalues are written in terms of the impedance poles of the three-port network, the result is

$$Z^0 = Z_0 + Z_3 \tag{12.4a}$$

$$Z^+ = Z_{+1} + Z_{-2} \tag{12.4b}$$

$$Z^- = Z_{-1} + Z_{+2} \tag{12.4c}$$

A typical pole of a three-port symmetrical isotropic junction is given below:

$$Z_n = \frac{j3\eta_r Z_r}{\pi}\left[\frac{\sin(n\psi)}{n\psi}\right]^2\left[\frac{J_{n-1}(kR_i)}{J_n(kR_i)} - n\left(\frac{1+\kappa/\mu}{kR_i}\right)\right]^{-1} \tag{12.5}$$

Z_r is the characteristic impedance of a typical stripline.

$$Z_r = 30\pi\ln\left(\frac{W+t+2H}{W+t}\right) \tag{12.6}$$

ψ is the coupling angle defined by

$$\sin\psi = \frac{W}{2r} \tag{12.7}$$

μ and κ are the diagonal and off-diagonal elements of the tensor permeability. The ratio κ/μ is known as the gyrotropy of the magnetic insulator.

Useful polynomial approximations for the Bessel functions for x between 0 and 3 and the recurrence formulae are shown in the table below, and are sufficient for computational purposes.

Polynomial approximations:

$$J_0(x) = 1 - 2.2499997\left(\frac{x}{3}\right)^2 + 1.2656208\left(\frac{x}{3}\right)^4 - 0.3163866\left(\frac{x}{3}\right)^6 + 0.0444479\left(\frac{x}{3}\right)^8 - 0.0039444\left(\frac{x}{3}\right)^{10} + 0.0002100\left(\frac{x}{3}\right)^{12}$$

$$J_1(x) = x\left[0.5 - 0.56249985\left(\frac{x}{3}\right)^2 + 0.21093573\left(\frac{x}{3}\right)^4 - 0.03954289\left(\frac{x}{3}\right)^6 + 0.00443319\left(\frac{x}{3}\right)^8 - 0.00031761\left(\frac{x}{3}\right)^{10} + 0.00001109\left(\frac{x}{3}\right)^{12}\right]$$

Recurrence formulae:

$$J_{n+1}(x) = \frac{2n}{x}J_n(x) - J_{n-1}(x)$$

$$J_{-n}(x) = (-1)^n J_n(x)$$

The characteristic equation for the frequencies of the resonator is obtained by resonating the degenerate impedance eigenvalues using suitable open-circuited transmission lines. The characteristic equation for the first two pairs of degenerate resonances is now established by a transverse resonance condition between the radial and uniform lines.

$$Z^{\pm} = j\eta_r Z_r \cot(kL) \tag{12.8}$$

L is the length of a typical open-circuited stub, defined by $L = R_O - R_i$. The phase constant (k) has the meaning in Eq. (12.2); W, t, and H are the linear dimensions of the uniform striplines.

The required condition is given by Eq. (12.9) for $n = 1$, $n = 2$.

$$\left(\frac{3\psi}{\pi}\right)\left(\frac{\sin\psi}{\psi}\right)^2\left[\frac{J_0(kR_i)}{J_1(kR_i)}-\left(\frac{1}{kR_i}\right)\right]^{-1}+\left(\frac{3\psi}{\pi}\right)\left(\frac{\sin 2\psi}{2\psi}\right)^2 \left[\frac{J_1(kR_i)}{J_2(kR_i)}-\left(\frac{2}{kR_i}\right)\right]^{-1} - \cot(kL) = 0 \tag{12.9}$$

The first two modes of the circular region have been retained in this formulation. The mode chart of the demagnetized wye resonator is depicted in Figure 12.8 for $n = 1$. The agreement between the calculations undertaken here and some finite element method (FEM) calculations is separately indicated in Figure 12.9.

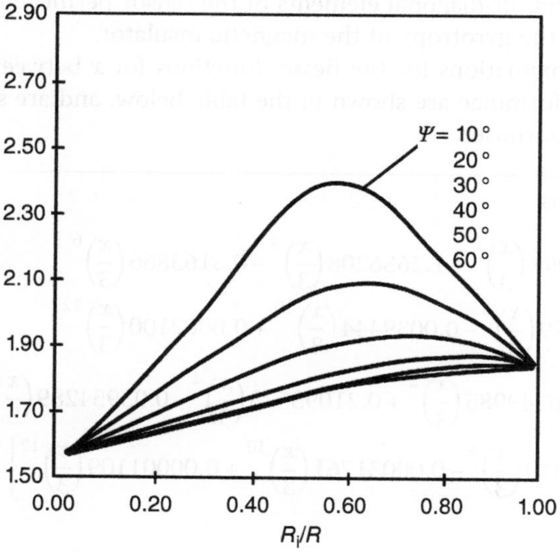

Figure 12.8 Cutoff wavenumber of the fundamental mode in a demagnetized wye resonator.

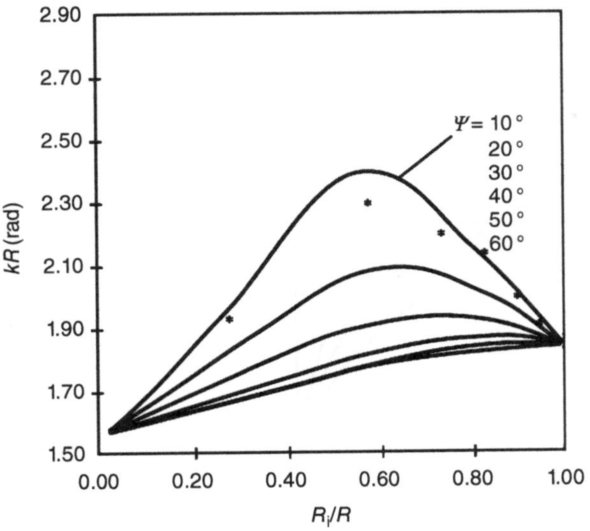

Figure 12.9 Comparison between closed form and FEM calculations of cutoff space of wye resonator.

12.5 The Gyromagnetic Cutoff Space

The split cutoff space of a gyromagnetic resonator is also readily established. The description of a typical pole of this sort of problem region is also a classic result in the literature. The corresponding cutoff numbers of the split gyromagnetic space are fixed by Eq. (12.10) for $n = +1$ and $n = -2$ and by Eq. (12.11) for $n = -1$ and $n = +2$.

$$\left(\frac{3\psi}{\pi}\right)\left(\frac{\sin\psi}{\psi}\right)^2\left[\frac{J_0(kR_i)}{J_1(kR_i)}-\left(\frac{1+\kappa/\mu}{kR_i}\right)\right]^{-1}+\left(\frac{3\psi}{\pi}\right)\left(\frac{\sin 2\psi}{2\psi}\right)^2 \left[\frac{J_{-3}(kR_i)}{J_{-2}(kR_i)}-2\left(\frac{1+\kappa/\mu}{kR_i}\right)\right]^{-1}-\cot(kL)=0 \tag{12.10}$$

$$\left(\frac{3\psi}{\pi}\right)\left(\frac{\sin\psi}{\psi}\right)^2\left[\frac{J_{-2}(kR_i)}{J_{-1}(kR_i)}+\left(\frac{1+\kappa/\mu}{kR_i}\right)\right]^{-1}+\left(\frac{3\psi}{\pi}\right)\left(\frac{\sin 2\psi}{2\psi}\right)^2 \left[\frac{J_1(kR_i)}{J_2(kR_i)}-2\left(\frac{1+\kappa/\mu}{kR_i}\right)\right]^{-1}-\cot(kL)=0 \tag{12.11}$$

μ and κ are the usual diagonal and off-diagonal entries of the tensor permeability. The ratio of κ and μ is known as the gyrotropy of the problem region. Figure 12.10 indicates the split mode chart with ψ_s = 0.20 rad obtained by

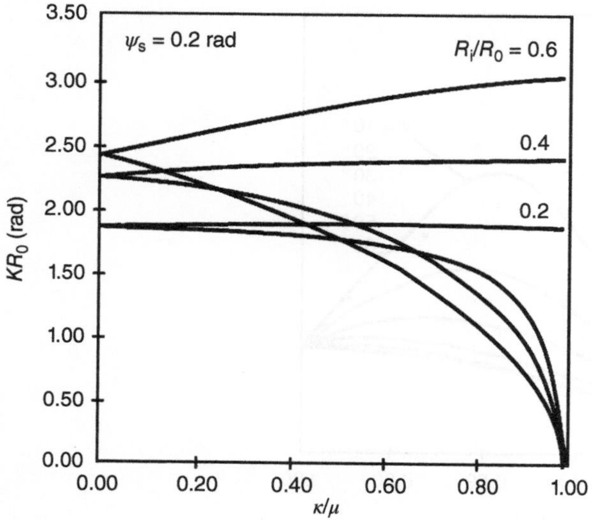

Figure 12.10 Split frequencies of planar gyromagnetic wye resonator in *kR* space using closed-form formulation (ψ = 0.20 rad, R_i/R_0 = 0.20, 0.40, and 0.60).

disregarding the $n = \pm 2$ modes. The opening between the split branches is essentially independent of the shape angle but deteriorates rapidly below about 0.15 rad. Figure 12.11 depicts another result with ψ_s = 0.40 rad.

12.6 TM Field Patterns of Triangular Planar Resonator

The TM-mode field patterns in a triangular-shaped demagnetized ferrite or dielectric resonator having no variation of the field patterns along the thickness of the resonator are given by

$$E_z = A_{m,n,l} T(x,y)$$

$$H_x = \frac{j}{\omega \mu_0 \mu_e} \frac{\partial E_z}{\partial y}$$

$$H_y = \frac{-j}{\omega \mu_0 \mu_e} \frac{\partial E_z}{\partial x}$$

$$H_z = E_x = E_y = 0$$

where $A_{m,n,l}$ is a constant. Figure 12.12 shows the geometry of the planar resonator discussed in this text.

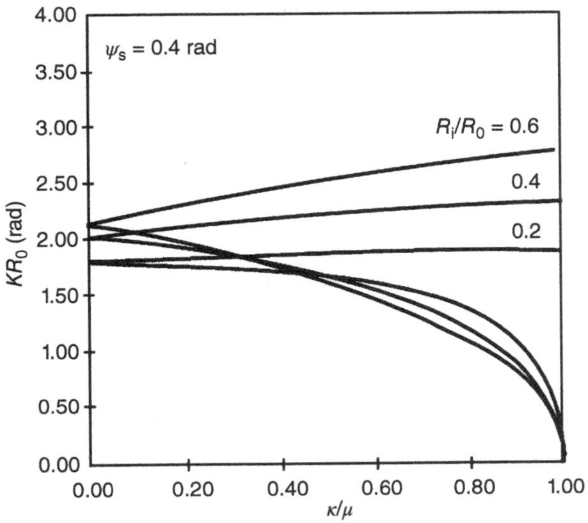

Figure 12.11 Split frequencies of planar gyromagnetic wye resonator in kR space using closed-form formulation ($\psi = 0.20$ rad, $R_i/R_0 = 0.20$, 0.40, and 0.60).

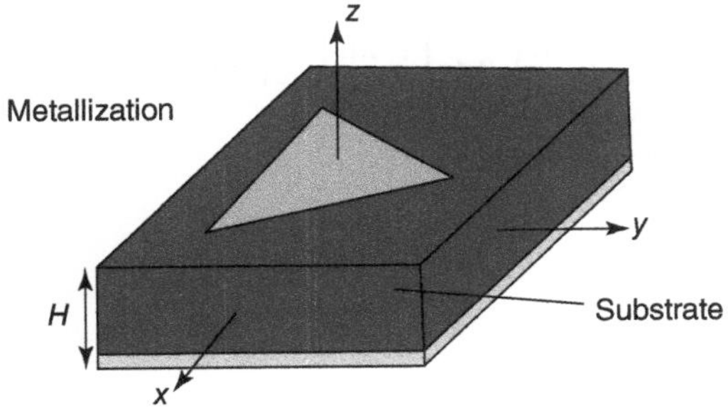

Figure 12.12 Schematic of microstrip triangular resonator.

For a magnetic boundary condition $T(x, y)$ may be obtained by duality from that of the TE mode with the following electric boundary conditions:

$$
\begin{aligned}
T(x,y) = {} & \cos\left[\left(\frac{2\pi x}{\sqrt{3}A}+\frac{2\pi}{3}\right)l\right]\cos\left[\frac{2\pi(m-n)y}{3A}\right] \\
& + \cos\left[\left(\frac{2\pi x}{\sqrt{3}A}+\frac{2\pi}{3}\right)m\right]\cos\left[\frac{2\pi(n-l)y}{3A}\right] \\
& + \cos\left[\left(\frac{2\pi x}{\sqrt{3}A}+\frac{2\pi}{3}\right)n\right]\cos\left[\frac{2\pi(l-m)y}{3A}\right]
\end{aligned}
$$

A is the length of the triangle side and $m + n + l = 0$.

E_z satisfies the wave equation:

$$\left(\frac{\partial^2}{\partial x^2} + \frac{\partial^2}{\partial y^2} + k^2_{m,n,l}\right) E_z = 0$$

where

$$k_{m,n,l} = \frac{4\pi}{3A}\sqrt{(m^2 + mn + n^2)}$$

The interchange of the three digits m, n, l leaves the cutoff number $k_{m,n,l}$ unchanged; similarly, the field patterns are retained, without rotation.

12.7 $TM_{1,0,-1}$ Field Components of Triangular Planar Resonator

The dominant mode in a planar triangular resonator is given with $m = 1$, $n = 0$, $l = -1$. The result is

$$E_z = A_{1,0,-1}\left[2\cos\left(\frac{2\pi x}{\sqrt{3}A} + \frac{2\pi}{3}\right)\cos\left(\frac{2\pi y}{3A}\right) + \cos\left(\frac{4\pi y}{3A}\right)\right]$$

$$H_x = -jA_{1,0,-1}\zeta_e\left[\cos\left(\frac{2\pi x}{\sqrt{3}A} + \frac{2\pi}{3}\right)\sin\left(\frac{2\pi y}{3A}\right) + \sin\left(\frac{4\pi y}{3A}\right)\right]$$

$$H_y = j\sqrt{3}A_{1,0,-1}\zeta_e\left[\sin\left(\frac{2\pi x}{\sqrt{3}A} + \frac{2\pi}{3}\right)\cos\left(\frac{2\pi y}{3A}\right)\right]$$

where

$$k_{1,0,-1} = \frac{4\pi}{3\mathrm{A}}$$

$$\zeta_e = \sqrt{\frac{\varepsilon_0\varepsilon_r}{\mu_0\mu_e}}$$

Figure 12.13 is a sketch of the magnetic and equipotential lines for the dominant $TM_{1,0,-1}$ mode in a triangular resonator.

12.8 Circulation Solutions

The cutoff numbers of the equilateral triangular planar resonator enters into the description of prism cavities met in the design of fixed and switched junction circulators. It fixes the relationship between the aspect ratio and the free space and waveguide wavelengths of the geometry. Figure 12.14 illustrates some

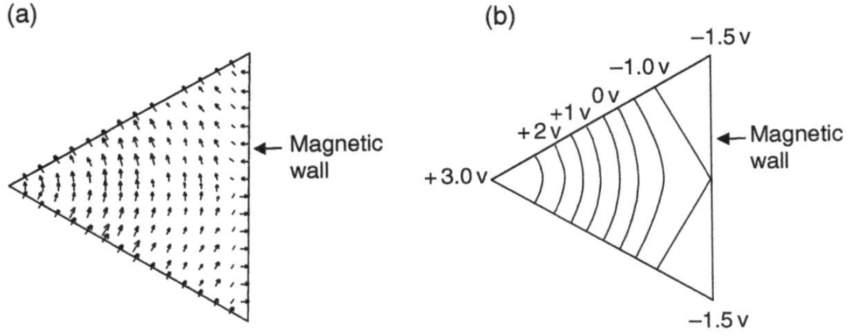

Figure 12.13 (a) Magnetic field pattern for dominant mode and (b) lines of equipotential ($A_{1,0,-1} = 1$).

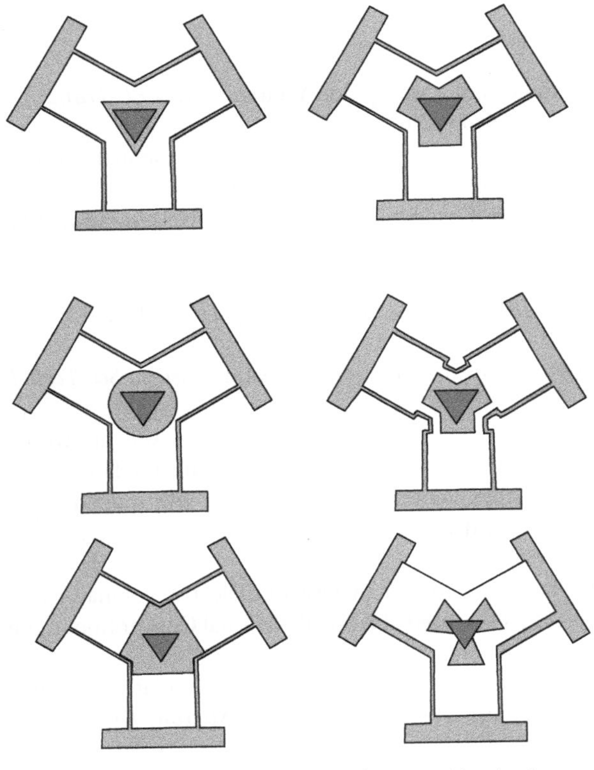

Figure 12.14 Schematic diagrams of waveguide circulators using quarter-wave-coupled triangular resonators.

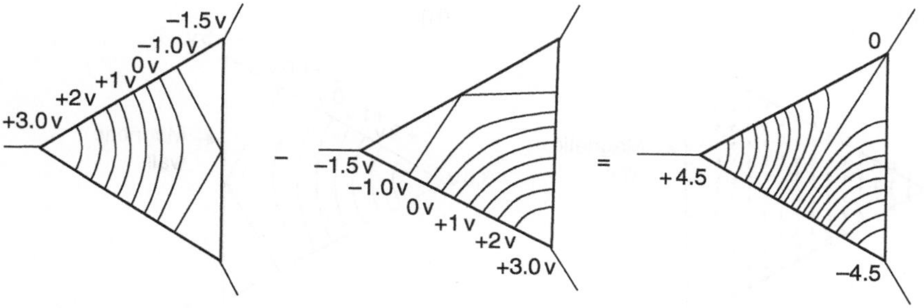

Figure 12.15 Standing wave solutions of waveguide circulator using a triangular resonator.

quarter-wave coupled arrangements. While it is difficult to visualize rotation in such resonators, circulation solutions may therefore be established by constructing a linear combination of isotropic resonators as shown in Figure 12.15.

Bibliography

Akaiwa, Y. (1974). Operation modes of a waveguide Y-circulator. *IEEE Trans. Microw. Theory Tech.* **MTT-22**: 954–959.

Helszajn, J. (1981). Standing wave solutions of planar irregular hexagonal and wye resonators. *IEEE Trans. Microw. Theory Tech.* **MTT-29**: 562–567.

Helszajn, J. and Nisbet, W.T. (1981). Circulators using planar wye resonator. *IEEE Trans. Microw. Theory Tech.* **MTT-29**: 689–699.

How, H., Fang, R.-M., Vittoria, C., and Schmidth, R. (1994). Design of six-port stripline junction circulators. *IEEE Trans. Microw. Theory Tech.* **MTT-42**: 1272–1275.

Ito, Y. and Yochuchi, H. (1969). Microwave junction circulators. *Fujitsu Sci. Tech. J.* 55–90.

Lyon, R.W. (1982). A finite element analysis of planar circulators using arbitrarily shaped resonators. *IEEE Trans. Microw. Theory Tech.* **MTT-30**: 1964–1974.

Nisbet, W.T. and Helszajn, J. (1980). Characterisation of planar wye shaped resonators for use in circulator hardware. Presented at MTT Symposium, Washington, DC.

Ogasawara, N. and Noguchi, T. (1974). Modal analysis of dielectric resonator of normal triangular cross-section. Presented at Annual National Convention of the IEEE, Japan (28 March 1974).

Riblet, G. (1980). Technique for broadbanding above resonance circulator junctions without the use of external matching networks. *IEEE Trans. Microw. Theory Tech.* **MTT-28**: 125–129.

Schelkunoff, S.A. (1943). *Electromagnetic Waves*, 393. New York: Van Nostrand.

Silvester, P. (1961). High order polynomial triangular finite elements for potential problems. *Int. J. Eng. Sci.* 7: 849–861.

13

The Turnstile Junction Circulator: First Circulation Condition

Joseph Helszajn

Heriot Watt University, Edinburgh, UK

13.1 Introduction

The original waveguide junction circulator is the turnstile geometry described in this chapter. It remains a classic commercial device to this day. Its geometry relies on a single half-wave long Faraday rotation section with its flat faces separated from the top and bottom walls of the rectangular waveguide by suitable gaps. Geometries using single and pairs of quarter-wave long resonators have also been described but are outside the remit of this chapter. The half-wave geometry may also be visualized, with respect to the symmetry plane of the resonator, as far as its midband frequency is concerned, as a pair of quarter-wave long resonators open-circuit at its open flat face and short-circuit at its symmetry plane. The adjustment of this class of junction is a classic eigenvalue problem, which is dealt with in some detail in Chapters 15 and 16. The text is restricted to a description of the specific engineering aspect entering into its operation.

The waveguide circulator dealt with here is often but not exclusively referred to as the turnstile arrangement. Its adjustment involves a two-step procedure. The first fixes the degeneracy between a pair of degenerate counterrotating modes and a quasi in-phase mode. The second amounts to replacing the dielectric resonator by a gyromagnetic one in order to remove the degeneracy between the counterrotating modes. The purpose of this chapter is to deal with the first condition.

The transition between the circular gyromagnetic waveguide and a typical rectangular waveguide feed may be catered for by introducing ideal transformers at the terminals of the counterrotating eigen-networks. These transformers do not enter into the adjustment of the first circulation condition of the junction.

Microwave Polarizers, Power Dividers, Phase Shifters, Circulators, and Switches, First Edition. Joseph Helszajn.

13.2 The Four-port Turnstile Junction Circulator

A classic nonreciprocal ferrite device, which relies on a Faraday rotation bit, is the four-port turnstile circulator. It relies for its operation on the junction of a 45° circular waveguide and the intersecting of two rectangular waveguides, as shown in Figure 13.1.

In order to appreciate the operation of this sort of circulator it is first of all necessary to understand that of the conventional turnstile junction. The schematic diagram in question is illustrated in Figure 13.1. This junction is a six-port network, having four rectangular waveguide ports and two circular waveguide ports. The scattering matrix of this circuit is given, in general, by

$$S = \left[\begin{array}{cccc:cc} \alpha & \gamma & \delta & \gamma & \varepsilon & 0 \\ \gamma & \alpha & \gamma & \delta & 0 & \varepsilon \\ \beta & \gamma & \alpha & \gamma & -\varepsilon & 0 \\ \gamma & \beta & \gamma & \alpha & 0 & -\varepsilon \\ \hdashline \varepsilon & 0 & -\varepsilon & 0 & \beta & 0 \\ 0 & \varepsilon & 0 & -\varepsilon & 0 & \beta \end{array}\right] \tag{13.1}$$

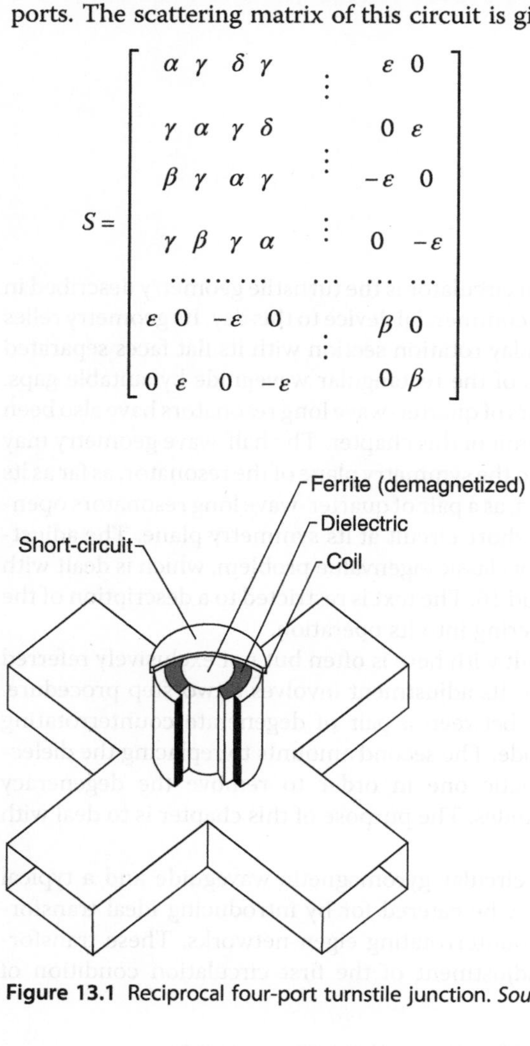

Figure 13.1 Reciprocal four-port turnstile junction. *Source:* After Allen (1956).

If the junction is matched,

$$\alpha = 0 \tag{13.2}$$

$$\beta = 0 \tag{13.3}$$

$$\delta = 0 \tag{13.4}$$

$$|\gamma| = 0 \tag{13.5}$$

$$\varepsilon = \frac{1}{\sqrt{2}} \tag{13.6}$$

The turnstile junction circulator is obtained from the conventional turnstile junction by introducing a 45° Faraday rotation bit in the round waveguide. This arrangement is shown in Figure 13.2.

The operation of the circulator in question may now be understood by considering a typical input wave at port 1. Such a wave produces no reflection at port 1, decouples ports 3 and 6, established equal in-phase waves at ports 2 and 4, and produces a component at port 5. The wave at port 5, upon traversing up and down, the 45° rotator section, is now aligned with port 6 at the plane of the rectangular waveguide. Such a wave decouples ports 1, 3, and 5 and produces out-of-phase waves at ports 2 and 4, which have equal amplitudes to those established by the original incident wave. The net effect is to produce a single output at port 2. Similar considerations indicate that a wave at port 2 is emergent at port 3 and so on in a cyclic manner.

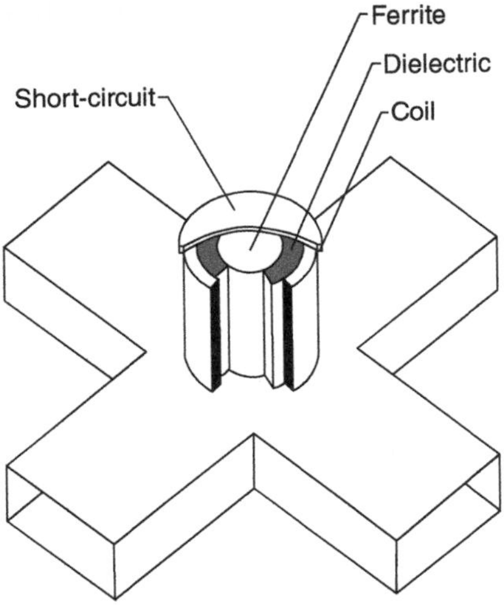

Figure 13.2 Nonreciprocal four-port turnstile junction. *Source:* After Allen (1956).

13.3 The Turnstile Junction Circulator

The original realization of the *H*-plane three-port turnstile junction circulator is depicted in Figure 13.3. It consists of a circular guide containing a longitudinally magnetized ferrite section at the junction of three rectangular waveguides. Figure 13.4 illustrates an *E*-plane configuration.

The operation of the turnstile junction may be understood by having recourse to superposition. It starts by decomposing single input waves at port 1 (say) into a linear combination of voltage settings at each port:

$$\begin{bmatrix} 1 \\ 0 \\ 0 \end{bmatrix} = \frac{1}{3}\begin{bmatrix} 1 \\ 1 \\ 1 \end{bmatrix} + \frac{1}{3}\begin{bmatrix} 1 \\ \alpha \\ \alpha^2 \end{bmatrix} + \frac{1}{3}\begin{bmatrix} 1 \\ \alpha^2 \\ \alpha \end{bmatrix} \tag{13.7}$$

where

$$\alpha = \exp(120j) \tag{13.8}$$

$$\alpha^2 = \exp(240j) \tag{13.9}$$

A scrutiny of the first, so-called in-phase generator setting indicates that it produces an electric field along the axis of the circular waveguide, which does not couple into it. The reflected waves at the three ports of the junction are therefore in this instance unaffected by the details of the gyromagnetic waveguide. A scrutiny of the second and third so-called counterrotating generator settings, indicates, however, that these establish counterrotating circularly polarized alternating magnetic fields at the open face of the circular gyromagnetic waveguide, which readily propagate. Since a characteristic of such a

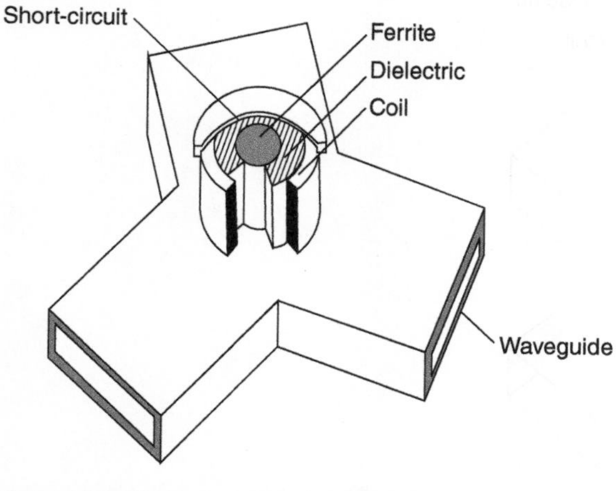

Figure 13.3 *H*-plane turnstile junction circulator. *Source:* After Schaug-Patterson (1958).

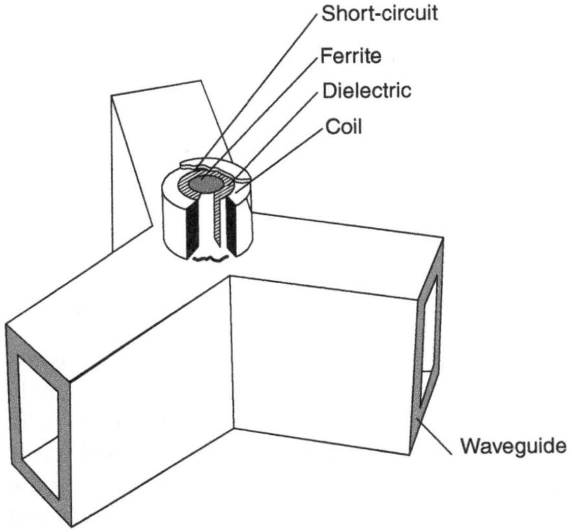

Figure 13.4 Schematic diagram of three-port *E*-plane turnstile circulator using single turnstile resonator.

waveguide is that is has different scalar permeabilities under the two arrangements it provides one practical means of removing the degeneracy between the reflected waves associated with these two generator settings.

A typical reflected wave at any port is constructed by adding the individual ones due to each possible generator setting. A typical term is realized by taking the product of a typical incident wave and a typical reflection coefficient:

$$\begin{bmatrix} b_1 \\ b_2 \\ b_3 \end{bmatrix} = \frac{\rho_0}{3}\begin{bmatrix} 1 \\ 1 \\ 1 \end{bmatrix} + \frac{\rho_-}{3}\begin{bmatrix} 1 \\ \alpha \\ \alpha^2 \end{bmatrix} + \frac{\rho_+}{3}\begin{bmatrix} 1 \\ \alpha^2 \\ \alpha \end{bmatrix} \tag{13.10}$$

An ideal circulator is now defined as

$$\frac{\rho_0 + \rho_- + \rho_+}{3} = 0 \tag{13.11}$$

$$\frac{\rho_0 + \alpha\rho_- + \alpha^2\rho_+}{3} = -1 \tag{13.12}$$

$$\frac{\rho_0 + \alpha^2\rho_- + \alpha\rho_+}{3} = 0 \tag{13.13}$$

To adjust this, and other circulators, requires a 120° phase difference between the reflection coefficients of the three different ways it is possible to excite the three rectangular waveguides. One solution is

$$\rho_+ = \exp\left[-2j\left(\theta_1 + \theta_+ + \frac{\pi}{2}\right)\right] \tag{13.14}$$

$$\rho_- = \exp\left[-2j\left(\theta_1 + \theta_- + \frac{\pi}{2}\right)\right] \tag{13.15}$$

$$\rho_0 = \exp[-2j\theta_0] \tag{13.16}$$

provided that

$$\theta_1 = \theta_0 = \frac{\pi}{2} \tag{13.17}$$

$$\theta_+ = -\theta_- = \frac{-\pi}{6} \tag{13.18}$$

The required phase angles of the three reflection coefficients are established by adjusting the length of the demagnetized ferrite section so that the angle between the in-phase and counterrotating reflection coefficients is initially 180°. The degenerate phase angles of the counter rotation reflection coefficient are separated by 120° by magnetizing the ferrite region, thereby producing the ideal phase angles of the circulator. These two steps represent the necessary and sufficient conditions for the adjustment of this class of circulator. Figure 13.5 depicts the eigenvalue problem of the junction.

13.4 Scattering Matrix

The design of any junction circulator is incomplete without a formulation of its scattering matrix. The bilinear relation between the eigenvalues of the junction and its scattering matrix is defined in the usual way by

$$S_{11} = \frac{\rho_0 + \rho_+ + \rho_-}{3} \tag{13.19a}$$

$$S_{21} = \frac{\rho_0 + \alpha\rho_+ + \alpha^2\rho_-}{3} \tag{13.19b}$$

$$S_{31} = \frac{\rho_0 + \alpha^2\rho_+ + \alpha\rho_-}{3} \tag{13.19c}$$

where

$$\alpha = \exp(120j) \tag{13.20a}$$

$$\rho_\pm = \frac{\xi_0 - Y_\pm}{\xi_0 + Y_\pm} \tag{13.20b}$$

$$\rho_0 = \frac{\xi_0 - Y_0}{\xi_0 + Y_0} \tag{13.20c}$$

The scattering matrix of the first circulation condition is defined by the eigenvalue diagram in Figure 13.6 as

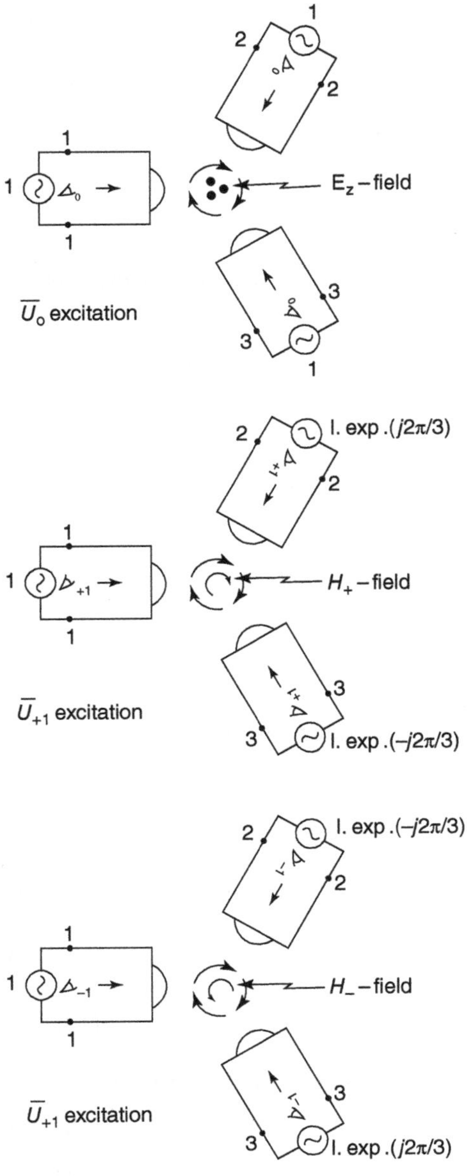

Figure 13.5 Eigen-solutions of the three-port circulator.

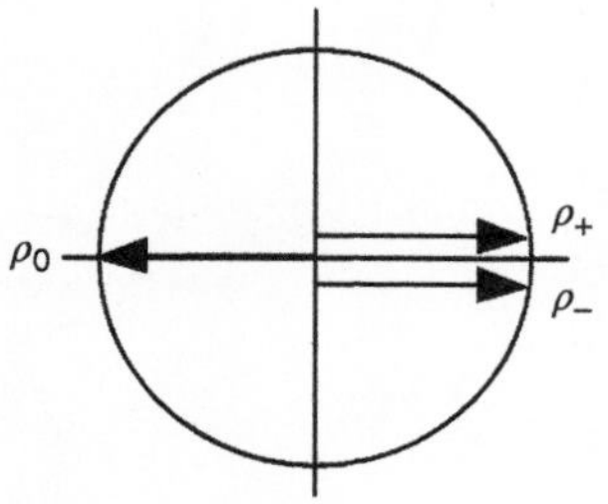

Figure 13.6 Eigenvalue diagram of reciprocal three-port junction for maximum power transfer.

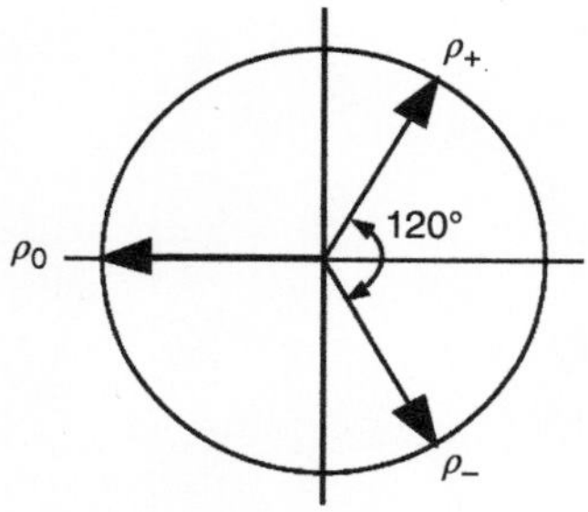

Figure 13.7 Eigenvalue diagram of ideal three-port junction circulator.

$$S_{11} = \frac{1}{3} \tag{13.21a}$$

$$S_{21} = \frac{2}{3} \tag{13.21b}$$

$$S_{31} = \frac{2}{3} \tag{13.21c}$$

That of the second circulation condition is defined by that in Figure 13.7.

$$S_{11} = 0 \tag{13.22a}$$

$$S_{21} = 1 \tag{13.22b}$$

$$S_{31} = 0 \tag{13.22c}$$

The above scattering parameters satisfy the unitary condition without ado

$$\bar{S} \cdot \bar{S}^* - \bar{I} = 0 \tag{13.23}$$

13.5 Frequencies of Cavity Resonators

The basic resonator met in the design of turnstile geometries is a half-wave long arrangement with an ideal or open magnetic sidewall and similar flat ones. The governing equation is the dispersion condition in an infinite long waveguide with a similar contour:

$$\left(\frac{2\pi}{\lambda_g}\right)^2 = \left(\frac{2\pi}{\lambda_0}\right)^2 \varepsilon_f - \left(\frac{2\pi}{\lambda_c}\right)^2 \quad (13.24)$$

where λ_0 is the free-space wavelength, λ_c is the cutoff wavelength and is the unknown of the problem, and λ_g is the wavelength of the waveguide.

The length of the cavity is denoted as L for a quarter-wave long cavity and $2L$ for half-wave geometry:

$$\lambda_g = 4L. \quad (13.25)$$

The unknown of the problem is λ_c. For a cylindrical waveguide,

$$\frac{2\pi}{\lambda_c} = \frac{1.84}{R} \quad (13.26)$$

For a triangular one,

$$\frac{2\pi}{\lambda_c} = \frac{4\pi}{3A} \quad (13.27)$$

R is the radius of the cylindrical resonator and A is the side of the triangular one. Introducing this condition in the dispersion relationship and rearranging gives

$$\left(\frac{2\pi}{\lambda_0}\right)^2 \varepsilon_f = \left(\frac{2\pi}{2L}\right)^2 + \left(\frac{2\pi}{\lambda_c}\right)^2. \quad (13.28)$$

No closed form cutoff numbers exist for other cross-sections such as a clover leaf waveguide. A numerical procedure is therefore mandatory to deal with the general problem. One means of doing so is to recognize that the general solution coincides with that of the planar resonator revealed by letting λ_g or L move to infinity. This approach avoids the need to have recourse to a three-dimensional solver.

13.6 Effective Dielectric Constant of Open Dielectric Waveguide

The characterization of the counterrotating frequencies of the open turnstile resonator is incomplete without catering for the open sidewalls. This may be done by constructing a one-to-one correspondence between an open waveguide

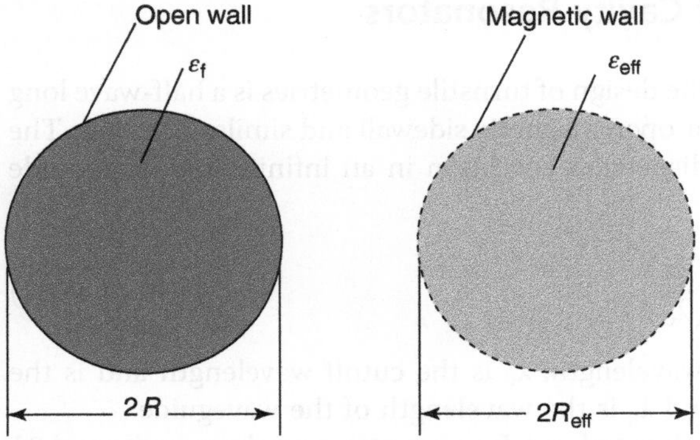

Figure 13.8 Equivalence between open dielectric waveguide with imperfect magnetic walls and equivalent waveguide model with idealized walls.

with a dielectric constant ε_f and an equivalent waveguide with an ideal magnetic wall but with an effective dielectric constant ε_{eff}.

Equivalent waveguide models are often employed to represent inhomogeneous waveguides or transmission lines. One possible equivalence between an open dielectric waveguide with a relative dielectric constant ε_f and radius R propagating the HE_{11} mode and a circular waveguide with magnetic wall boundary conditions with an effective dielectric constant ε_{eff} and effective radius R_{eff} propagating the TM_{11} mode is depicted in Figure 13.8. ε_{eff} and R_{eff} may be determined from knowledge of the phase constant of the inhomogeneous waveguide and the complex power flow. If only the phase constant is required, or if only matching between similar cross-sections is required, then an approximate waveguide model in terms of an effective dielectric constant is sufficient and will be adopted here in the first instance.

The characteristic equation for the propagation constant β of the open dielectric waveguide is a standard result and is reproduced here for completeness sake:

$$\left[\frac{J_n'(u)}{uJ_n(u)}+\frac{K_n'(w)}{wK_n(w)}\right]\left[\frac{\varepsilon_1}{\varepsilon_2}\frac{J_n'(u)}{uJ_n(u)}+\frac{K_n'(w)}{wK_n(w)}\right]=n^2\left[\frac{1}{u^2}+\frac{1}{w^2}\right]\left[\frac{\varepsilon_1}{\varepsilon_2}\frac{1}{u^2}+\frac{1}{w^2}\right] \tag{13.29}$$

ε_1 is the dielectric constant of the ferrite region, ε_2 is that of the surrounding material and

$$u^2=\left(k_0^2\varepsilon_1-\beta^2\right)R^2 \tag{13.30}$$

$$w^2=\left(\beta^2-k_0^2\varepsilon_2\right)R^2 \tag{13.31}$$

where R is the radius of the ferrite, k_0 is given in Eq. (13.33), $J_n(u)$ is the Bessel function of the first kind of order n, and $K_n(w)$ is the modified Bessel function of the second kind representing outward traveling waves.

The effective dielectric constant of the equivalent waveguide model with an idealized magnetic wall may now be evaluated from knowledge of the phase constant β of the open waveguide by making use of the following relationship:

$$\beta^2 = k_0^2 \varepsilon_{\text{eff}} - k_c^2 \tag{13.32}$$

where

$$k_0 = \frac{2\pi}{\lambda_0} \tag{13.33}$$

$$k_c = \frac{1.84}{R} \tag{13.34}$$

Figure 13.9 illustrates the relationship between the effective dielectric constant ε_{eff} and k_0R for ε_f equal to 10, 12.5, and 15.

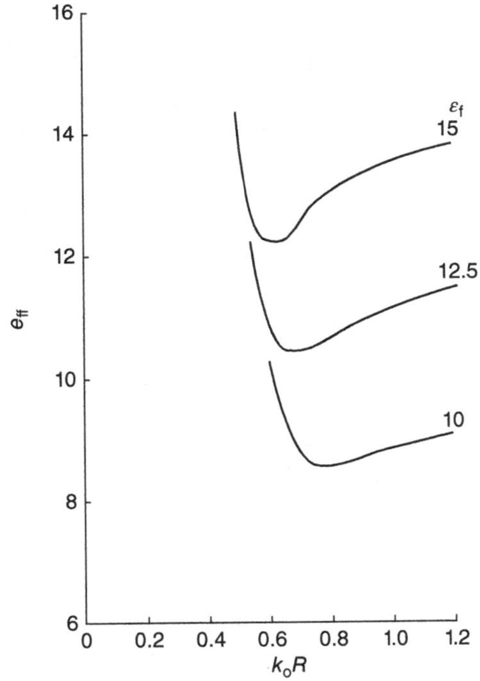

Figure 13.9 Effective dielectric constant of open dielectric waveguide. *Source:* Reprinted from Helszajn and Sharp (1986) with permission.

This result indicates that for $\varepsilon_f = 15$, say, $\varepsilon_{eff} = 12.2$ for $k_0R = 0.60$, 12.6 for $k_0R = 0.70$, 13.1 for $k_0R = 0.80$, and 13.4 for $k_0R = 0.90$. The origin of this discrepancy may be separately understood by evaluating the power flow through the waveguide P_i and outside it P_0. Such a calculation indicates that for $k_0R = 0.80$ and $\varepsilon_f = 16$, say, P_i/P_0 is of the order of 25.

13.7 The Open Dielectric Cavity Resonator

The cavity resonator in the design of a junction circulator is a half-wave long geometry with open flat faces separated by dielectric spacers from top and bottom pistons. The arrangement considered here is illustrated in Figure 13.10.

The calculation of the degenerate counterrotating frequencies of the open resonator is well understood. A mode-matching procedure must separately satisfy the boundaries between regions 1 and 2, 1 and 3, and 1 and 4 as defined in Figure 13.11a. The boundary between 1 and 3 is usually neglected in that there is little or no field in region 3. The one between regions 1 and 4 may be catered for by replacing the open dielectric resonator by an equivalent closed one with an effective dielectric constant ε_{eff}. The boundary between 1 and 2 is met by forming a transverse resonance condition between the two. The dual geometry obtained by placing an image or electric wall at the symmetry plane of Figure 13.11a is indicated in Figure 13.11b. Its circuit topology is illustrated in Figure 13.12.

The characteristic equation is

$$\frac{\varepsilon_{eff}k_0}{\beta_0}\cot(\beta_0 L) - \frac{\varepsilon_d k_0}{\alpha_0}\coth(\alpha_0 S) = 0 \tag{13.35}$$

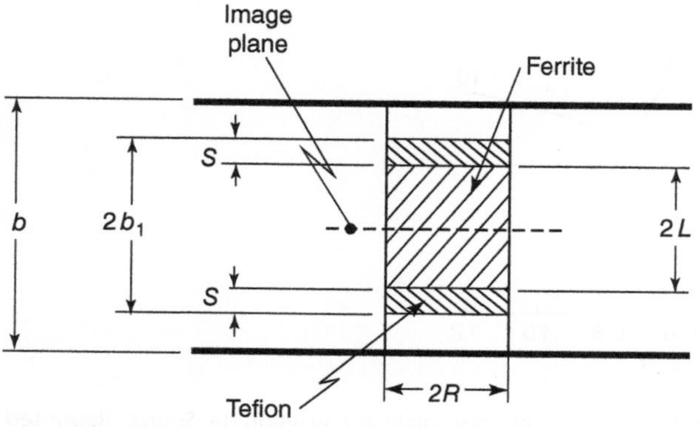

Figure 13.10 Schematic diagram of a practical gyromagnetic resonator.

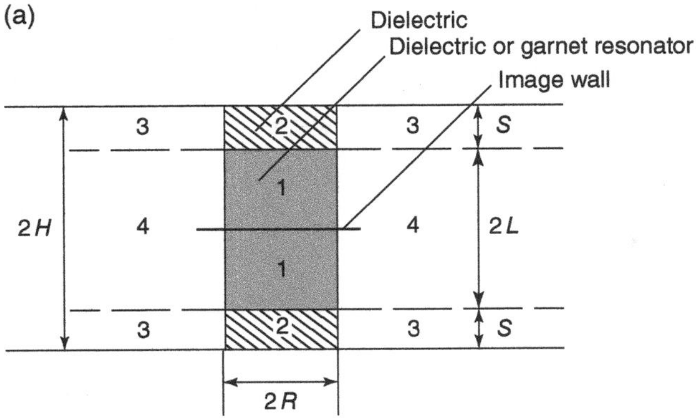

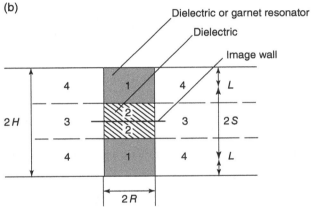

Figure 13.11 (a) Schematic diagram of half-wave long open ferrite or dielectric resonator loaded by image wall. (b) Schematic diagram of coupled quarter-wave long ferrite or dielectric resonators loaded by image wall.

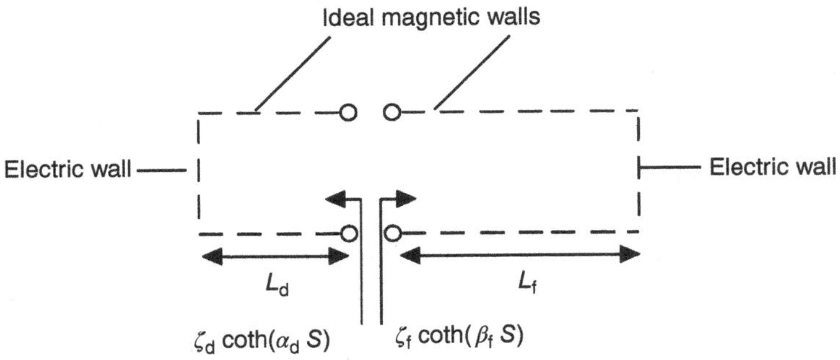

Figure 13.12 Circuit topology of closed composite gyromagnetic resonator.

where L is the half space of the resonator, S is the thickness of the dielectric gap, and ε_{eff} is the effective dielectric constant of the gyromagnetic resonator; it is obtained from either measurement or calculation. ε_d is the relative dielectric constant of the spacers, and

$$\left(\frac{\beta_0}{k_0}\right)^2 = \varepsilon_{eff} - \left(\frac{k_c}{k_0}\right)^2 \tag{13.36}$$

$$\left(\frac{\alpha_0}{k_0}\right)^2 = \left(\frac{k_c}{k_0}\right)^2 - \varepsilon_d \tag{13.37}$$

The condition in Eq. (13.35) may, for computational purposes, be written as

$$\varepsilon_{eff}\left(\frac{k_0}{\beta_0}\right)\cot\left[\left(\frac{\beta_0}{k_0}\right)\left(\frac{L}{A}\right)k_0A\right] - \varepsilon_d\left(\frac{k_0}{\alpha_0}\right)\coth\left[\left(\frac{\alpha_0}{k_0}\right)\left(\frac{L}{A}\right)\left(\frac{1-q_\pm}{q_\pm}\right)k_0A\right] = 0 \tag{13.38}$$

where $q_\pm$ is the gap factor, defined by

$$q_\pm = \frac{L}{L+S} \tag{13.39}$$

The factor $(1 - q_\pm)/q_\pm$ that appears in the characteristic equation may also be written in terms of the gap-resonator ratio S/L:

$$\frac{1-q_\pm}{q_\pm} = \frac{S}{L} \tag{13.40}$$

The characteristic equation in Eq. (13.38) fixes the frequencies of the degenerate eigenvalues of the circulator. The gap S is here not, in practice, an independent variable but is determined by the frequency of the in-phase eigen-network. Figure 13.13a and b show quarter-wave-coupled turnstile junctions using single and pairs of quarter-wave long resonators, open-circuited at one flat face and short-circuited at the other.

13.8 The In-phase Mode

The dielectric gap in the design of a three-port junction circulator fixes the so-called in-phase mode of the junction. A suitable mode with a magnetic wall at the origin and an electric wall on the boundary is a quasi-planar geometry supporting a $TM_{0,1}$ mode. Its cutoff number is given by

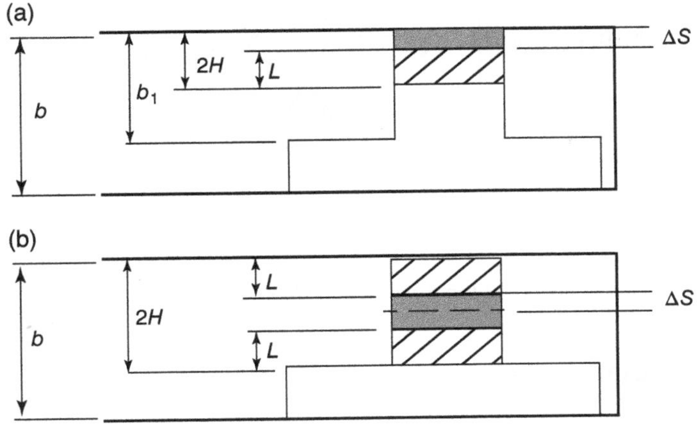

Figure 13.13 Quarter-wave coupled turnstile junctions using single (a) and pair (b) of quarter-wave long resonators.

$$k_0\sqrt{\varepsilon_{\text{eff}}}R = 2.40 \tag{13.41}$$

The resonator here is a two-plate arrangement with dielectric constant layers equal to ε_{f} and ε_{d}. A quasi-static approximation is obtained by taking the equivalent series capacitance, which leads to

$$\frac{L+S_0}{\varepsilon_0\varepsilon_{\text{eff}}} = \frac{L}{\varepsilon_0\varepsilon_{\text{f}}} + \frac{S_0}{\varepsilon_0\varepsilon_{\text{d}}} \tag{13.42}$$

After simplification, the equivalent permittivity may be written as

$$\varepsilon_{\text{eff}} = \frac{\varepsilon_{\text{d}}\varepsilon_{\text{f}}}{q_0\varepsilon_{\text{f}} + (1-q_0)\varepsilon_{\text{d}}} \tag{13.43}$$

where q_0 is the gap factor, given by

$$q_0 = \frac{L}{L+S_0} \tag{13.44}$$

The connection between q_0 and k_0R is readily understood.

The exact model of the in-phase resonator is obtained by introducing a triplet of electric walls at the symmetry plane of the junction. The arrangement is here depicted in Figure 13.14.

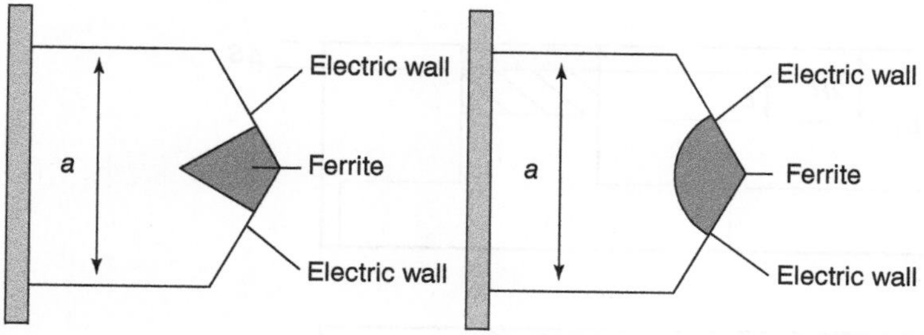

Figure 13.14 In-phase eigen-network of the three-port junction circulator.

13.9 First Circulation Condition

The first of the two circulation conditions coincides with that for which the curve between q_{eff} and k_0R and $q_\pm$ versus k_0R for parametric values of the aspect ratio R/L and A/L of the resonator intersect. The points of the first relationship are compatible with a short-circuit boundary condition at the terminals of the resonator; the second with an open-circuit at the same terminals. The required condition is satisfied, provided

$$q_0 = q_\pm = q \tag{13.45}$$

$$q = \frac{L}{L+S} \tag{13.46}$$

where S is the gap between the open flat face of the resonator and the top wall of the waveguide.

Bibliography

Aitken, F.M. and Mclean, R. (1963). Some properties of the waveguide Y circulator. *Proc. Inst. Electr. Eng.* **110** (2): 256–260.

Akaiwa, Y. (1974). Operation modes of a waveguide Y-circulator. *IEEE Trans. Microw. Theory Tech.* **MTT-22**: 954–959.

Akaiwa, Y. (1978). A numerical analysis of waveguide H-plane Y-junction circulators with circular partial-height ferrite post. *J. Inst. Electron. Commun. Eng. Jpn.* **E61**: 609–617.

Allen, P.J. (1956). The turnstile circulator. *IRE Trans. Microw. Theory Tech.* **MTT-4**: 223–227.

Auld, B.A. (1959). The synthesis of symmetrical waveguide circulators. *IRE Trans. Microw. Theory Tech.* **MTT-7**: 238–246.

Denlinger, E.J. (1974). Design of partial-height ferrite waveguide circulators. *IEEE Trans. Microw. Theory Tech.* **MTT-22**: 810–813.

Green, J.J. and Sandy, F. (1974). Microwave characterization of partially magnetized ferrites. *IEEE Trans. Microw. Theory Tech.* **MTT-22**: 645–651.

Hauth, W. (1981). Analysis of circular waveguide cavities with partial-height ferrite insert. *11th Proceedings of European Microwave Conference*, Amsterdam, the Netherlands (7–11 September 1981), pp. 383–388.

Helszajn, J. (1974). Common waveguide circulator configurations. *Electron. Eng.* **46**: 66–68.

Helszajn, J. (1999). Adjustment of degree-2 H-plane waveguide turnstile circulator using prism resonator. *Microw. Eng. Eur.* (July): 35–48.

Helszajn, J. and Sharp, J. (1983). Resonant frequencies, Q-factor, and susceptance slope parameter of waveguide circulators using weakly magnetized open resonators. *IEEE Trans. Microw. Theory Tech.* **MTT-31**: 434–441.

Helszajn, J. and Sharp, J. (1985). Adjustment of in-phase mode in turnstile junction circulators. *IEEE Trans. Microw. Theory Tech.* **MTT-33** (4): 339–343.

Helszajn, J. and Sharp, J. (1986). Dielectric and permeability effects in HE_{111} open demagnetised ferrite resonators. *IEE Proc., Pt. H* **133** (4): 271–275.

Helszajn, J. and Sharp, J. (2005). Verification of first circulation condition of turnstile waveguide circulators using a finite element solver. *IEEE Trans. Microw. Theory Tech.* **MTT-53** (7): 2309–2316.

Helszajn, J. and Tan, F.C.F. (1975a). Mode charts for partial-height ferrite waveguide circulators. *Proc. Inst. Electr. Eng.* **122** (1): 34–36.

Helszajn, J. and Tan, F.C.F. (1975b). Design data for radial waveguide circulators using partial-height ferrite resonators. *IEEE Trans. Microw. Theory Tech.* **MTT-23**: 288–298.

Helszajn, J. and Tan, F.C.F. (1975c). Susceptance slope parameter of waveguide partial-height ferrite circulators. *Proc. Inst. Electr. Eng.* **122** (72): 1329–1332.

Hogan, C.L. (1952). The ferromagnetic Faraday effect at microwave frequencies and its applications – the microwave gyrator. *Bell. Syst. Tech.* **31**: 1–31.

Montgomery, C., Dicke, R.H., and Purcel, E.M. (1948). *Principles of Microwave Circuits*, Ch. 12. New York: McGraw-Hill Book Co. Inc.

Owen, B. (1972). The identification of modal resonances in ferrite loaded waveguide junction and their adjustment for circulation. *Bell Syst. Tech. J.* **51** (3): 595–627.

Owen, B. and Barnes, C.E. (1970). The compact turnstile circulator. *IEEE Trans. Microw. Theory Tech.* **MTT-18**: 1096–1100.

Rado, G.T. (1953). Theory of the microwave permeability tensor and faraday effect in non-saturated ferromagnetic materials. *Phys. Rev.* **89**: 529.

Schaug-Patterson, T. (1958). Novel Design of a 3-port Circulator. Norwegian Defence Research Establishment Report, Rpt No. R-59 (January).

Schlömann, E. (1970). Microwave behaviour of partially magnetized ferrites. *J. Appl. Phys.* **41** (1): 204–214.

14

The Turnstile Junction Circulator: Second Circulation Condition

Joseph Helszajn[1] *and Mark McKay*[2]

[1] *Heriot Watt University, Edinburgh, UK*
[2] *Honeywell, Edinburgh, UK*

14.1 Introduction

A classic junction circulator is the turnstile junction based on the topologies in Figure 14.1. Its first circulation condition has been the subject of a large number of numerical and experimental calculations. Its second circulation condition, however, has been usually experimentally tackled by replacing the dielectric resonator by a gyromagnetic one. The purpose of this chapter is to deduce the second condition in closed form. This is done by having recourse to the properties of a quarter-wave long circular or triangular gyromagnetic waveguide together with the introduction of ideal transformers at the terminals of the counterrotating eigen-networks of the junction. The chapter includes closed-form descriptions of the quality factor, gyrator conductance, and susceptance slope parameter of the one-port complex gyrator circuit of the junction. This is done in terms of the properties of a quarter-wave long gyromagnetic waveguide open-circuited at one flat face and short-circuited at the other together with the turns-ratio of an ideal transformer at the terminals of the counterrotating eigen-networks of the junction. A means of experimentally deducing the turns-ratio of the ideal transformer from a measurement of the susceptance slope parameter is given special attention. The notion of ideal transformers at the terminals of the eigen-networks of any junction has been introduced in Akaiwa (1974).

The literature of the turnstile junction circulator may be divided into one group which deals with its phenomenological operation, another which deals with its numerical description and still another which mainly deals with its experimental adjustment.

Microwave Polarizers, Power Dividers, Phase Shifters, Circulators, and Switches,
First Edition. Joseph Helszajn.

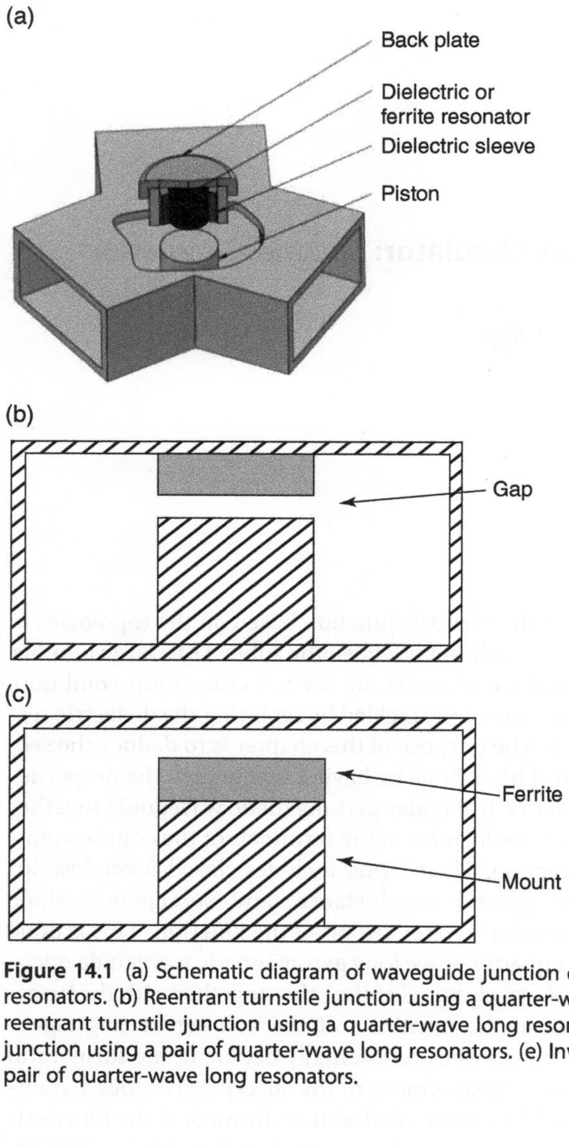

Figure 14.1 (a) Schematic diagram of waveguide junction circulators using single turnstile resonators. (b) Reentrant turnstile junction using a quarter-wave long resonator. (c) Inverted reentrant turnstile junction using a quarter-wave long resonator. (d) Reentrant turnstile junction using a pair of quarter-wave long resonators. (e) Inverted turnstile junction using a pair of quarter-wave long resonators.

14.2 Complex Gyrator of Turnstile Circulator

The purpose of this section is to reiterate the first and second circulation conditions in a slightly different way. This is done as a preamble to dealing with the second circulation condition of the turnstile circulator. The elements entering

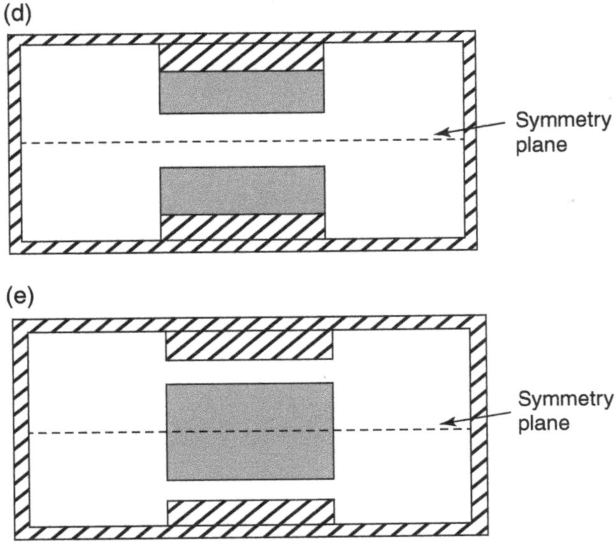

Figure 14.1 (Continued)

into its description are the susceptance slope parameter (b'), the gyrator conductance (g), and the quality factor (Q_L). The two circulation conditions are here deduced from the imaginary and real parts of the one-port gyrator circuit instead of the scattering parameters of the junction.

An important concept that enters into the description of a junction circulator is its complex gyrator immitance:

$$y_{in} = \frac{Y_{in}}{Y_0} = \frac{y_+ + y_-}{2} - j\sqrt{3}\left(\frac{y_+ - y_-}{2}\right), \quad y_0 = \infty \tag{14.1}$$

This equation is exact provided the in-phase eigen-network is separately idealized by a short-circuit boundary condition. The imaginary part of the complex gyrator circuit, the so-called first circulation condition, fixes its midband frequency. The real part, the so-called second condition, fixes its gyrator conductance. The definition of the complex gyrator circuit is illustrated in Figure 14.2.

The first circulation condition is obtained by reducing the gyrator conductance to zero,

$$\frac{y_+ + y_-}{2} = 0, \quad y_0 = \infty \tag{14.2a}$$

$$-j\sqrt{3}\left(\frac{y_+ - y_-}{2}\right) = 0, \quad y_0 = \infty \tag{14.2b}$$

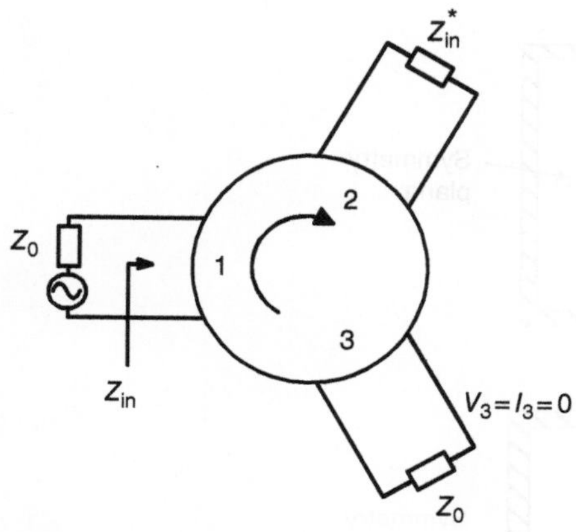

Figure 14.2 Definition of complex gyrator circuit.

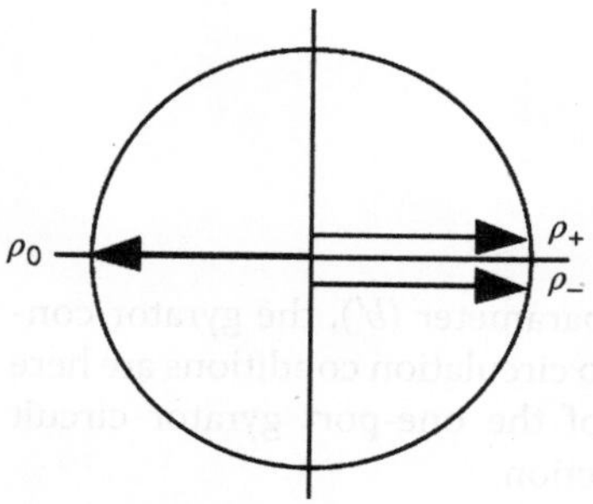

Figure 14.3 Eigenvalue diagram of reciprocal three-port junction for maximum power transfer.

The assumption here is a vanishingly small but not zero gyrotropy. This gives

$$y_+ = y_- = y_1, \quad y_0 = \infty \tag{14.3a}$$

$$y_1 = 0, \quad y_0 = \infty \tag{14.3b}$$

The corresponding reflection coefficients are

$$\rho_+ = \rho_- = 1 \tag{14.4a}$$

$$\rho_0 = -1 \tag{14.4b}$$

The classic eigenvalue diagram obtained here is indicated in Figure 14.3. The second circulation condition is defined by

$$\frac{y_+ + y_-}{2} = 0, \quad y_0 = \infty \tag{14.5a}$$

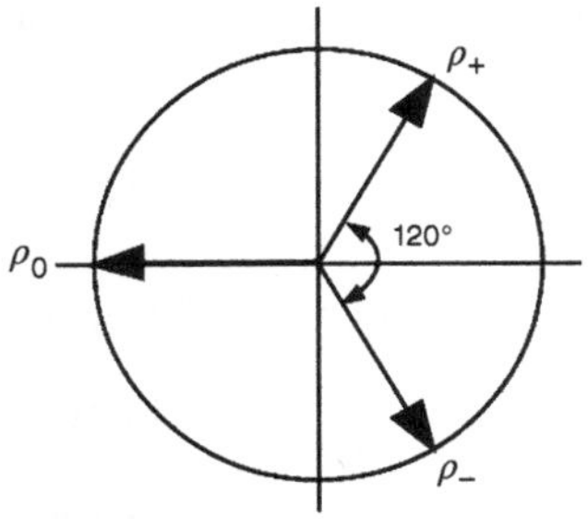

Figure 14.4 Eigenvalue diagram of ideal three-port junction circulator.

and

$$-j\sqrt{3}\left(\frac{y_+ - y_-}{2}\right) = 1, \quad y_0 = \infty \tag{14.5b}$$

The corresponding admittance eigenvalues are here given by

$$y_+ = \frac{-j\sqrt{3}}{2}, \quad y_0 = \infty \tag{14.6a}$$

$$y_- = \frac{j\sqrt{3}}{2}, \quad y_0 = \infty \tag{14.6b}$$

Application of the above two conditions fixes the gyrotropy of the junction. The reflection angles ϕ_0 and $\phi_\pm$ are here displaced by 120°, as shown in Figure 14.4. The reflection eigenvalues are

$$\rho_\pm = 1 \cdot \exp\left(\pm j\frac{\pi}{3}\right) \tag{14.7a}$$

and

$$\rho_0 = -1 \tag{14.7b}$$

14.3 Susceptance Slope Parameter, Gyrator Conductance, and Quality Factor

A complete description of the one-port complex gyrator circuit requires knowledge of the normalized susceptance slope parameter b' of the imaginary part of the gyrator admittance,

$$b' = \left(\frac{\omega_0}{2}\right)\left(\frac{\partial b}{\partial \omega}\right)_{\omega = \omega_0} \tag{14.8}$$

where b is the susceptance. The parameter b' is a measure of the bandwidth of the circuit. The result here is

$$b' = \left(\frac{\omega_0}{2}\right)\left(\frac{y_- - y_+}{\omega_+ - \omega_-}\right), \quad y_0 = \infty \tag{14.9}$$

The above may be understood by recognizing that $y_\pm = 0$ at $\omega = \omega_\pm$. This gives

$$b = \frac{y_-}{2}, \quad \text{at} \quad \omega = \omega_+ \tag{14.10a}$$

$$b = \frac{y_+}{2}, \quad \text{at} \quad \omega = \omega_- \tag{14.10b}$$

Writing the difference between the admittance eigenvalues in terms of the gyrator conductance separately gives the classic relationship:

$$g = \sqrt{3}b'\left(\frac{\omega_+ - \omega_-}{\omega_0}\right), \quad y_0 = \infty \tag{14.11}$$

A further relationship that enters into the description of a one-port G-STUB circuit is its quality factor Q_L, given by

$$Q_L = \frac{b'}{g} \tag{14.12}$$

The result here is

$$\frac{1}{Q_L} = \sqrt{3}\left(\frac{\omega_+ - \omega_-}{\omega_0}\right), \quad y_0 = \infty \tag{14.13}$$

The split frequencies of the junction coincide with the 9.5 dB points in the frequency response at one port with the other two terminated with matched loads.

This quantity fixes the gain-bandwidth product of the complex gyrator circuit as is universally understood

$$(2\delta_0)(\text{RL})Q_L = \text{constant} \tag{14.14}$$

RL is the return loss (dB) and $2\delta_0$ is the normalized bandwidth. The constant on the right-hand side of this condition is determined by the nature and degree of any matching network.

14.4 Propagation in Gyromagnetic Waveguides

The nature of the split phase constant $\beta_\pm$ in a longitudinally magnetized open and closed circular waveguide are classic topics in the literature. Characteristic equations for the calculations of the split phase constants in a gyromagnetic waveguide with either electric or magnetic walls are available in the literature. The characteristic equation of the partially filled gyromagnetic circular waveguide with an electric wall is also dealt with in the literature. A closed-form solution of the magnetic wall geometry based on perturbation theory and one in a related isotropic waveguide are available in the literature. The agreement between the former description and the exact solution is excellent. A characteristic equation of the open anisotropic gyromagnetic waveguide has historically been avoided in this sort of problem because of its complexity and the simpler approximate formulation based on perturbation theory has been used instead. The split phase constants obtained in this way are given by

$$\beta_{\pm}^2 = \left(\frac{\omega_0}{c}\right)^2 \varepsilon_f(\mu \mp C_{11}\kappa) - k_c^2 \tag{14.15}$$

where ε_f is the dielectric constant of the magnetic insulator. The constant C_{11} embodies the variation of the alternating magnetic field over the cross-section of the waveguide. For a cylindrical resonator, the value k_c of the isotropic waveguide and the constant C_{11} are given by

$$k_c = \frac{1.84}{R} \tag{14.16}$$

$$C_{11} = \frac{2}{(1.84)^2 - 1} \tag{14.17}$$

For a prism resonator, k_c and C_{11} are given by

$$k_c = \frac{4\pi}{3A} \tag{14.18}$$

$$C_{11} = \frac{\sqrt{3}}{\pi} \tag{14.19}$$

The agreement between the exact and perturbation solution in the case of a circular gyromagnetic waveguide with an ideal magnetic wall is illustrated in Figure 14.5.

This sort of waveguide displays both split propagation constants and split cutoff frequencies. The split cutoff frequencies are determined with $\beta_{\pm}$ given by

$$\beta_{\pm} = 0 \tag{14.20}$$

14.5 Eigen-network of Turnstile Circulator

The counterrotating eigen-networks are short-circuited unit elements (UEs) or stubs, where admittances are pure imaginary numbers.

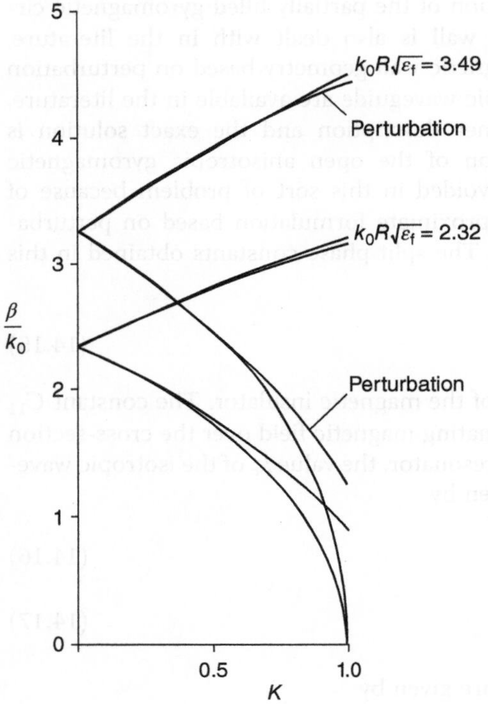

Figure 14.5 Comparison between perturbation and numerical solutions in the case of a circular gyromagnetic waveguide with an ideal magnetic wall.

$$y_{\pm} = \frac{Y_0}{\xi_0} = jn^2 p\xi_{\pm} \cot(\theta_0 + \theta_{\pm}) \tag{14.21}$$

where θ_0 is the insertion phase angle of the isotropic waveguide, $\theta_{\pm}$ are the split phase angles of the gyromagnetic waveguide, n is the turns-ratio. The ideal transformers entering into the descriptions of the counterrotating networks represent the discontinuity between the rectangular waveguide feeds and the circular waveguide Faraday rotation section. ξ_0 and $\xi_{\pm}$ are wave-admittance, written as

$$\xi_0 = \sqrt{\frac{\varepsilon_0}{\mu_0}} \tag{14.22a}$$

$$\xi_{\pm} = \sqrt{\frac{\varepsilon_f}{\mu \mp \kappa}} \tag{14.22b}$$

$p = 1$ in the case of a turnstile junction based on a quarter-wave long resonator short-circuited at one end and open-circuited at the other. $p = 1/2$ in the case of one using a half-wave resonator open-circuited at both flat faces. Expanding the above equation near θ_0 equal to $\pi/2$ gives

$$y_{\pm} = -jn^2 p\xi_{\pm} \tan\theta_{\pm} \tag{14.23}$$

The required development proceeds by identifying the reciprocal and nonreciprocal phase angles of the gyromagnetic waveguide.

$$\beta_0 \pm \Delta\beta_{\pm} = \sqrt{k_0^2 \varepsilon_f (\mu \mp C_{11}\kappa) - k_c^2} \tag{14.24}$$

The diagonal element μ of the tensor permeability is one in a saturated magnetic insulator. The split propagation constants are

$$\Delta\beta_{\pm} = \beta_{\pm} - \beta_0, \ \text{rad mm}^{-1} \tag{14.25}$$

where

$$\beta_0 = \sqrt{k_0^2 \varepsilon_f - k_c^2} \tag{14.26a}$$

$$\beta_0 + \Delta\beta_{\pm} = \sqrt{\beta_0^2 \mp (k_0^2 \varepsilon_f C_{11}\kappa)} \tag{14.26b}$$

The length L of the resonator is

$$\theta_0 = \beta_0 L = \frac{\pi}{2} \tag{14.27}$$

The split phase angles of the eigen-networks are

$$\theta_+ = \Delta\beta_+ L \tag{14.28a}$$

$$\theta_- = \Delta\beta_- L \tag{14.28b}$$

The eigen-networks of the reentrant and inverted reentrant circulators using a single quarter-wave long resonator are depicted in Figure 14.6a. The ones using pairs of quarter-wave long resonators are indicated in Figure 14.6b. The frequencies of all four arrangements are identical but the susceptance slope parameters of the single quarter-wave arrangements are twice those of the other two.

14.6 The Quality Factor of the Turnstile Circulator

A closed-form derivation of the quality factor of the turnstile circulator is readily deduced. It starts by writing the split frequencies of the closed resonator below:

$$k_+^2 \varepsilon_f - k_c^2 - k_0^2 \varepsilon_f C_{11}\kappa = \left(\frac{\pi}{2L}\right)^2 \tag{14.29a}$$

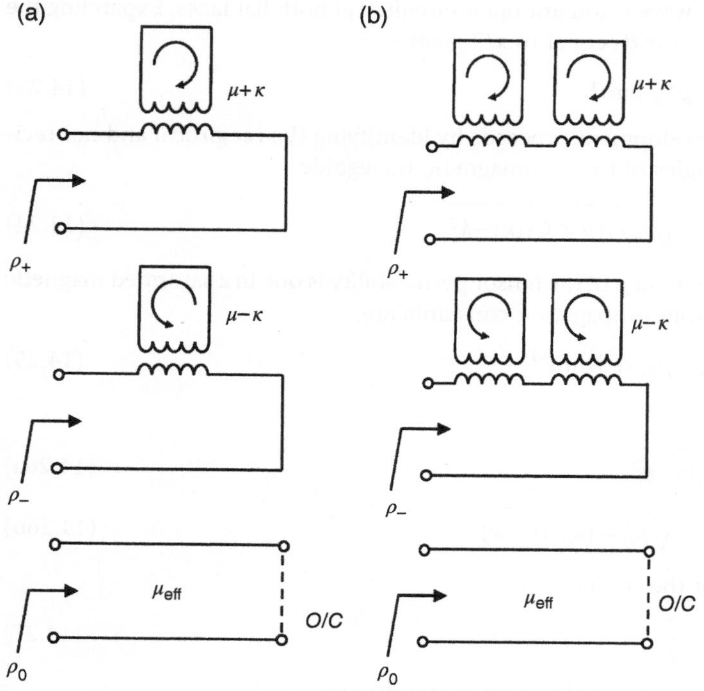

Figure 14.6 Eigen-networks of reentrant and inverted reentrant turnstile circulators using (a) single quarter-wave long resonator and (b) doublets of quarter-wave long resonators.

$$k_-^2 \varepsilon_f - k_c^2 + k_0^2 \varepsilon_f C_{11} \kappa = \left(\frac{\pi}{2L}\right)^2 \tag{14.29b}$$

The assumption here is that the frequency variation of the nonreciprocal term may be neglected compared to that of the reciprocal one. The required derivation continues by forming the difference between the above two equations:

$$\left(k_+^2 - k_-^2\right)\varepsilon_f - 2k_0^2 \varepsilon_f C_{11} \kappa = 0 \tag{14.30}$$

The result is

$$\frac{k_+ - k_-}{k_0} = \frac{\omega_+ - \omega_-}{\omega_0} = C_{11}\kappa \tag{14.31}$$

provided

$$k_+ + k_- \approx 2k_0 \tag{14.32}$$

The result, in keeping with this class of junction, is a property of the gyrotropy only.

14.7 Susceptance Slope Parameter of Turnstile Junction

The susceptance slope parameter of the junction is constructed by forming the split admittances ($y_\pm$) at the split frequencies ($\omega_\pm$). The development assumes that the nonreciprocal angles ($\theta_\pm$) of the gyromagnetic waveguide may be neglected compared to the reciprocal parts. It is assumed further that the split frequencies are not very far from the degenerate midband frequency. Introducing these assumptions in a typical admittance function gives the required result.

$$b' = \left(\frac{\omega_0}{2}\right)\left(\frac{y_+ - y_-}{\omega_+ - \omega_-}\right) = n^2 p \xi_{\text{eff}} \left(\frac{4}{\pi}\right)\left(\frac{k_0}{\beta}\right) \tag{14.33}$$

where

$$\xi_{\text{eff}} = \sqrt{\frac{\varepsilon_{\text{f}}}{\mu_{\text{eff}}}} \tag{14.34a}$$

and

$$\frac{1}{\mu_{\text{eff}}} = \frac{1}{\mu_+} + \frac{1}{\mu_-} \tag{14.34b}$$

The gyrator conductance is the product of the susceptance slope parameter and the split frequencies without ado.

The susceptance slope parameter here is that of a regular quarter-wave short-circuited isotropic waveguide with an effective permeability μ_{eff}.

One way to extract the turns-ratio of the transformer of the counterrotating eigen-networks is by satisfying the immittance eigenvalues $y_\pm$ at $g = 1$. Another means of doing so is to have recourse to the definition of b'.

Bibliography

Aitken, F.M. and Mclean, R. (1963). Some properties of the waveguide Y circulator. *Proc. Inst. Electr. Eng.* **110** (2): 256–260.

Akaiwa, Y. (1974). Operation modes of a waveguide Y-circulator. *IEEE Trans. Microw. Theory Tech.* **MTT-22**: 954–959.

Akaiwa, Y. (1978). A numerical analysis of waveguide H-plane Y-junction circulators with circular partial-height ferrite post. *J. Inst. Electron. Commun. Eng. Jpn.* **E61**: 609–617.

Allen, P.J. (1956). The turnstile circulator. *IRE Trans. Microw. Theory Tech.* **MTT-4**: 223–227.

Auld, B.A. (1959). The synthesis of symmetrical waveguide circulators. *IRE Trans. Microw. Theory Tech.* **MTT-7**: 238–246.

Clarricoats, P.J.B. (1964). Some properties of circular waveguides containing ferrites. *Proc. IEE* **104** (Part B, Suppl. 6): 286.

Clarricoats, P.J.B. (1969). A perturbation method for circular waveguides containing ferrites. *Proc. IEE* **106** (Part B): 335.

Denlinger, E.J. (1974). Design of partial-height ferrite waveguide circulators. *IEEE Trans. Microw. Theory Tech.* **MTT-22**: 810–813.

Gamo, H. (1953). The Faraday rotation of waves in a circular waveguide. *J. Phys. Soc. Jpn.* **8**: 176–182.

Green, J.J. and Sandy, F. (1974). Microwave characterization of partially magnetized ferrites. *IEEE Trans. Microw. Theory Tech.* **MTT-22**: 645–651.

Hauth, W. (1981). Analysis of circular waveguide cavities with partial-height ferrite insert. *Proceedings of European Microwave Conference*, Amsterdam, the Netherlands (7–11 September 1981), pp. 383–388.

Helszajn, J. (1973). Microwave measurement techniques for junction circulators. *IEEE Trans. Microw. Theory Tech.* **MTT-21**: 347–351.

Helszajn, J. (1974). Common waveguide circulator configurations. *Electron. Eng.* **46**: 66–68.

Helszajn, J. (1976). A unified approach to lumped element, stripline and waveguide junction circulators. *IEE Proc. Microw. Opt. Acoust.* **1** (1): 18–26.

Helszajn, J. (1994). Experimental evaluation of junction circulators: a review. *IEE Proc. Microw. Antennas Propag.* **141** (5): 351–358.

Helszajn, J. (1999). Adjustment of degree-2 H-plane waveguide turnstile circulator using prism resonator. *Microw. Eng. Eur.* (July): 35–48.

Helszajn, J. (2012). An FE algorithm for the adjustment of the first circulation condition of the turnstile waveguide circulator. *IEEE Trans. Microw. Theory Tech.* **MTT-60**: 3079–3087.

Helszajn, J. and Gibson, A.A.P. (1987). Mode nomenclature of circular gyromagnetic and anisotropic waveguides with magnetic and open walls. *Proc. IEE, Part H* **134** (6): 488–496.

Helszajn, J. and Sharp, J. (1983). Resonant frequencies, Q-factor, and susceptance slope parameter of waveguide circulators using weakly magnetized open resonators. *IEEE Trans. Microw. Theory Tech.* **MTT-31**: 434–441.

Helszajn, J. and Sharp, J. (1985). Adjustment of in-phase mode in turnstile junction circulators. *IEEE Trans Microw. Theory Tech.* **MTT-33** (4): 339–343.

Helszajn, J. and Sharp, J. (1986). Dielectric and permeability effects in HE_{111} open demagnetised ferrite resonators. *IEE Proc. Pt. H* **133** (4): 271–275.

Helszajn, J. and Sharp, J. (2005). Verification of first circulation condition of turnstile waveguide circulators using a finite element solver. *IEEE Trans Microw. Theory Tech.* **MTT-53** (7): 2309–2316.

Helszajn, J. and Tan, F.C.F. (1975a). Mode charts for partial-height ferrite waveguide circulators. *Proc. Inst. Electr. Eng.* **122** (1): 34–36.

Helszajn, J. and Tan, F.C.F. (1975b). Design data for radial waveguide circulators using partial-height ferrite resonators. *IEEE Trans. Microw. Theory Tech.* **MTT-23**: 288–298.

Helszajn, J. and Tan, F.C.F. (1975c). Susceptance slope parameter of waveguide partial-height ferrite circulators. *Proc. Inst. Electr. Eng.* **122** (72): 1329–1332.

Hogan, C.L. (1952). The ferromagnetic Faraday effect at microwave frequencies and its applications – the microwave gyrator. *Bell. Syst. Tech.* **31**: 1–31.

Kales, M.L. (1953). Modes in waveguides containing ferrites. *J. Appl. Phys.* **24**: 604–608.

Montgomery, C., Dicke, R.H., and Purcel, E.M. (1948). *Principles of Microwave Circuits*, Ch. 12. New York: McGraw-Hill Book Co. Inc.

Owen, B. (1972). The identification of modal resonances in ferrite loaded waveguide y-junctions and their adjustment for circulation. *Bell Syst. Tech. J.* **51** (3): 595–627.

Owen, B. and Barnes, C.E. (1970). The compact turnstile circulator. *IEEE Trans. Microw. Theory Tech.* **MTT-18**: 1096–1100.

Rado, G.T. (1953). Theory of the microwave permeability tensor and Faraday effect in non-saturated ferromagnetic materials. *Phys. Rev.* **89**: 529.

Riblet, G., Helszajn, J., and O'Donnell, B.C. (1979). Loaded Q-factors of partial height triangular resonators for use in waveguide circulators. *European Microwave Conference*, Brighton, UK (17–20 September 1979).

Schaug-Patterson, T. (1958). Novel Design of a 3-port Circulator. Norwegian Defence Research Establishment Report, Rpt No. R-59 (January).

Schlömann, E. (1970). Microwave behaviour of partially magnetized ferrites. *J. Appl. Phys.* **41** (1): 204–214.

Simon, J. (1965). Broadband strip-transmission line Y junction circulators. *IEEE Trans. Microw. Theory Tech.* **MTT-13**: 335–345.

Suhl, H. and Walker, L.R. (1952). Faraday rotation in guided waves. *Phys. Rev.* **86**: 122.

Suhl, H. and Walker, L.R. (1954). Topics in guided wave propagation through gyromagnetic media. *Bell Syst. Tech. J.* **33**: 579–659, 939–986, 1122–1194.

Van Trier, A.A. (1953). Guided electromagnetic waves in anisotropic media. *Appl. Sci. Res.* **3**: 305–371.

Waldron, R.A. (1958). Electromagnetic wave propagation in cylindrical waveguides containing gyromagnetic media. *J. Br. IRE* **18**: 597–612, 677–690, 733–746.

Waldron, R.A. (1960). Features of cylindrical waveguides containing gyromagnetic media. *J. Br. IRE* **20**: 695–706.

Waldron, R.A. (1962). Properties of ferrite-loaded cylindrical waveguides in the neighbourhood of cut-off. *Proc. IEE* **109** (Part B, Suppl. 21): 90–94.

Waldron, R.A. (1963). Properties of inhomogeneous cylindrical waveguides in the neighbourhood of cut-off. *J. Br. IRE* **25**: 547–555.

15

A Finite-Element Algorithm for the Adjustment of the First Circulation Condition of the H-plane Turnstile Waveguide Circulator

Joseph Helszajn

Heriot Watt University, Edinburgh, UK

15.1 Introduction

The three-port turnstile circulator may be visualized as a five-port network consisting of a cylindrical gyromagnetic waveguide having two orthogonal ports that are closed with a short-circuit piston at the junction of three H-plane rectangular waveguides. This sort of junction supports one in-phase eigen-network and a pair of degenerate or split counterrotating eigen-networks. The first of its two circulation conditions coincides with the maximum power transfer condition of the junction prior to the application of the gyrotropy. Its adjustment is an eigenvalue problem. It fixes all the physical parameters of the circulator, except for the gyrotropy. This paper describes an algorithm in conjunction with a finite-element solver for the adjustment of this class of junction. It is met, provided the in-phase and counterrotating eigen-networks of the junction are 90° long and the corresponding reflection angles differ by 180°. The algorithm introduced in this chapter may also be used to determine the split frequencies of the junction by replacing the demagnetized permeability of the resonator, one at a time, by appropriate scalar counterrotating permeabilities. An inverted reentrant turnstile junction in half-height WR75 waveguide is characterized by way of an example. A reentrant turnstile junction in standard WR75 waveguide is separately synthesized.

The chapter describes one universal algorithm for the solution of this class of device. There are altogether three classic geometries using a single cylindrical resonator and six employing a prism one. The possible configurations met in connection with the cylindrical arrangements are illustrated in Figure 15.1. The triplet of structures associated with each possible orientation of the prism resonator is understood without ado.

Microwave Polarizers, Power Dividers, Phase Shifters, Circulators, and Switches, First Edition. Joseph Helszajn.

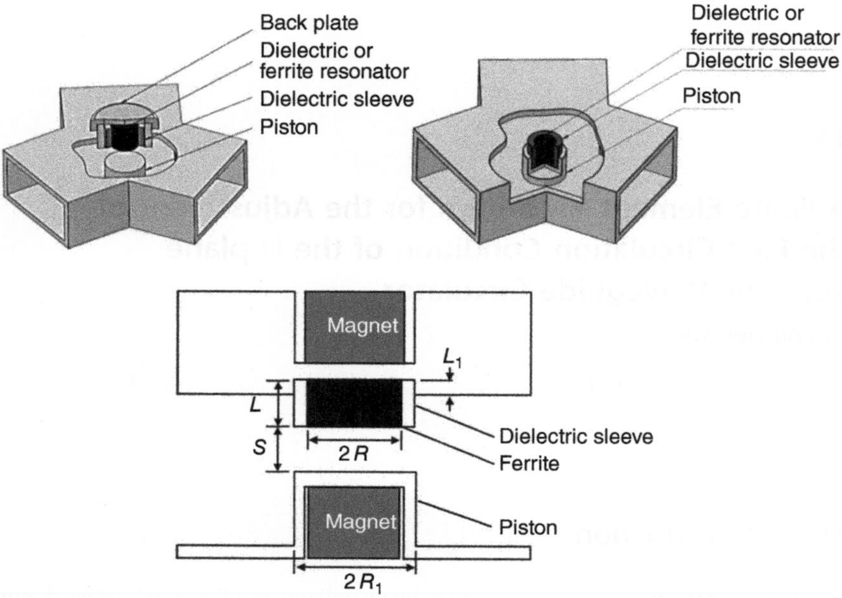

Figure 15.1 Schematic diagrams of waveguide junction circulators using single turnstile resonators.

The basic element is a quarter-wave-long open gyromagnetic resonator with a gap between one open flat face and one waveguide wall and one short-circuited flat face on the other waveguide wall. It determines the counterrotating eigen-networks of the junction. The in-phase eigen-network is a quasi-planar geometry that is fixed by the circulator composite structure made up of the ferrite and the air or dielectric gap with top and bottom electric walls. The solution of each geometry is separately fixed by the relative dielectric constant of the ferrite material and the position of the operating frequency with respect to the cutoff frequency of the waveguide. The permeability of any demagnetized ferrite must also be taken into account. The solution of this class of junction involves two independent and two dependent variables.

The dependent variables are the radial wavenumber k_0R and the gap factor of the junction (q_{eff}); the independent ones are the wavenumber of the specification (k_0) and the aspect ratio of the resonator, radius (R), and length (L), (R/L). A Finite Element (FE) solver is obviously essential in order to deal with the various configurations met with this junction. The algorithm developed in this paper may also be used to determine the split frequencies of the resonator by replacing the demagnetized permeability one at a time by counterrotating scalar quantities.

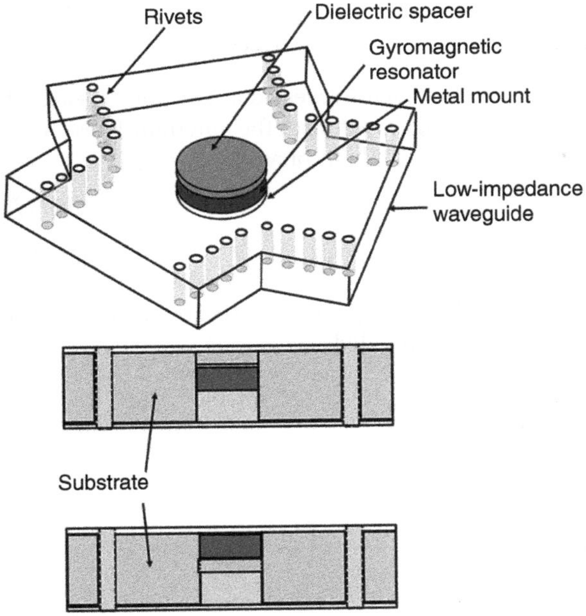

Figure 15.2 Schematic diagram of integrated substrate waveguide circulator using turnstile resonators.

The *H*-plane waveguide turnstile junction circulator is only one geometry that relies for its operation on a turnstile resonator. Figure 15.2 illustrates an integrated surface waveguide that relies on the same principles. The second circulation condition of this type of junction is established by replacing the dielectric by a gyromagnetic one. This chapter includes some data on this condition.

15.2 Bandpass Frequency of a Turnstile Junction

The required midband relationships between the variables entering into the first circulation condition of the junction are satisfied, provided the in-phase and counterrotating eigenvalues are 90° long and 180° out of phase.

$$s_0 = -1 \tag{15.1a}$$

$$s_+ = s_- = 1 \tag{15.1b}$$

The eigenvalues are, for the purpose of calculations, related to the scattering matrix of the junction in the usual way.

$$s_0 = S_{11} + 2S_{21} \tag{15.2a}$$

$$s_+ = s_- = S_{11} - S_{21} \tag{15.2b}$$

The eigenvalues are the reflection coefficients revealed at any port by each of the three possible generator settings or eigenvectors of the junction. These have unit amplitude and differ from each other only in phase.

$$s_0 = 1.\exp - j2\theta_0 \tag{15.3a}$$

$$s_\pm = 1.\exp - j2\left(\theta_\pm + \frac{\pi}{2}\right) \tag{15.3b}$$

The eigenvalue diagram at the pass band frequency of a reciprocal junction is indicated in Figure 15.3a. It is satisfied, provided

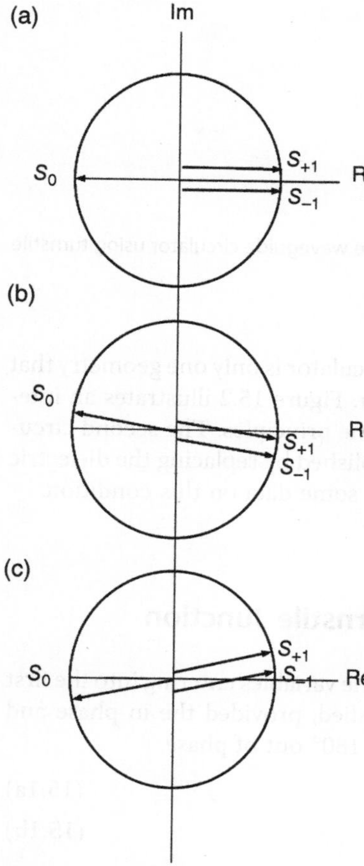

Figure 15.3 (a–c) Eigenvalue diagrams of reciprocal three-port junction.

$$\theta_0 = \theta_{\pm} = \frac{\pi}{2} \tag{15.4}$$

θ_0 and $\theta_{\pm}$ are the electrical lengths of the in-phase and counterrotating eigen-networks, respectively. The $\pi/2$ term in the degenerate counterrotating reflection coefficients is associated with the short-circuit boundary condition that exists at the axis of the problem region. The reflection eigenvalues are also sometimes expressed in terms of the reflection angles ϕ_0 and $\phi_{\pm}$:

$$s_0 = 1.\exp - j\phi_0 \tag{15.5a}$$

$$s_{\pm} = 1.\exp - j\phi_{\pm} \tag{15.5b}$$

The reflection coefficient associated with this diagram is $S_{11} = 1/3$. Two other eigenvalue diagrams in the vicinity of the required solution are separately shown in Figure 15.3b and c.

The amplitudes of the reflection coefficients are also in these instances equal to that of the ideal solution but the reflection angles do not, however, correspond to that of the pass band.

15.3 In-phase and Counterrotating Modes of Turnstile Junction

The degenerate counterrotating modes entering into the adjustment of a reciprocal turnstile junction are approximately specified by a pair of HE_{11} modes in an open dielectric waveguide supporting an open magnetic wall at one flat face and a short-circuit at the other.

Its adjustment involves a gap factor:

$$q_{\pm} = \frac{L}{L + S_{\pm}} \tag{15.6a}$$

The symmetric mode is a quasi-planar TM_{010} one with top and bottom electric walls and an open sidewall. It does not propagate along the axis of the resonator. Its adjustment involves a gap factor:

$$q_0 = \frac{L}{L + S_0} \tag{15.6b}$$

The unknowns of the problem region are the aspect ratio (R/L), the radius of the resonator (R), and the length of the gap (s) between its open face and the opposite waveguide wall. The latter two quantities are usually expressed in terms of the radial wavenumber (k_0R) and a gap factor (q_{eff}):

$$q_{\text{eff}} = \frac{L}{L + S} \tag{15.7}$$

All the other quantities including the wavenumber k_0 entering into the description of the junction are independent variables and are specified as a preamble to the optimization subroutine.

An approximation of the first circulation condition of a degree-2 junction may be established by replacing $s_{\pm} = 1$ by $s_{\pm} = -1$ and $s_0 = -1$ by $s_0 = 1$ in the degree-1 solution. This condition is usually satisfied by introducing suitable quarter-wave long or alternate line impedance transformers at each port. The variables in addition to those met in connection with the degenerate junction are the electrical angle(s) and the impedance(s) of a typical transformer.

15.4 Reference Plane

The reference plane of the solution is obtained here and elsewhere by replacing the resonator region by a metal plug. The assumption here, if nothing else, is the extent of any leakage by the open apertures on either side of the post into the output waveguides. Measurements indicate that this leakage is of the order of 2% in each output waveguide. It furthermore assumes that the electrical planes of both the in-phase and counterrotating eigenvalues have a common surface with that of the mechanical boundary of the open resonator. The phase constant associated with the nonuniform radial region connecting the dielectric resonator to a typical rectangular waveguide is dealt with by separately extremizing this problem.

A property of such a region is that an open-circuit is not mapped into a short-circuit over the same length that a short-circuit is mapped into an open-circuit. One consequence of this feature is that the angle between the reflection angles at the resonator terminals is not preserved at the terminals of a typical waveguide. Furthermore, the frequency of the 9½ dB points in the return loss of the junction no longer coincides with the real axis of the Smith chart but resides instead either side of it. This effect is of issue in the design of degree-2 junctions. The organization of the junction has also a significant influence on the fringing field so that each possible structure must be, strictly speaking, separately evaluated. Figure 15.4a and b indicate the calibration process for both reflection and transmission parameters of the junction at its reference terminals.

15.5 FE Algorithm

The adjustment of the reciprocal turnstile junction is met when the in-phase and degenerate counterrotating eigen-networks are degenerate. This condition is satisfied, provided:

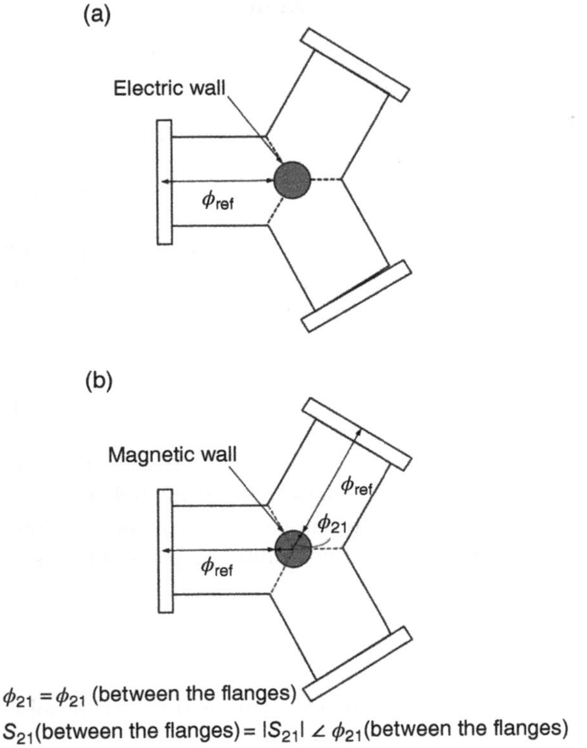

Figure 15.4 Construction of reference terminals of resonator: (a) reflection coefficient and (b) transmission coefficient.

$$k_0R^0 = k_0R^{\pm} = k_0R \quad \frac{R}{L} = \text{constant}, \quad k_0 = \text{constant} \tag{15.8a}$$

$$q_0 = q_{\pm} = q_{\text{eff}} \quad \frac{R}{L} = \text{constant}, \quad k_0 = \text{constant} \tag{15.8b}$$

The first condition fixes the radius of the resonator, R, and thereafter its length, L. The gap, S, between the open face of the resonator and the image wall of the junction is fixed by the second condition in terms of the gap factor defined in the last section.

The design procedure, in the absence of fringing, is straight forward. The in-phase eigenvalue fixes the gap factor and gives a relationship between S and L. The absolute values of these two quantities are then fixed without ado by the counterrotating ones. In the presence of fringing, however, the situation is more complicated in that the two conditions are now coupled.

One general approach commences by constructing in-phase and counterrotating polynomials connecting q_0 and $q_\pm$ to k_0R, which satisfy $s_0 = -1$ and $s_\pm = 1$ for parametric values of R/L.

$$q_0 = F(k_0R), \quad \frac{R}{L} = \text{constant}, \quad s_0 = -1 \tag{15.9a}$$

$$q_\pm = F'(k_0R), \quad \frac{R}{L} = \text{constant}, \quad s_\pm = 1 \tag{15.9b}$$

The preceding polynomials may be constructed by having recourse to an FE procedure or some other numerical method. The two polynomials are thereafter equated and combined into a single characteristic equation involving the variable k_0R for discrete values of R/L.

$$q_\pm - q_0 = 0, \quad \frac{R}{L} = \text{constant}, \quad s_\pm = 1, \quad s_0 = -1 \tag{15.10}$$

The roots of this equation for parametric values of R/L provide the link between the latter variable and k_0R. The calculation is completed once the actual gap factor q_{eff} is evaluated in terms of the same roots. This may be done by having recourse to either of the two original polynomial representations of the first circulation condition:

$$q_{\text{eff}} = q_\pm = q_0 = 0, \quad \frac{R}{L} = \text{constant}, \quad s_\pm = 1, \quad s_0 = -1 \tag{15.11}$$

It is convenient, for the purpose of engineering to assemble the specific solutions in polynomial form:

$$k_0R = P\left(\frac{R}{L}\right), \quad s_\pm = 1, \quad s_0 = -1 \tag{15.12a}$$

$$q_{\text{eff}} = Q(k_0R), \quad s_\pm = 1, \quad s_0 = -1 \tag{15.12b}$$

15.6 FE Adjustment

The normalized variables introduced here are desirable in order to store universal data. Absolute quantities such as R, L, and S are, however, necessary in order to initialize any numerical procedure. A moot point in the organization of any calculations is the description of the geometry in terms of realistic physical variables. The initial choices adopted here rely on historic experimental data on an inverted reentrant turnstile junction in standard WR90 waveguide. An inverted reentrant turnstile junction in half-height WR75 waveguide is, however, adopted by way of illustration of the proposed algorithm in this work. Taking $R/L = 2.0$ produces initial values for k_0R and q_{eff} of 0.8 and 0.85, respectively.

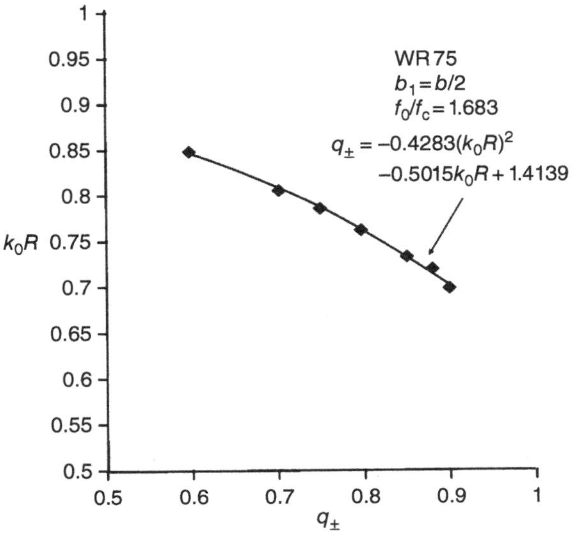

Figure 15.5 k_0R versus $q_\pm$ of inverted reentrant turnstile junction in half-height WR75 waveguide for counterrotating mode (f_0/f_c = 1.683, R/L = 2.0).

Figures 15.5 and 15.6 show the connections between k_0R and q_0 and $q_\pm$ for a resonator with an aspect ratio $R/L = 2.0$ for a reciprocal inverted reentrant turnstile junction using a dielectric resonator with a relative dielectric constant of 15.0 in half-height WR75 waveguide at a frequency of f_0/f_c = 1.683.

The respective polynomial solutions are

$$q_0 = -5.4122(k_0R)^2 + 6.3056(k_0R) - 0.8604, \quad \frac{R}{L} = 2.0 \tag{15.13a}$$

$$q_\pm = -0.4283(k_0R)^2 - 0.5015(k_0R) + 1.4139, \quad \frac{R}{L} = 2.0 \tag{15.13b}$$

Repetitive recourse to the reference plane of the junction is avoided by varying q_{eff} for parametric values of k_0R at k_0 rather than the other way round. Figure 15.7 illustrates a typical flow chart for the evaluation of either quantity.

A typical calculation amounts to partitioning the k_0R interval into m segments and the q_{eff} one into n segments. A typical regular grid is produced with m = 6 and n = 4 implying 24 problem drawings and six calibration steps or drawings each of which involves replacing the resonator geometry by a metal pillar of the same radius. One process proceeds by constructing the polynomial relationship between q_{eff} and both ϕ_0 and $\phi_\pm$ for any specific values of k_0R.

These polynomials are then employed to solve for q_0 at θ_0=90° and $q_\pm$ at $\theta_\pm$=90° or equivalently $\phi_0 = \pi$ and $\phi_\pm = 0$. The in-phase and counterrotating

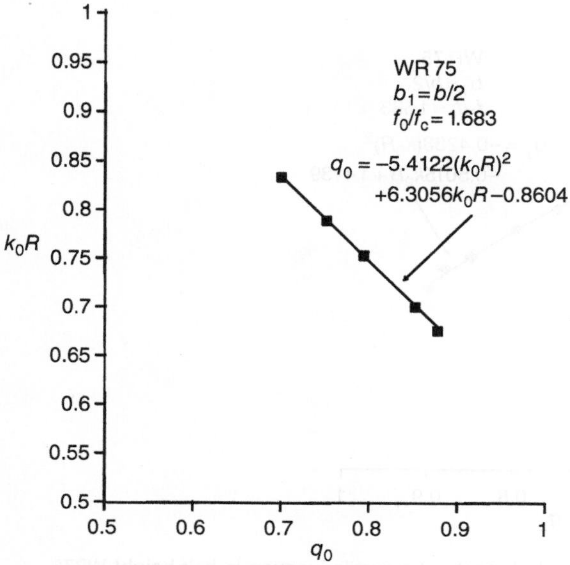

Figure 15.6 k_0R versus q_0 of inverted reentrant turnstile junction in half-height WR75 waveguide for in-phase mode ($f_0/f_c = 1.683$, $R/L = 2.0$).

eigenvalues of the geometry are typically located in the space defined by $0.60 \le k_0R \le 0.90$ and $0.60 \le q_{\text{eff}} \le 0.80$.

The characteristic equation from which k_0R may be deduced is

$$-5.4122(k_0R)^2 + 6.3056(k_0R) - 0.8604 + 0.4283(k_0R)^2 + 0.5015(k_0R) - 1.4139 = 0, \quad \frac{R}{L} = 2.0 \tag{15.14}$$

The root of this characteristic equation is k_0R equals 0.783. The corresponding value of q_{eff} is 0.758.

A scrutiny of the graphical solution in Figure 15.8 indicates that the angle between the two polynomials is, at the intersection point, relatively small. This suggests that the failure to accurately reproduce the boundary conditions of the problem region is not, in practice, as important as once supposed. A flow chart of this process is indicated in Figure 15.9.

Both the adopted contour of the reference plane of the junction and the characterization of the geometry of the resonator are extremized in the same way.

The solution produced by the proposed algorithm is unique to the wavenumber employed in the calculation. This remark may be understood by recognizing

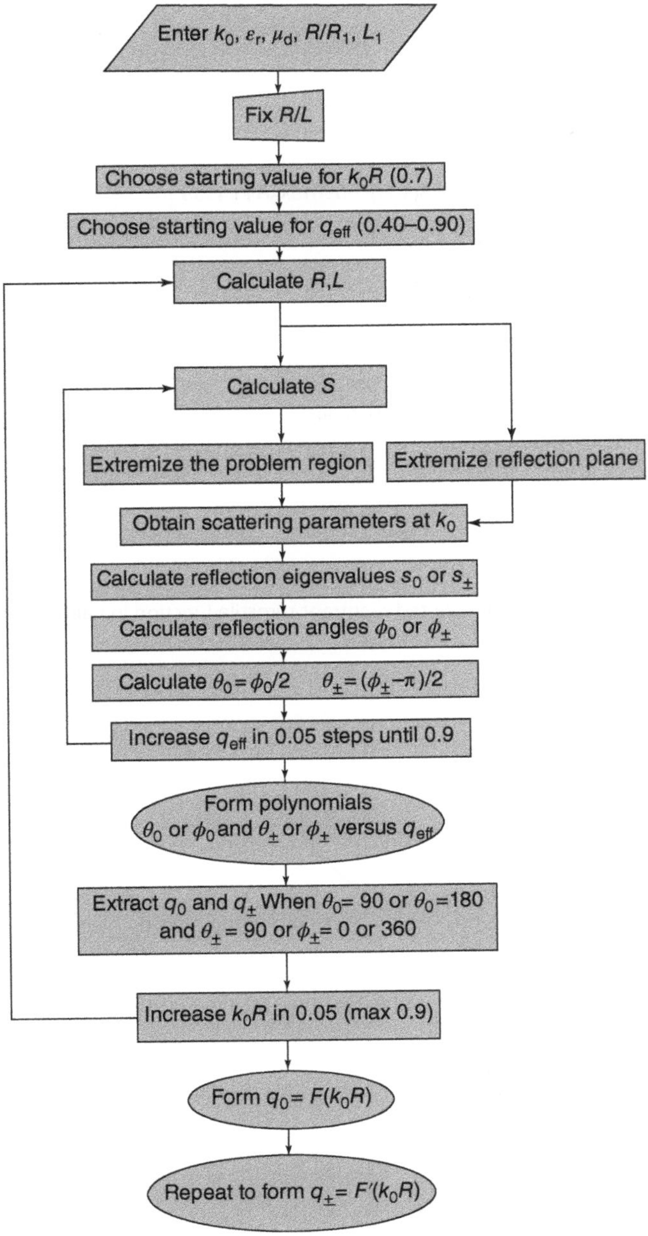

Figure 15.7 Flowchart for the solution of either $s_0 = -1$ or $s_\pm = 1$ at the resonator edge.

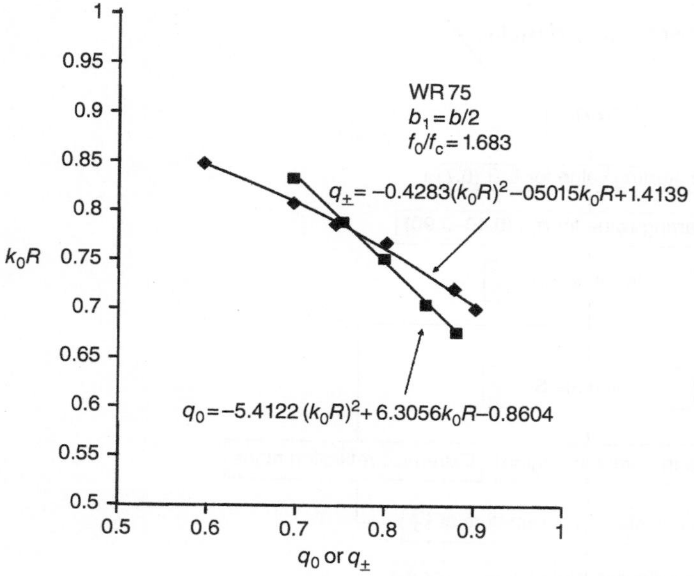

Figure 15.8 First circulation condition of inverted reentrant turnstile junction in half-height WR75 waveguide ($f_0/f_c = 1.683$, $R/L = 2.0$).

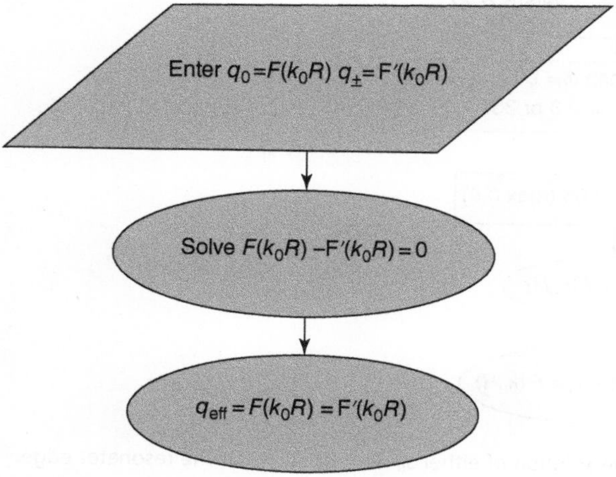

Figure 15.9 Flowchart algorithm for first circulation condition.

that the same combination of the product k_0R but at a different value of k_0 will produce a different R–L–S envelope in conjunction with a perturbation in the existing fringing field of the geometry.

The shortcoming of the approximate closed form solution is that, of course, it does not account for any fringing field in its description. In order to accurately scale any existing design, it is therefore essential to respect all the parameters entering into its description and to recalculate the geometry whenever the frequency of the device or the cutoff frequency of the waveguide or the details of the junction are modified.

One way to verify the robustness of any solution is to resort to experiment or analysis. Figure 15.10 depicts a Smith chart representation of the structure under consideration. It accurately reproduces the pass band frequency of the junction as asserted. A scrutiny of this result suggests that the frequency variation of the in-phase eigenvalue may be neglected compared to those of the counterrotating eigenvalues. It also indicates that the frequency variation of S_{11} is more or less that of $s_{\pm}$.

The permeability met in connection with a demagnetized magnetic insulator must be accurately accounted for separately. It is related to the magnetization of the material and the frequency by (Schlömann 1970)

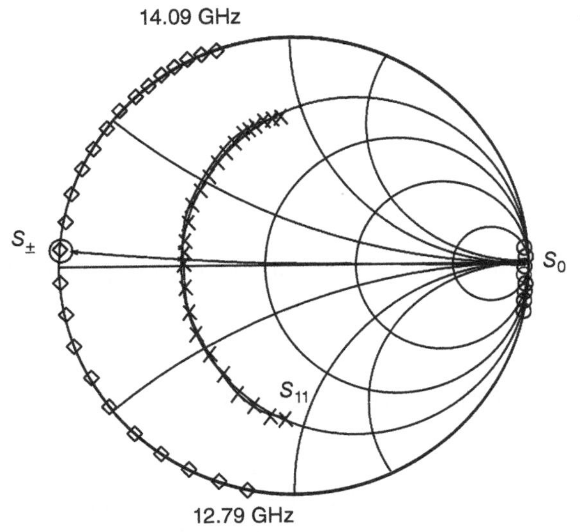

Figure 15.10 Smith chart of the first circulation condition of inverted reentrant turnstile junction in half-height WR75 waveguide ($f_0/f_c = 1.683$, $R/L = 2.0$).

$$\mu_d = \frac{1}{3} + \frac{2}{3}\left[1 - \left(\frac{\gamma M_0}{\omega}\right)^2\right]^{\frac{1}{2}} \tag{15.15}$$

γ is the gyromagnetic ratio 2.21×10^5 (rad s^{-1} per A m^{-1}), M_0 is the saturation magnetization (A m^{-1}), and ω is the radian frequency (rad s^{-1}).

Most experimental data in the literature on this class of junction has actually been restricted to ones using demagnetized gyromagnetic resonators. This effect is separately dealt with in Schlömann (1970), Rado (1953), Green and Sandy (1974), and Helszajn and Sharp (1986, 2011).

15.7 The Reentrant Turnstile Junction in Standard WR75 Waveguide

The robustness of the algorithm introduced for the adjustment of the three-port turnstile circulator in this chapter has been separately experimentally verified by fabricating one junction in standard waveguide based on an existing simulation (Hauth 1981). Its Smith chart solution is reproduced in Figure 15.11 for completeness sake.

The arrangement under consideration is a reentrant instead of inverted reentrant turnstile geometry. A full-height waveguide assembly has been chosen for this purpose in order to avoid the complication of making transitions between half- and full-height waveguide.

Its details are summarized by

$$k_0 = 0.278\,\text{rad}\,\text{mm}^{-1}$$

$$\frac{R}{L} = 2.0$$

$$k_0 R = 0.858$$

$$q_{\text{eff}} = 0.536$$

Figure 15.12 compares the calculated and measured frequency responses of the solution in question.

15.8 Susceptance Slope Parameter of Degree-1 Junction

The susceptance slope parameter of a degree-1 demagnetized junction, which is a measure of its bandwidth, may be extracted by constructing its frequency response about its center frequency at one typical port with the other two terminated in matched loads. One formulation of this parameter, in the case for

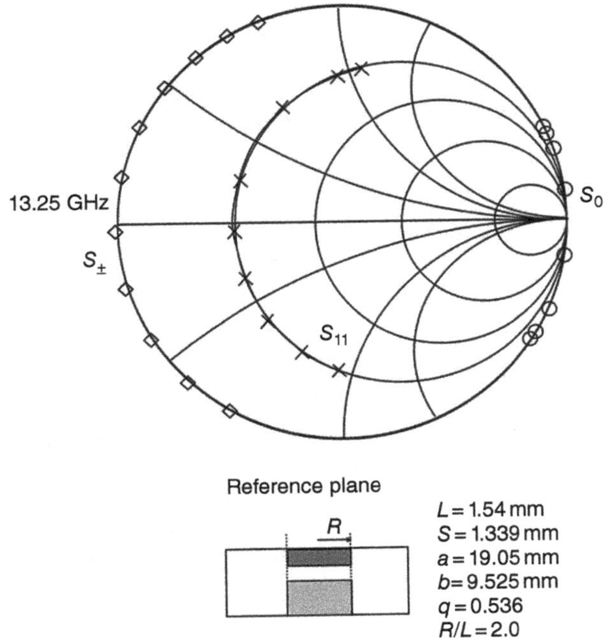

Figure 15.11 Smith chart of reentrant turnstile junction in standard WR75 waveguide ($f_0/f_c = 1.683$, $R/L = 2.0$).

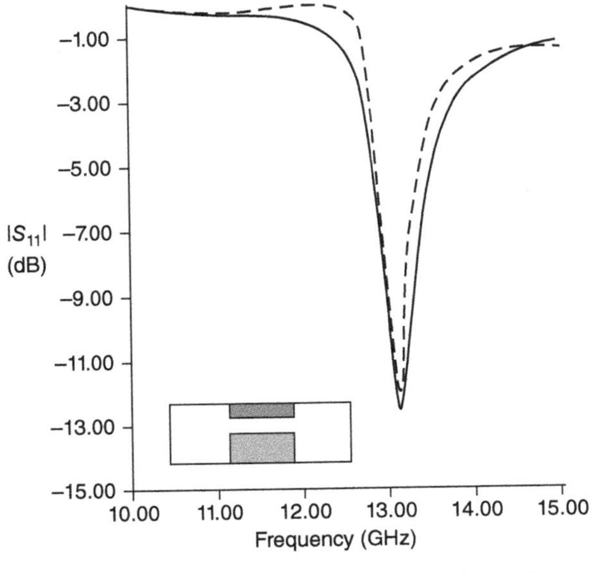

Figure 15.12 Comparison between calculated (---) and experimental (––) frequency response of reentrant turnstile junction in WR75 waveguide at each port.

which the frequency response of the in-phase eigen-network can be neglected compared to those of the degenerate ones is (Helszajn and Sharp 2003)

$$b' = \frac{B'}{Y_0} = \frac{(2/3)\{[(\mathrm{VSWR})^2 - 2.5(\mathrm{VSWR}) + 1]/2(\mathrm{VSWR})\}^{1/2}}{2\delta_0} \tag{15.16}$$

where

$$2\delta_0 = \frac{\omega_2 - \omega_1}{\omega_0} \tag{15.17}$$

$\omega_{1,2}$ are band edge frequencies and ω_0 is the midband frequency. VSWR is the voltage standing wave ratio at $\omega_{1,2}$. The factor 2/3 in Eq. (15.16) connects the susceptance slope parameter of the reciprocal junction to that of the complex gyrator circuit of the corresponding circulator. The susceptance slope parameter of a turnstile junction is in practice dependent on the aspect ratio of the resonator. The normalized value obtained with half-height waveguide in the example shown in Figure 15.13 is 14. The corresponding value in standard WR75 is half that displayed by the half-height waveguide.

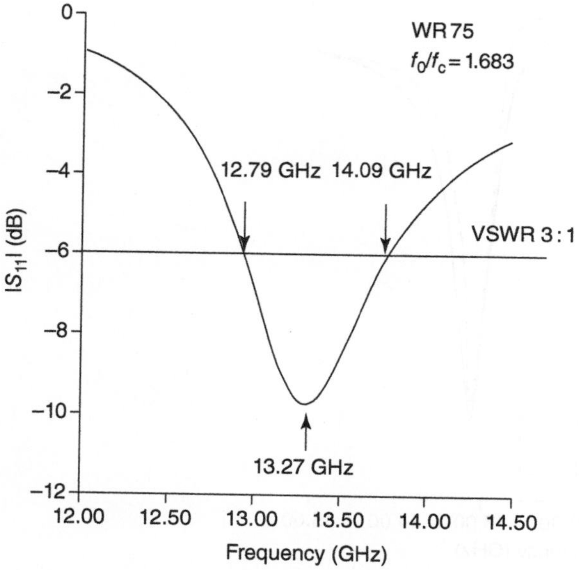

Figure 15.13 Frequency response of first circulation condition of inverted reentrant turnstile junction in half-height WR75 waveguide ($f_0/f_c = 1.683$, $R/L = 2.0$).

15.9 Split Frequencies of Gyromagnetic Resonators

The solver introduced in connection with the first circulation condition of the turnstile circulator may also be employed to approximately investigate, one at a time, the split frequencies $\omega_\pm$ of the gyromagnetic resonator on either side of the isotropic one ω_0. A knowledge of these frequencies allows the quality factor Q_L of the gyrator circuit of the magnetized junction to be evaluated without recourse to a magnetic solver. This may be done by replacing, one at a time, the demagnetized permeability μ_d of the magnetic insulator by counterrotating partially magnetized scalar permeabilities $\mu_\pm$:

$$\mu_\pm = \mu_p \mp C_{11}\kappa_p \tag{15.18}$$

C_{11} is a constant met in connection with the characteristic equation of the cutoff space of a planar resonator with top and bottom electric walls and a magnetic sidewall. One useful approximation for a cylindrical resonator is (Helszajn and Tan 1975b):

$$C_{11} = \frac{2}{(k_c R)^2 - 1}, \quad 0 \le \kappa_p \le 0.50 \tag{15.19}$$

where

$$k_c R = 1.84$$

and

$$k_c = k_0\sqrt{\varepsilon_f \mu_{eff}} \tag{15.20}$$

$$\mu_{eff} = \frac{\mu_p^2 - (C_{11}\kappa_p)^2}{\mu_p} \tag{15.21}$$

The calculations in this section assume a saturated magnetic insulator for which

$$\mu_p = 1 \tag{15.22}$$

$$\kappa_p = \frac{\gamma M_0}{\mu_0} \tag{15.23}$$

The split frequencies $\omega_\pm$ coincide with $s_\pm = 1$ at the resonator terminals. Figure 15.14 shows the eigenvalue diagrams at $\omega = \omega_0$.

One experimental procedure from which the split frequencies of the resonator can be extracted is obtained by determining the frequencies at which the return loss of the terminated junction passed through 9½ dB. This result is readily recognized both here and elsewhere by examining the connection between the reflection coefficient S_{11} and the eigenvalues of the junction.

$$S_{11} = \frac{s_0 + s_+ + s_-}{3} \tag{15.24}$$

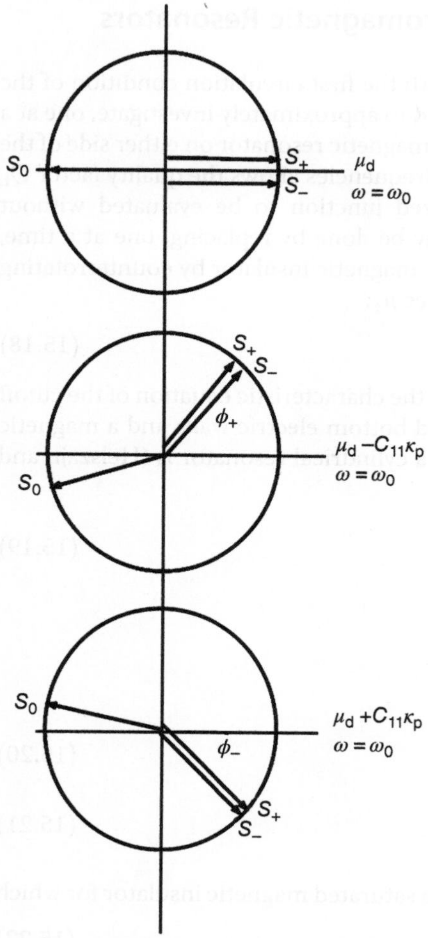

Figure 15.14 Eigenvalue diagrams associated with scalar counterrotating permeabilities.

The quality factor is given in terms of ω_0 and $\omega_\pm$ by

$$\frac{1}{Q_L} = \sqrt{3}\left(\frac{\omega_+ - \omega_-}{\omega_0}\right) \tag{15.25}$$

This quantity fixes the gain bandwidth product of the complex gyrator circuit as is universally understood.

$$(2\delta_0)(\mathrm{RL})Q_L = \text{constant} \tag{15.26}$$

RL is the return loss (dB), $2\delta_0$ is the normalized bandwidth. The constant on the right-hand side of this condition is determined by the nature and degree of any matching network.

The frequency responses corresponding to $\mu_d = 1.0$ and $\mu_\pm = 0.7, 1.3$ are indicated in Figure 15.15. The split frequencies in this picture correspond to the eigenvalue diagrams in Figure 15.3b and c.

The condition associated with these two split frequencies are (Helszajn and Sharp 1986)

$$\frac{1}{Z^\pm} = Z^0 \tag{15.27}$$

Either condition ensures that the reflection angles are 180° out of phase but neither guarantees that the eigen-networks are commensurate. The prevailing condition depends on whether the in-phase eigen-network is larger or smaller than 90°. In a uniform transmission line the above relationship should hold at any plane from the load.

Figure 15.16 depicts the frequency response of the return loss at port 1 with the other two ports terminated in matched loads, in half-height WR75 waveguide, for parametric values of the gyrotropy (κ). Each magnetization step is associated with a different effective permeability (μ_{eff}), the effect of which is to perturb the midband frequency of the junction. In keeping with the

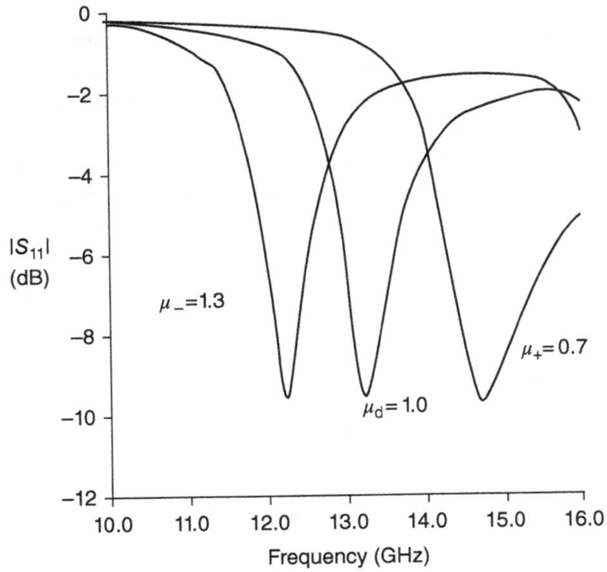

Figure 15.15 9½ dB frequencies corresponding to $\mu_d = 1.0$ and $\mu_\pm = 0.7, 1.3$.

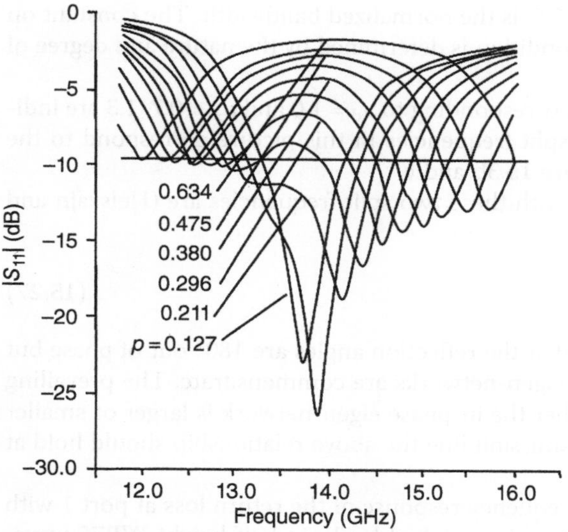

Figure 15.16 Return loss at port 1 of inverted reentrant turnstile circulator in half-height WR75 waveguide for parametric values of normalized magnetization in saturated magnetic insulator.

numerical data, the permeability is lowered as the magnetization is increased thereby increasing the frequency. A saturation magnetization $\mu_0 M_0$ equal to 0.1600 T corresponds with κ equal to 0.338 at 13.25 GHz. The corresponding normalized split frequency $(\omega_+ - \omega_-)/\omega_0$ is equal to 0.196. The value obtained by having recourse to the split cutoff space of a planar gyromagnetic resonator is 0.220. The quality factor Q_L associated with this gyromagnetic splitting is equal to 2.63. It is compatible with the realization of communication circulators with commercial specifications.

The exact split frequencies of the junction may also be readily calculated using a suitable commercial solver.

References

Aitken, F.M. and Mclean, R. (1963). Some properties of the waveguide Y circulator. *Proc. Inst. Electr. Eng.* **110** (2): 256–260.

Akaiwa, Y. (1974). Operation modes of a waveguide Y-circulator. *IEEE Trans. Microw. Theory Tech.* **MTT-22**: 954–959.

Akaiwa, Y. (1978). A numerical analysis of waveguide H-plane Y-junction circulators with circular partial-height ferrite post. *J. Inst. Electron. Commun. Eng. Jpn.* **E61**: 609–617.

Auld, B.A. (1959). The synthesis of symmetrical waveguide circulators. *IRE Trans. Microw. Theory Tech.* **MTT-7**: 238–246.

Denlinger, E.J. (1974). Design of partial-height ferrite waveguide circulators. *IEEE Trans. Microw. Theory Tech.* **MTT-22**: 810–813.

Green, J.J. and Sandy, F. (1974). Microwave characterization of partially magnetized ferrites. *IEEE Trans. Microw. Theory Tech.* **MTT-22**: 645–651.

Hauth, W. (1981). Analysis of circular waveguide cavities with partial-height ferrite insert. *11th Proceedings of European Microwave Conference*, Amsterdam, the Netherlands (7–11 September 1981), pp. 383–388.

Helszajn, J. (1974). Common waveguide circulator configurations. *Electron. Eng.* (94): 66–68.

Helszajn, J. (1999). Adjustment of degree-2 H-plane waveguide turnstile circulator using prism resonator. *Microw. Eng. Eur.* (July): 35–48.

Helszajn, J. and Sharp, J. (1983). Resonant frequencies, Q-factor, and susceptance slope parameter of waveguide circulators using weakly magnetized open resonators. *IEEE Trans. Microw. Theory Tech.* **MTT-31**: 434–441.

Helszajn, J. and Sharp, J. (1985). Adjustment of in-phase mode in turnstile junction circulators. *IEEE Trans. Microw. Theory Tech.* **MTT-33** (4): 339–343.

Helszajn, J. and Sharp, J. (1986). Dielectric and permeability effects in HE_{111} open demagnetised ferrite resonators. *IEE Proc.* **133** (Pt. H, 4): 271–275.

Helszajn, J. and Sharp, J. (2003). Frequency response of quarter-wave coupled reciprocal stripline junctions. *Microw. Eng. Eur.* (March/April): 29–35.

Helszajn, J. and Sharp, J. (2005). Verification of first circulation condition of turnstile waveguide circulators using a finite element solver. *IEEE Trans. Microw. Theory Tech.* **MTT-53** (7): 2309–2316.

Helszajn, J. and Sharp, J. (2011). Fringing effects in re-entrant and inverted re-entrant turnstile waveguide junctions using cylindrical resonators. *IET Mirow, Ant. Prop.* **5** (9): 1109–1115.

Helszajn, J. and Tan, F.C.F. (1975a). Mode charts for partial-height ferrite waveguide circulators. *Proc. Inst. Elec. Eng.* **122** (1): 34–36.

Helszajn, J. and Tan, F.C.F. (1975b). Design data for radial waveguide circulators using partial-height ferrite resonators. *IEEE Trans. Microw. Theory Tech.* **MTT-23**: 288–298.

Helszajn, J. and Tan, F.C.F. (1975c). Susceptance slope parameter of waveguide partial-height ferrite circulators. *Proc. Inst. Electr. Eng.* **122** (72): 1329–1332.

Montgomery, C., Dicke, R.H., and Purcel, E.M. (1948). *Principles of Microwave Circuits*. New York, Ch. 12: McGraw-Hill Book Co. Inc.

Owen, B. (1972). The identification of modal resonances in ferrite loaded waveguide junction and their adjustment for circulation. *Bell Syst. Tech. J.* **51** (3): 595–627.

Owen, B. and Barnes, C.E. (1970). The compact turnstile circulator. *IEEE Trans. Microw. Theory Tech.* **MTT-18**: 1096–1100.
Rado, G.T. (1953). Theory of the microwave permeability tensor and Faraday effect in non-saturated ferromagnetic materials. *Phys. Rev.* **89**: 529.
Schaug-Patterson, T. (1958). Novel Design of a 3-port Circulator. Norwegian Defence Research Establishment Report, Rpt No. R-59 (January).
Schlömann, E. (1970). Microwave behaviour of partially magnetized ferrites. *J. Appl. Phys.* **41** (1): 204.

16

The E-plane Waveguide Wye Junction: First Circulation Conditions

Joseph Helszajn[1] *and Marco Caplin*[2]

[1] *Heriot Watt University, Edinburgh, UK*
[2] *Apollo Microwaves Ltd, Dorval, Quebec, Canada*

16.1 Introduction

The *E*- and *H*-plane three-port waveguide wye turnstile junction circulators are essential devices in most telecommunication systems. Whereas the *H*-plane geometry has been given much attention in the literature, the *E*-plane has received less attention. The purpose of this chapter is to investigate the first circulation condition of the latter geometry. The transmission parameters of the scattering matrix are here all negative. The first circulation condition of the *E*-plane circulator is characterized by either a pass or stop band depending on the indices of the axial and radial mode of the resonator.

The adjustment of this class of junction is a classic eigenvalue problem. The degenerate counterrotating eigenvalues are fixed by having recourse to Dicke's theorem. It states that these coincide with the position of a piston at one port, which will decouple a second one from an input port. The nondegenerate eigenvalue is then obtained by making use of the equality between the trace of the square matrix containing the eigenvalues along its main diagonal of the matrix and that of the scattering one. The first of these conditions coincides with the definition of the characteristic planes of the junction. The junction described in this chapter uses a pair of turnstile resonators on each narrow wall of the waveguide. It may be visualized as a seven-port network consisting of an *E*-plane junction of three rectangular waveguides and *H*-plane circular waveguides at its axis with two orthogonal ports. The required three-port is obtained by closing the circular waveguides with a short-circuit piston. Figure 16.1 depicts the *E*-plane arrangement under consideration.

Microwave Polarizers, Power Dividers, Phase Shifters, Circulators, and Switches, First Edition. Joseph Helszajn.

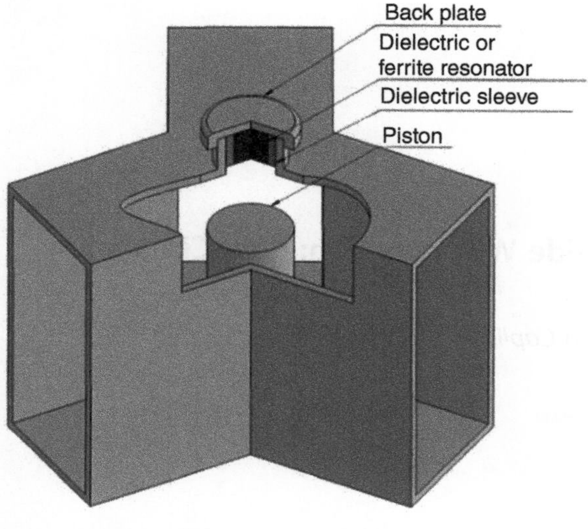

Figure 16.1 Schematic diagram of *E*-plane junction.

16.2 Scattering Matrix of Reciprocal *E*-plane Three-port *Y*-junction

The scattering matrix of the three-port *E*-plane wye junction differs from its *H*-plane counterpart in that it contains negative elements. Two negative terms are present in the description of the *E*-plane tee junction. The scattering matrix at the pass band characteristic plane of the *E*-plane wye junction may be written as

$$\bar{S} = \begin{bmatrix} S_{11} & -S_{21} & -S_{21} \\ -S_{21} & S_{11} & -S_{21} \\ -S_{21} & -S_{21} & S_{11} \end{bmatrix} \tag{16.1}$$

This matrix satisfies the unitary condition. It separately has symmetry about the main diagonal in keeping with the reciprocity condition, as well as the phase condition between ports 1 and 3 of the *E*-plane tee junction and similar phase relations with the other two pairs of adjacent ports in a cyclic manner. The origin of the negative elements may be understood by scrutinizing the field pattern of the junction across the narrow dimension of the waveguide with the generator setting at a side port of the junction. The mapping between the tee and wye junctions is indicated in Figure 16.2.

The nature of the scattering matrix adopted in this work may be tested by forming the unitary condition:

$$SS^{\mathrm{T}} = \mathrm{I} \tag{16.2a}$$

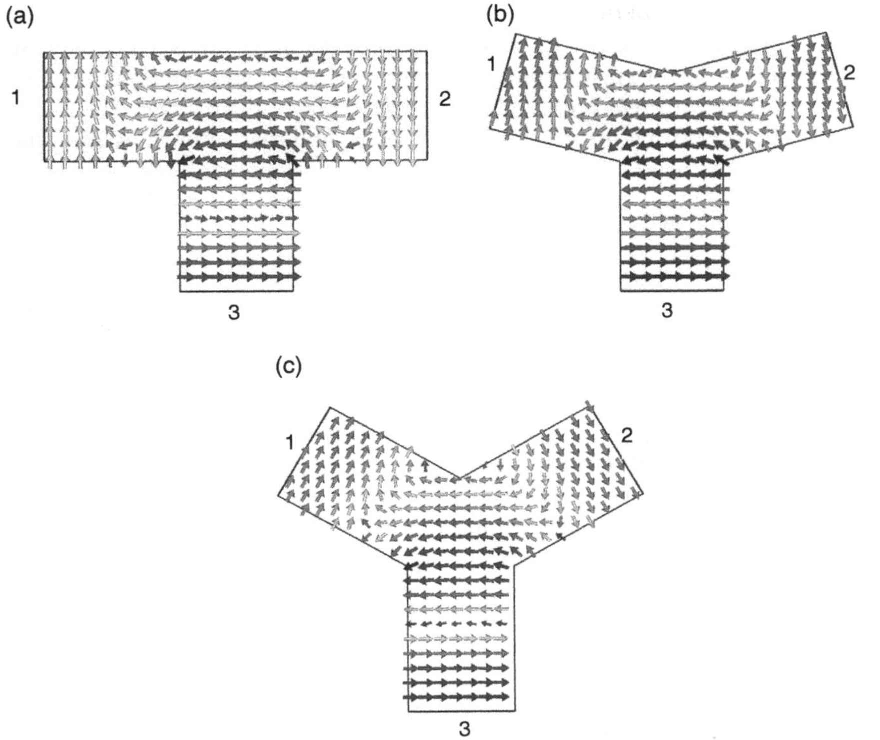

Figure 16.2 (a–c) Electrical field pattern across the narrow dimension of an *E*-plane waveguide wye junction with the generator sitting at port 3.

One property of the matrix is that this junction cannot be satisfied with

$$S_{11} = 0 \tag{16.2b}$$

Introducing this statement into the unitary condition scattering matrix gives

$$2S_{21} = 1 \tag{16.2c}$$

$$S_{21} = 0 \tag{16.2d}$$

The reference planes of the junction are not unique. Another possibility is to place it at the stop band characteristic plane 90° away from that in Eq. (16.1), that is,

$$\bar{S} = \begin{bmatrix} -S_{11} & S_{21} & S_{21} \\ S_{21} & -S_{11} & S_{21} \\ S_{21} & S_{21} & -S_{11} \end{bmatrix} \tag{16.3}$$

The reference plane of the scattering matrix in Eq. (16.1) corresponds with that of a piston that will produce a pass band between the other two ports. The plane in Eq. (16.3) is that of a piston which will produce a stop band between the other two ports, herewith referred to as a stop band characteristic plane. The possibility of characterizing the junction at one plane and thereafter translating the result by 90° to the other is understood. The converse operation is equally valid.

The scattering matrix of the *H*-plane junction is reproduced below for completeness sake. It is written as

$$\bar{S} = \begin{bmatrix} \pm S_{11} & \pm S_{21} & \pm S_{21} \\ \pm S_{21} & \pm S_{11} & \pm S_{21} \\ \pm S_{21} & \pm S_{21} & \pm S_{11} \end{bmatrix} \tag{16.4}$$

The positive sign is compatible with the scattering matrix in Eq. (16.1). The negative one is compatible with that in Eq. (16.3). The two planes differ in that the first displays a shunt STUB-G complex gyrator circuit whereas the second presents a series STUB-R load. The former solution is that met in most commercial circulators and is the one adopted here.

Equations (16.5) and (16.6) indicate the scattering matrices of the first stop band circulation condition at the pass and stop band characteristic planes of the junction.

$$\bar{S} = \begin{bmatrix} S_{11} & 0 & 0 \\ 0 & S_{11} & 0 \\ 0 & 0 & S_{11} \end{bmatrix} \tag{16.5}$$

$$\bar{S} = \begin{bmatrix} -S_{11} & 0 & 0 \\ 0 & -S_{11} & 0 \\ 0 & 0 & -S_{11} \end{bmatrix} \tag{16.6}$$

The above two matrices are compatible with Figure 16.3c and d in the next section as well as Eq. (16.7) in the same section.

16.3 Reflection Eigenvalue Diagrams of Three-port Junction Circulator

The adjustment of any *m*-port junction possible circulator is an eigenvalue problem. The first and second circulation conditions of a three-port junction circulator are depicted in Figure 16.3.

The left column in these illustrations corresponds to the first so-called circulation condition of the circulator and the right column represents the second.

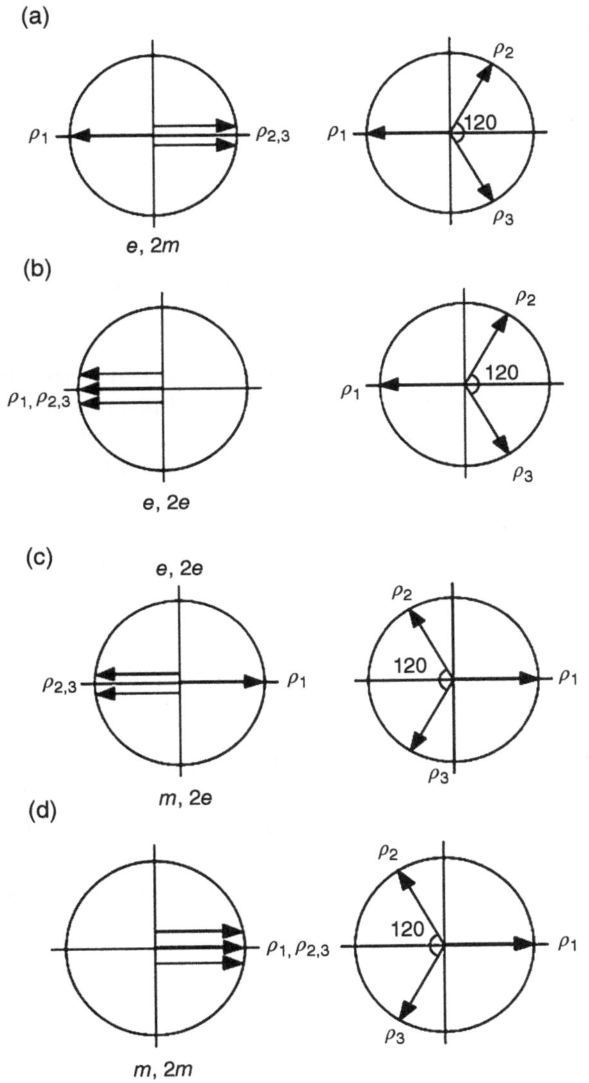

Figure 16.3 (a–d) Eigenvalue diagrams of the three-port symmetrical junction circulator (90° away from axis of junction).

The relationship between the eigenvalues of the junction and the scattering parameter is given in the usual way by

$$S_{11} = \frac{\rho_1 + 2\rho_{2,3}}{3} \tag{16.7a}$$

$$S_{21} = \frac{\rho_1 - \rho_{2,3}}{3} \tag{16.7b}$$

For a reciprocal junction

$$\rho_2 = \rho_3 = \rho_{2,3}. \tag{16.7c}$$

ρ_1 in Eq. (16.7) is the nondegenerate eigenvalue and ρ_2 and ρ_3 are the degenerate counterrotating eigenvalues. The first four first-circulation conditions illustrated in the left-hand side of Figure 16.3 are $S_{11} = 1/3, -1/3, -1$, and 1, respectively. The second circulation condition is in each instance satisfied, provided $S_{11} = 0$. This places the eigenvalues 120° apart on the eigenvalue diagram, as shown on the right column of Figure 16.3.

The eigenvalue diagram applicable in any situation is dictated by both the choice of reference planes and whether the junction is an *E*- or *H*-plane arrangement. The stop band characteristic planes are displaced by 90° from each other. The two eigenvalue diagrams in Figure 16.3a and b have pass band frequency responses at the midband frequency of the junction. The two in Figure 16.3c and d have stop bands there. Their diagrams are actually not unique. All four eigenvalues may be reduced to a single stop band solution by introducing a thin metal wire along the symmetry axis of the junction.

16.4 Eigen-networks

The eigen-networks of the three-port junction circulator, in its most simple form, are one-port open- or short-circuit unit elements (UEs), which display the reflection eigenvalues of the junction. The possible eigen-networks of the problem region are separately indicated in Figure 16.4.

The in-phase eigenvector $\bar{U}_1$ and circularly polarized counterrotating ones $\bar{U}_{2,3}$ are written as

$$\bar{U}_1 = \frac{1}{\sqrt{3}} \begin{bmatrix} 1 \\ 1 \\ 1 \end{bmatrix} \tag{16.8a}$$

$$\bar{U}_2 = \frac{1}{\sqrt{3}} \begin{bmatrix} 1 \\ \alpha \\ \alpha^2 \end{bmatrix} \tag{16.8b}$$

$$\bar{U}_3 = \frac{1}{\sqrt{3}} \begin{bmatrix} 1 \\ \alpha^2 \\ \alpha \end{bmatrix} \tag{16.8c}$$

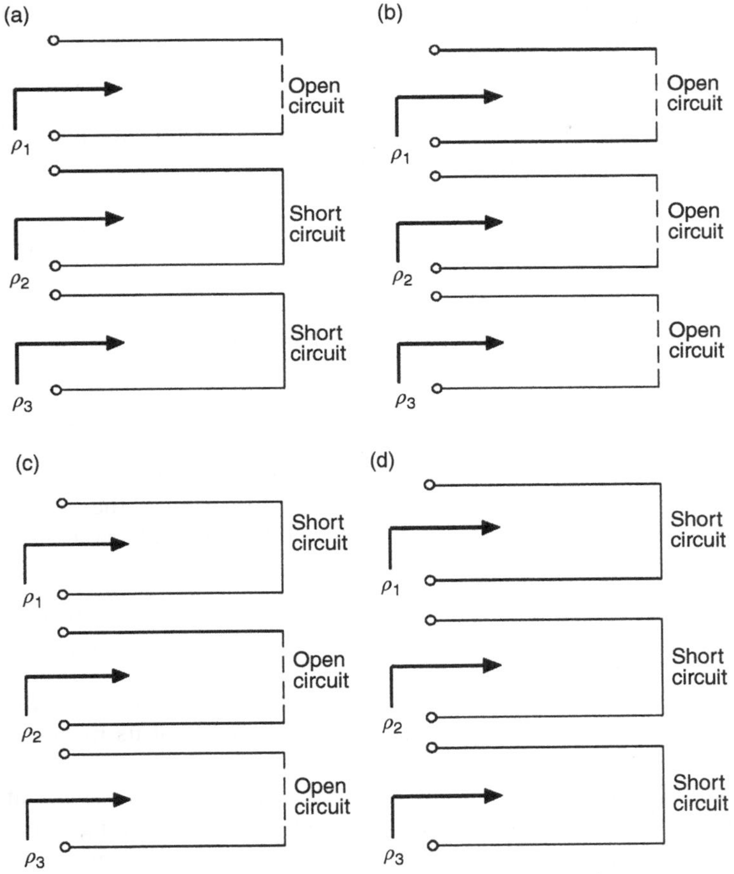

Figure 16.4 Eigen-networks of the *H*-plane circulator (a, b), and that of the *E*-plane circulator (c, d).

where

$$\alpha = 1\exp(j120)$$

ϕ_1 and $\phi_{2,3}$ are reflection angles given by

$$\phi_1 = 2\theta_1 + \frac{p\pi}{2}, \text{rad} \tag{16.9a}$$

$$\phi_{2,3} = 2\theta_{2,3} + \frac{p\pi}{2}, \text{rad} \tag{16.9b}$$

p is equal to 1 for an open-circuited stub and 2 for a short-circuited one. θ_1 and $\theta_{2,3}$ are the electrical lengths of the eigen-networks, respectively.

The eigen-networks are revealed by application of the eigenvectors one at a time. The reflection eigenvalues satisfy

$$\rho_i \bar{U}_i = \bar{S}\bar{U}_i \tag{16.10}$$

The roots of the characteristic equation at the pass band characteristic plane of the junction described in Eq. (16.1) are obtained from

$$\begin{vmatrix} S_{11}-\rho_i & -S_{21} & -S_{21} \\ -S_{21} & S_{11}-\rho_i & -S_{21} \\ -S_{21} & -S_{21} & S_{11}-\rho_i \end{vmatrix} = 0 \tag{16.11}$$

with $i = 1,2,3$. The eigenvalues at the pass band characteristic plane are

$$\rho_1 = S_{11} - 2S_{21} \tag{16.12a}$$

$$\rho_{2,3} = S_{11} + S_{21} \tag{16.12b}$$

It is separately deduced in terms of Dicke's formulation in Section 16.6.

The reflection eigenvalues at the stop band characteristic plane of the scattering matrix in Eq. (16.3) are displaced from that of Eq. (16.1) by π radians, as written as

$$\rho_1 = -S_{11} + 2S_{21} \tag{16.13a}$$

$$\rho_{2,3} = -S_{11} - S_{21} \tag{16.13b}$$

The nature of the *H*-plane eigen-networks may be established in terms of the electric fields of the waveguide, that of the *E*-plane in terms of its magnetic fields.

The two reciprocal arrangements also differ in the insertion phase angles at the characteristic planes, when injecting from one port and measuring at the other two. These are spaced 360° apart in the case of the *H*-plane geometry and 180° apart in that of the *E*-plane one.

16.5 Pass Band and Stop Band Characteristic Planes

Two classic planes separated by 90° are met in the description of wye junctions. One plane is that of a piston at one port which will produce a pass band between the other two. The second is that of a similar piston at the same port, which will produce a stop band. The latter planes are henceforth denoted as the stop band characteristic planes of the junction and the pass band characteristic one. Each arrangement displays the degenerate reflection eigenvalues of the junction at the respective planes of the piston.

The calculation of the first circulation condition of any junction circulator may in principle be undertaken at one or the other of its classic planes. The difference between the two is that the complex gyrator circuit at the first characteristic plane is that of a series STUB-R complex gyrator circuit at its terminals, whereas the second establishes a shunt STUB-G one. The complex gyrator circuit encountered in the design of junction circulators is usually the former circuit. Its reference plane is, therefore, adopted in this paper.

An important property of a piston at a characteristic plane is that it establishes similar planes at the other ports of the junction. The reflection coefficient at the stop band characteristic plane is given by

$$\rho_{\text{in}} = -1 \tag{16.14}$$

The reflection coefficient at the pass band characteristic plane, 90° away from the first one, is

$$\rho_{\text{in}} = 1 \tag{16.15}$$

The degenerate reflection eigenvalues are obtained by decoupling one output port from a second one using a variable short-circuit at the third one.

16.6 The Dicke Eigenvalue Solution

The eigenvalues of the symmetrical reciprocal *E*-plane wye junction may be obtained by forming the roots of the characteristic equation. These may also be deduced by recognizing that the degenerate eigenvalues of the junction coincide with the reflection coefficients at a typical input port with one output port decoupled from it by a piston at the other output port, as proposed by Dicke. The third eigenvalue may then be deduced by having recourse to the trace of the diagonal eigenvalue matrix and that of the scattering matrix.

The development of the degenerate eigenvalues at the stop band characteristic plane is dealt with here by way of an example. It starts by forcing the relationship between the incident and reflected waves of the junction, given by

$$\begin{bmatrix} b_1 \\ b_2 \\ b_3 \end{bmatrix} = \begin{bmatrix} -S_{11} & S_{21} & S_{21} \\ S_{21} & -S_{11} & S_{21} \\ S_{21} & S_{21} & -S_{11} \end{bmatrix} \begin{bmatrix} a_1 \\ a_2 \\ a_3 \end{bmatrix} \tag{16.16}$$

where

$$b_1 = -S_{11}a_1 + S_{21}a_2 \tag{16.17a}$$

$$b_2 = S_{21}a_1 - S_{11}a_2 \tag{16.17b}$$

$$0 = S_{21}a_1 + S_{21}a_2 \tag{16.17c}$$

with

$$a_3 = b_3 = 0 \tag{16.18a}$$

$$a_2 = -b_2 \tag{16.18b}$$

The reflection coefficient at port 1 is now given without ado by

$$\rho_{\text{in}} \equiv \frac{b_1}{a_1} = -S_{11} - S_{21} \tag{16.19}$$

and

$$\rho_{2,3} = \rho_{\text{in}} \tag{16.20}$$

The third eigenvalue is deduced by making use of the equality between the trace of the square diagonal matrix containing the eigenvalues of the problem and that of the square scattering matrix, that is,

$$\rho_1 + 2\rho_{2,3} = -3S_{11} \tag{16.21}$$

or

$$\rho_1 = -S_{11} + 2S_{21} \tag{16.22}$$

This result is identical to that derived by having recourse to the eigenvalue problem.

The equality between the reflection coefficient at the pass band characteristic plane and the degenerate eigenvalues at the same plane is readily understood. This may be demonstrated without ado by recognizing that the pass and stop bands' characteristic planes are interlaced and separated by 90°. The result is in keeping with the scattering matrix in Eq. (16.1). The reflection coefficient is here +1 instead of −1.

16.7 Stop Band Characteristic Plane

One means of testing the description of the scattering matrix adopted here is to see whether it satisfies the boundary conditions at the characteristic planes established previously.

A typical calculation of the scattering parameters at the characteristic planes produced in an unloaded *E*-plane wye junction (Figure 16.5) in WR75 at a frequency of 13.25 GHz gives

$$S_{11} = 0.3476\angle -190.75^\circ$$

$$S_{21} = 0.6630\angle -173.25^\circ$$

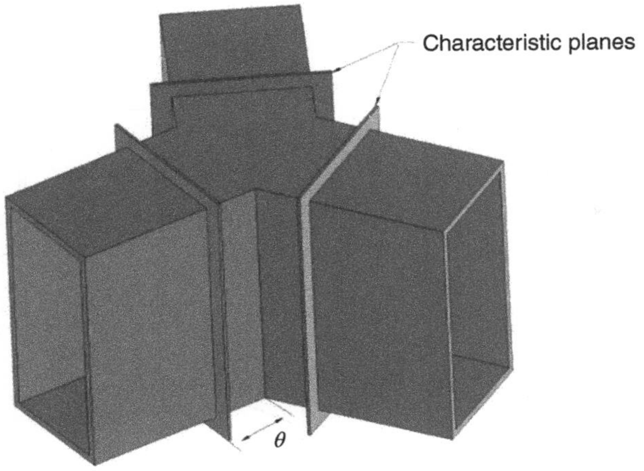

Figure 16.5 Schematic diagram of *E*-plane wye junction showing the positions of the characteristic planes.

The electrical length θ measured from the opening of the rectangular waveguide to the characteristic plane is

$$\theta = \frac{2\pi}{\lambda_g} l = 1.75\,\text{rad} = 100.5^\circ \tag{16.23}$$

Introducing these parameters in the definition of ρ_{in} gives

$$\rho_{in} = -S_{11} - S_{21} = 0.9999\angle -179.24^\circ \tag{16.24}$$

as asserted.

16.8 The *E*-plane Geometry

The *E*-plane junction consists of one or two quarter-wave long gyromagnetic prism or cylindrical resonators on one or both narrow waveguide walls at the junction of the three rectangular waveguides. Figures 16.6 and 16.7 illustrate two possibilities. Another resonator structure is a truncated cylindrical geometry. The geometry dealt with here is that of a junction containing a pair of cylindrical resonators. The reference terminals of the junction are the openings of the regular waveguides. The junction, in the case of the geometry using disk resonators, is defined in terms of its physical parameters by the radius of the resonator R, by its length L, by the gap S between the open face of a typical resonator and the symmetrical plane of the waveguide. The junction is completely fixed by the aspect ratio R/L of the resonator, by the radial wave number

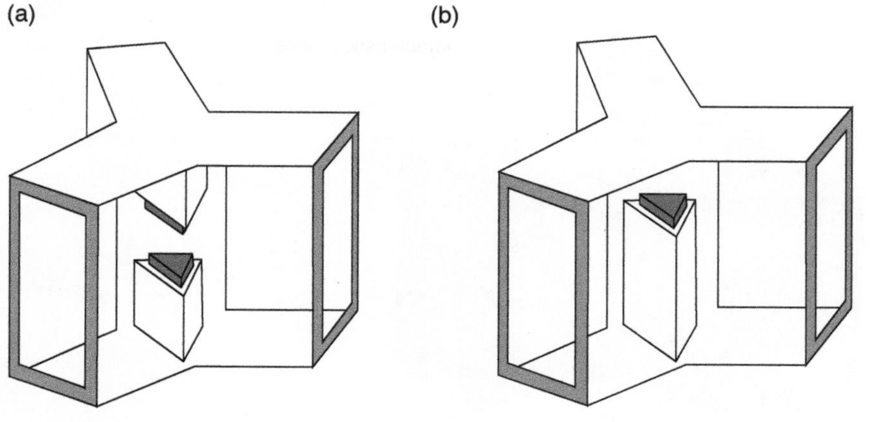

Figure 16.6 Schematic diagram of *E*-plane prism circulator using (a) coupled quarter-wave long resonators and (b) single quarter-wave long resonator.

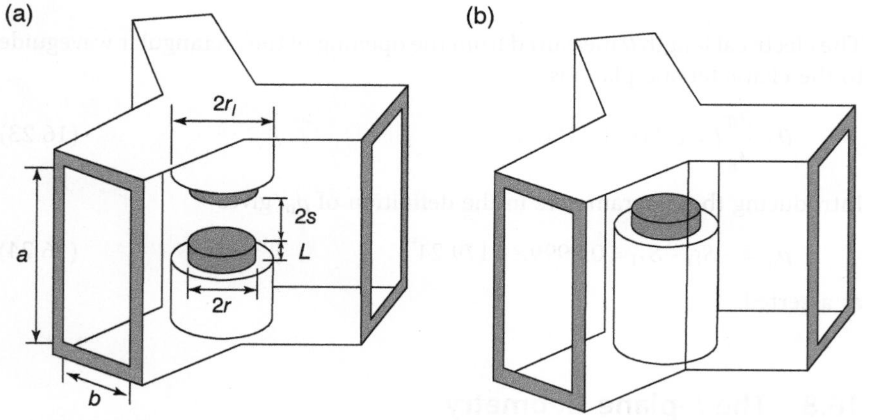

Figure 16.7 Schematic diagram of *E*-plane cylindrical circulator using (a) coupled quarter-wave long resonators and (b) single quarter-wave long resonator.

k_0R, and by the gap–resonator ratio S/L. The wavenumber k_0 completes its description.

A property of the junction under consideration is that it has two different neighboring solutions. One, a so-called small gap, has a pass band about its mid-band frequency. The other, a so-called large-gap solution, has a stop band. The prevailing solution depends upon whether the nondegenerate eigenvector has an electric or magnetic wall at the terminals of the junction. A feature of the small gap solution is that its in-phase eigen-network may be tuned to display a short-circuit boundary condition, which would otherwise be an open-circuited one, at its input terminals. This property of an evanescent UE is demonstrated in the next section. The degenerate eigen-networks support quasi-HE_{11} field patterns, which propagate up and down the length of the resonator. It may be adjusted to exhibit an open-circuit at the same terminals. The overall arrangement is, therefore, compatible with the eigenvalue diagram in Figure 16.3a and is the problem dealt with in this paper.

16.9 First Circulation Condition

The adjustment of the first circulation condition of the eigenvalue diagrams in Figure 16.3a demands that the eigen-networks are commensurate and that the reflection coefficients are either in-phase or out-of-phase at the terminals of the junction,

$$\rho_1 = -1 \text{ or } +1, k_0, \frac{R}{L} = \text{constant} \tag{16.25a}$$

$$\rho_{2,3} = +1 \text{ or } -1, k_0, \frac{R}{L} = \text{constant} \tag{16.25b}$$

The demagnetized junction displays a pass band whenever

$$\rho_1 = -1 \tag{16.26a}$$

$$\rho_{2,3} = +1 \tag{16.26b}$$

It displays stop band conditions at the same possible pairs of terminals whenever

$$\rho_1 = +1 \tag{16.27a}$$

$$\rho_{2,3} = +1 \tag{16.27b}$$

A typical demagnetised three-port E plane junction may either display a pass or stop band. Tables 16.1 and 16.2 indicate, in keeping with Dicke's theorem, that

Table 16.1 $\phi_1 = 2\theta_1 + \pi$, s/c.

θ_1	ϕ_1	ρ_1
$\frac{\pi}{2}$	2π	$+1$
π	3π	-1
$\frac{3\pi}{2}$	4π	$+1$
2π	5π	-1

Table 16.2 $\phi_{2,3} = 2\theta_{2,3}$, o/c.

$\theta_{2,3}$	$\phi_{2,3}$	$\rho_{2,3}$
$\frac{\pi}{2}$	π	−1
π	2π	+1
$\frac{3\pi}{2}$	3π	−1
2π	4π	+1

for a specified even mode stepping the odd one by 90° exchanges a typical pass or stop band into a stop and pass band.

The required solution is

$$k_0 R_1 = k_0 R_{2,3} = k_0 R \tag{16.28a}$$

$$\frac{S_1}{L} = \frac{S_{2,3}}{L} = \frac{S_{eff}}{L} \tag{16.28b}$$

The boundary condition may also be expressed in terms of the gap factor. The latter notation is the one adopted in this paper.

$$q_1 = q_{2,3} = q_{eff}. \tag{16.29}$$

These are written as

$$q_1 = \frac{L}{L + S_1} \tag{16.30a}$$

$$q_{2,3} = \frac{L}{L + S_{2,3}} \tag{16.30b}$$

$$q_{eff} = \frac{L}{L + S_{eff}} \tag{16.30c}$$

k_0 is the radian wavenumber (rad m^{-1}), written as

$$k_0 = \frac{2\pi}{\lambda_0}, \text{rad m}^{-1}$$

and λ_0 (m) is the free-space wavelength. The work undertaken here is at 13.25 GHz in WR75 waveguide for which k_0 = 0.278 rad mm^{-1}.

16.10 Calculations of Eigenvalues

The required eigenvalue diagram of the small gap solution is dealt with here. The eigenvalues are given in Eq. (16.12). The intersection between the two conditions, $q_1(k_0R)$ and $q_{2,3}(k_0R)$, is the required solution.

The adjustments of ϕ_1 and $\phi_{2,3}$ are the objects of the optimization. Writing ϕ_1 and $\phi_{2,3}$ in terms of the scattering parameters gives

$$\rho_1 = 1\exp - j\phi_1 = S_{11} - 2S_{21} \tag{16.31a}$$

$$\rho_{2,3} = 1\exp - j\phi_{2,3} = S_{11} + S_{21} \tag{16.31b}$$

In general,

$$\rho_1 = 1\exp - j\phi_1 \tag{16.32a}$$

$$\rho_{2,3} = 1\exp - j\phi_{2,3} \tag{16.32b}$$

The required angles of the reflection eigenvalues are here equal to

$$\phi_1 = \pi, \text{rad}$$

The in-phase eigenvalue of the junctions may be deduced by having recourse to the appropriate linear combination of the scattering matrix of the junction. The approach utilized here, however, is to construct its one-port eigen-network. This network is obtained by partitioning the junction into equal regions using a triplet of electric walls. The circuit obtained thus is depicted in Figure 16.8.

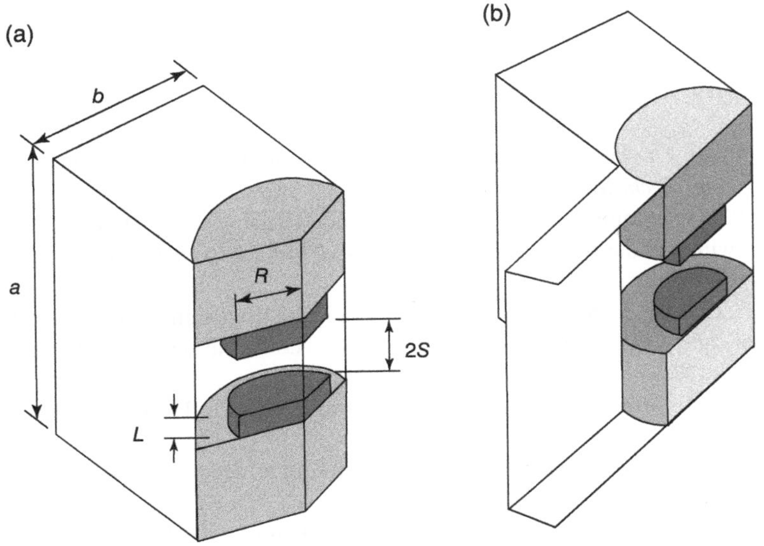

Figure 16.8 Schematic diagram of (a) the in-phase eigen-network and (b) the counterrotating eigen-network.

The calibration plane in the design of the junction circulator is not unique. Distinct boundary conditions produce equally valid solutions but with different combinations of radian wave numbers and gap–resonator ratios.

Figure 16.9 is a Smith chart of a typical plot.

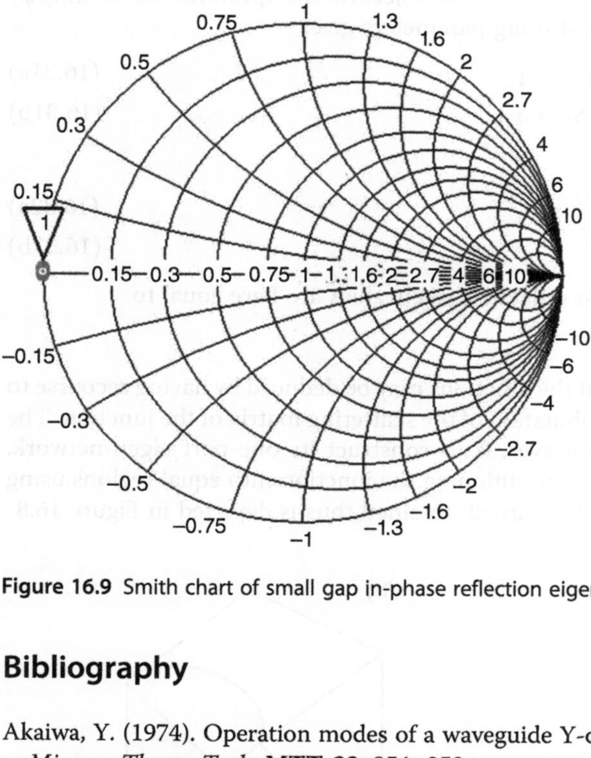

Figure 16.9 Smith chart of small gap in-phase reflection eigenvalue ρ_1.

Bibliography

Akaiwa, Y. (1974). Operation modes of a waveguide Y-circulator. *IEEE Trans. Microw. Theory Tech.* **MTT-22**: 954–959.

Akaiwa, Y. (1978). A numerical analysis of waveguide H-plane Y-junction circulators with circular partial-height ferrite post. *J. Inst. Electron. Commun. Eng. Jpn.* **E61**: 609–617.

Allanson, J.T., Cooper, R., and Cowling, T.G. (1946). The theory and experimental behavior of right-angled junctions in rectangular–section waveguides. *IEE Proc.* **93** (23): 177–187.

Altman, J.L. (1964). *Microwave Circuits*, Van Nostrand Series in Electronics and Communications. New York: Van Nostrand.

Auld, B.A. (1959). The synthesis of symmetrical waveguide circulators. *IRE Trans. Microw. Theory Tech.* **MTT-7**: 238–246.

Buchta, G. (1966). Miniaturized broadband E-tee circulator at X-band. *Proc. IEEE* **54**: 1607–1608.

Caplin, M., D'Orazio, W., and Helszajn, J. (1999). First circulation condition of E-plane circulator using single-prism resonator. *IEEE Microw. Guid. Wave Lett.* **9** (3): 99–101.

Casanueva, A., Leon, A., Mediavilla, A., and Helszajn, J. (2013). Characteristic planes of microstrip and unilateral finline tee-junctions. *Progress in Electromagnetics Research Symposium Proceedings*, Stockholm, Sweden (12–15 August 2013), pp. 173–179.

Chait, H.N. and Curry, R.T. (1959). Y-circulator. *J. Appl. Phys.* **30**: 152.

Davis, L.E. and Longley, S.R. (1963). E-plane three-port X-band waveguide circulators. *IEEE Trans. Microw. Theory Tech.* **MTT-11**: 443–445.

DeCamp, E.E. Jr. and True, R.M. (1971). 1-MW four-port E-plane junction circulator. *IEEE Trans. Microw. Theory Tech.* **MTT-19**: 100–103.

Goebel, U. and Schieblock, C. (1983). A unified equivalent circuit representation of H and E-plane junction circulators. *European Microwave Conference*, Nurnberg, Germany (3–8 September 1983), pp. 803–808.

Helszajn, J. (1970). The adjustment of the m-port single-junction circulator. *IEEE Trans. Microw. Theory Tech.* **MTT-18**: 705–711.

Helszajn, J. (1987). Complex gyrator of an evanescent mode E-plane junction circulator using H-plane turnstile resonators. *IEEE Trans. Microw. Theory Tech.* **MTT-35**: 797–806.

Helszajn, J. (2012). An FE algorithm for the adjustment of the first circulation condition of the turnstile waveguide circulator. *IEEE Trans. Microw. Theory Tech.* **MTT-60**: 3079–3087.

Helszajn, J. (2016). The electrically symmetrical E-plane waveguide tee junction at the Dicke and Altman planes. *IEEE Trans. Microw. Theory Tech.* **MTT-64**: 1–9.

Helszajn, J. and Cheng, S. (1990). Aspect ratio of open resonators in the design of evanescent mode E-plane circulators. *IEE Proc. Microw. Antennas Propag.* **137**: 55–60.

Helszajn, J. and McDermott, M. (1972). Mode charts for E-plane circulators. *IEEE Trans. Microw. Theory Tech.* **MTT-20**: 187–188.

Helszajn, J. and Sharp, J. (1986). Dielectric and permeability effects in HE1,1,1/2 open demagnetized ferrite resonators. *IEE Proc.* **133** (Part H): 271–276.

Helszajn, J., Caplin, M., Frenna, J., and Tsounis, B. (2014). Characteristic planes and scattering matrices of E and H-plane waveguide tee junction. *IEEE Microw. Wireless Compon. Lett.* **24** (4): 209–211.

Helszajn, J., Carignan, L.-P., and Sharp, J. (2016). Calibration and adjustments of degree-1 and -2 re-entrant turnstile circulators using a partially recessed quarter-wave long dielectric resonator. *IET Microw. Antennas Propag.* **10**: 1–7.

McGrown, J.W. and Wright, W.H. Jr. (1967). A high power, Y-junction E-plane circulator. *G-MTT International Microwave Symposium Digest*, Boston, MA (8–11 May 1967), pp. 85–87.

Montgomery, C.G., Dicke, R.H., and Purcell, E.M. (1948). *Principles of Microwave Circuits*, MIT Radiation Laboratory Series, vol. **VIII**, 432. New York: McGraw-Hill Book Co.

Omori, S. (1968). An improved E-plane waveguide circulator. *G-MTT International Microwave Symposium Digest*, Detroit, MI (20–22 May 1968), pp. 228–236.

Owen, B. (1972). The identification of modal resonances in ferrite loaded waveguide junction and their adjustment for circulation. *Bell Syst. Tech. J.* **51** (3): 595–627.

Owen, B. and Barnes, C.E. (1970). The compact turnstile circulator. *IEEE Trans. Microw. Theory Tech.* **MTT-18**: 1096–1100.

Schaug-Patterson, T. (1958). Novel Design of a 3-port Circulator. Norwegian Defence Research Establishment Report, Rpt No. R-59 (January).

Solbach, K. (1972). Equivalent E-plane circulators. *IEEE Trans. Microw. Theory Tech.* **MTT-20**: 187–188.

Wright, W. and McGowan, J. (1968). High-power, Y-junction E-plane circulators. *IEEE Trans. Microw. Theory Tech.* **MTT-16**: 557–559.

Yoshida, S. (1959). E-type T circulator. *Proc. IRE* **47**: 208.

17

Adjustment of Prism Turnstile Resonators Latched by Wire Loops

Joseph Helszajn[1] *and William D'Orazio*[2]

[1] *Heriot Watt University, Edinburgh, UK*
[2] *Apollo Microwaves Ltd, Dorval, Quebec, Canada*

17.1 Introduction

An important microwave component is a switched circulator using a single wire loop embedded in a gyromagnetic resonator. The operation of the switch relies on the two remnant states of the hysteresis loop of a magnetic insulator. This is done using suitable current pulses. The resonator in Goodman (1965) is a half-wave long tri-toroidal geometry with open flat faces and that in Passaro and McManus (1966) and Katoh et al. (1980) is a half-long prism configuration. The operation of each arrangement has been interpreted in terms of a tri-toroidal geometry consisting of an inner core magnetized in one sense and a triplet of ribs magnetized in the other. The purpose of this chapter is to investigate the operation of a prism resonator using a single irregular hexagonal wire loop in terms of the ratio of the inner and one typical rib of the outer tri-toroidal geometry. The results obtained here are expressed in terms of a shape factor defined by the ratio of the inner core of the tri-toroidal resonator and one typical outer rib. This quantity is bracketed between 1 and typically 10. A shape factor of infinity corresponds to a homogeneous resonator.

Figure 17.1 depicts the schematic diagram of the arrangement considered here. It may be visualized as an inverted reentrant junction together with a doublet of quarter-wave long resonators with the flat face of one typical resonator closed by a virtual electric wall and the other separated from the in-phase piston by a suitable gap. This type of junction, however, is usually described in terms of a half-wave long geometry on its axis and this convention is retained here.

Microwave Polarizers, Power Dividers, Phase Shifters, Circulators, and Switches, First Edition. Joseph Helszajn.

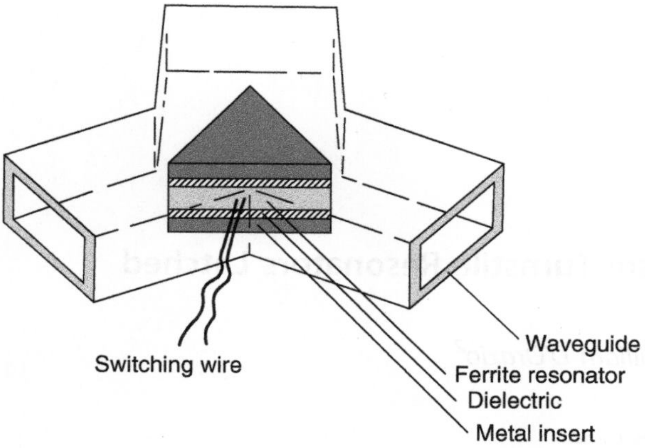

Figure 17.1 Schematic diagram of a junction circulator using a latched prism resonator.

The axial direct magnetization produced by a wire carrying current has equal magnitudes but opposite polarities on either side of the wire loop and decays away from it in a classic, predictable manner. The distribution of the alternating radio frequency magnetic field also varies on either side of the wire, but with unequal intensities. Unless a full three-dimensional solver is available, the experimental procedure introduced here is a practical solution to the adjustment of the switch resonator under consideration.

A knowledge of the quality factor of the resonator together with its susceptance slope parameter is sufficient to define the complex gyrator circuit of the circulator. The dependent parameter is here the gyrator conductance. The quality factor of the complex gyrator is a property of the gyrotropy of the resonator and is established in this chapter. The susceptance slope parameter is a property of the geometry of the resonator.

17.2 The Prism Resonator

The prism resonator under consideration consists of a half-wave long gyromagnetic waveguide with open flat faces separated by dielectric spacers from top and bottom triangular platforms. The tri-toroidal prism resonator in Figure 17.2 is defined by its wave number k_0, the overall length of the half-wave long prism resonator ($2L_0$), the side dimension of the prism (A), its aspect ratio A/L_0, the side of the subsidiary triangle (L), and its gap factor q_{eff}. The latter quantity is set by half the length of the resonator (L_0) and that of a typical dielectric spacer (S).

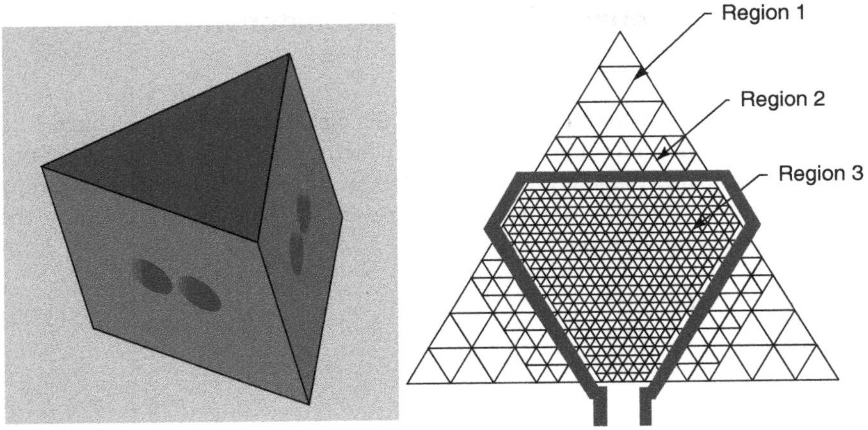

Figure 17.2 Geometry of tri-toroidal prism resonator.

The relative dielectric constants of the two regions ε_{d} and ε_{f} complete its characterization. The relative dielectric constant of the spacers used here is $\varepsilon_{\mathrm{d}} = 2.1$.

The thickness and dielectric constant of the spacers are not independent variables but are both fixed by the in-phase eigen-network of the junction. The degenerate counterrotating eigen-networks are essentially set by the linear dimensions of the prism and its dielectric constant. It approximately determines the midband frequency of the demagnetized junction. The exact solution is an eigenvalue problem as outlined in chapter 15.

The physical geometry under consideration is summarized below.

$$k_0 L = 0.445$$

$$\frac{A}{L} = 4.448$$

$$\frac{S}{L} = 0.338$$

$$\varepsilon_{\mathrm{d}} = 2.1$$

$$\varepsilon_{\mathrm{f}} = 15$$

$$k_0 = 0.0792\,\mathrm{rad\,mm^{-1}}$$

The gyrotropy is

$$\kappa = \frac{\gamma M_0}{\omega_0} = 0.68 \tag{17.1}$$

The waveguide size is WR229 and the saturation magnetization $\mu_0 M_0$ of the inner region of the magnetic insulator is 0.1200 T.

17.3 Split Frequency of Cavity Resonator with Up or Down Magnetization

The most important quantity entering into the description of either a fixed field or latched junction circulator is the quality factor (Q_L) of its complex gyrator circuit. This quantity is solely determined by the difference between the opening of the counterrotating frequencies of the resonator. The quality factor of the closed gyromagnetic resonator is

$$\frac{1}{Q_L} = \sqrt{3}\left(\frac{\omega_+ - \omega_-}{\omega_0}\right) \tag{17.2}$$

and

$$\left(\frac{\pi}{2L_0}\right)^2 = \left(\frac{\omega_\pm}{c}\right)^2 \varepsilon_f(\mu \pm C_{1,1}\kappa) - \left(\frac{4\pi}{3A}\right)^2 \tag{17.3}$$

μ and κ are the relative diagonal and off-diagonal elements of the tensor permeability. The gyromagnetic constant is (Akaiwa 1974)

$$C_{1,1} = \frac{\sqrt{3}}{\pi} \tag{17.4}$$

The factor $C_{1,1}$ accounts for the fact that the counterrotating magnetic fields in the gyromagnetic waveguide are only circularly polarized on its axis. This quantity also enters into the description of the split cutoff numbers or frequencies of the related planar resonator with top and bottom electric walls and a magnetic sidewall. $4\pi/3A$ is the cutoff number of the dominant $TM_{1,0}$ mode in the triangular waveguide.

In a saturated material, μ and κ are given by (Green and Sandy 1974)

$$\mu = 1 \tag{17.5a}$$

$$\kappa = \frac{\omega_m}{\omega} \tag{17.5b}$$

where

$$\omega_m = \gamma M_r \tag{17.6}$$

M_r is the remnant magnetization of the garnet resonator (A m^{-1}), γ is the gyromagnetic ratio (2.21 × 105 rad s^{-1} per A m^{-1}), and ω is the radian frequency (rad s^{-1}).

Figure 17.3 indicates some results for two different materials at 4 GHz. The magnetic flux density in this illustration is that at one of the two flat faces of the prism.

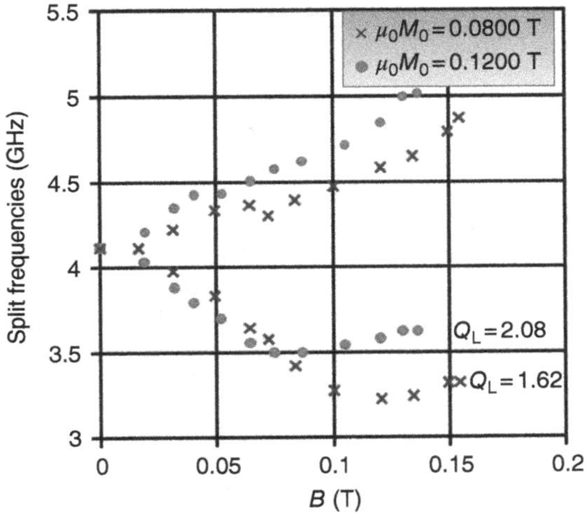

Figure 17.3 Experimental split frequencies of prism resonators ($\mu_0 M_0 = 0.0800$ T and 0.1200 T).

One of the two materials employed in Figure 17.3 is a garnet one with a magnetization ($\mu_0 M_0$) of 0.0800 T, a squareness (M_r/M_0) equal to 0.80, a relative dielectric constant (ε_f) of 15.3, and μ_0 is the free space permeability ($4\pi \times 10^{-7}$ H m^{-1}). The side dimension of the resonator (A) is 25.0 mm and its overall length ($2L_0$) is 11.24 mm.

The relative dielectric constant of the region between the open faces of the resonator and the waveguide or image walls is 2.1; its thickness (S) is 1.9 mm. The gap factor q_{eff} is 0.75. Its quality factor, Q_L, at the knee of the lower branch of the two split frequencies is equal to 2.08. The other material utilized in obtaining the data in Figure 17.3 has a magnetization ($\mu_0 M_0$) of 0.1200 T, a squareness of 0.78, and a relative dielectric constant of 15.1. Its quality factor Q_L is equal to 1.62.

17.4 Quality Factor of Gyromagnetic Resonator with Up and Down Magnetization

The physical variables of the in-phase eigen-network do not explicitly appear in the description of the complex gyrator circuit, provided it is separately idealized. The development of a circulator or switch using an internally latched resonator differs only from that of a conventional one in that the degree of splitting that is realizable between the degenerate counterrotating modes is in this situation somewhat less than is otherwise the case. This situation may be expressed in terms of a splitting factor (k_f) by writing the former relationship as (Helszajn and Sharp 2012)

$$\frac{1}{Q_{\text{eff}}} = \sqrt{3}k_{\text{f}}\left(\frac{\omega_+ - \omega_-}{\omega_0}\right) \tag{17.7}$$

where

$$k_{\text{f}} = \frac{\omega'_+ - \omega'_-}{\omega_+ - \omega_-} \tag{17.8}$$

and ω'_+ and ω'_- are the split frequencies of the resonator with its inner region magnetized along the positive z direction and its outer one in the opposite direction. The factor k_{f} may therefore be experimentally determined by evaluating the preceding relationship for the homogeneous and inhomogeneous circuits, respectively.

A scrutiny of the network problem indicates that values for Q_{eff} between ½ and 2½ are optimum for the design of quarter-wave coupled devices. An experimental or theoretical knowledge of this latter quantity is therefore essential for design and its evaluation is the main endeavor of this chapter.

Figure 17.2 indicates a typical discretization.

17.5 Shape Factor of Tri-toroidal Resonator

The ratio of the two oppositely magnetized regions is defined by a shape factor q:

$$q = \frac{\text{Surface area of the inner region magnetized along the} + z\text{ direction}}{\text{Surface area of one typical outer region magnetized along the} - z\text{ direction}} \tag{17.9}$$

If the side dimension of the overall triangular resonator is taken as A and those of the outer regions as L, then

$$q = \frac{A^2 - 3L^2}{L^2} \tag{17.10}$$

The condition $q = 1$ corresponds to that for which the cross-sectional areas of the two regions are equal.

Six specific geometries are investigated in this chapter. These are defined by $L_0 = 5.62$ mm, $A = 25$ mm, $L = 8.33$ mm, $L = 9.13$ mm, $L = 10.2$ mm, $L = 10.4$ mm, $L = 11.4$ mm, and $L = 12.5$ mm. The corresponding ratios of the magnetized surface areas are $q = 6.01$, $q = 4.50$, $q = 3.01$, $q = 2.78$, $q = 1.81$, and $q = 1.00$. The wire used is a 24 SWG solid copper wire. Table 17.1 summarizes the data. The general relationship obtained experimentally here between q and k_{f} is depicted in Figure 17.4.

Table 17.1 Shape factor q of the six geometries investigated in this work.

A (mm)	L (mm)	q
25	8.33	6.01
25	9.13	4.50
25	10.2	3.01
25	10.4	2.78
25	11.4	1.81
25	12.5	1.00

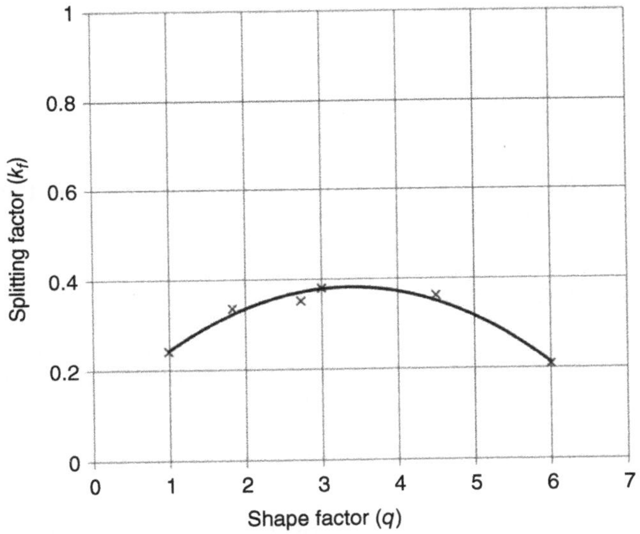

Figure 17.4 Experimental connection between q and k_f in prism resonator with up and down magnetization.

The condition q = 1.0 is that for which the flux density in the core is three times that in a typical rib. The maximum splitting factor in this instance is given by $k_f = 0.4$. It is compatible with a quality factor $Q_{eff} = 4.6$. k_f in Figure 17.4 represents the split frequencies in the resonator at saturation produced by a current-carrying wire divided by that produced by a uniform magnetic field.

Figure 17.5 is a photograph of the experimental hardware.

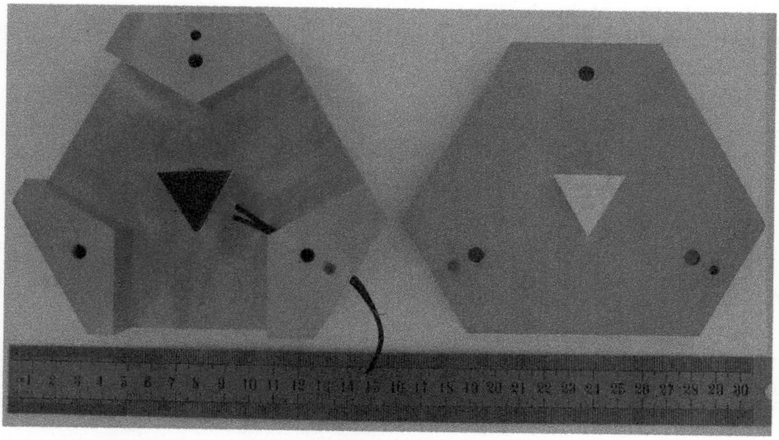

Figure 17.5 Photograph of switch resonator.

17.6 Squareness Ratio

Another quantity that enters into the description of latched devices is the squareness of the hysteresis loop of the magnetic insulator. It is defined here by

$$R = \frac{\text{Splitting with field removed}}{\text{Splitting with field applied}} \tag{17.11}$$

Figure 17.6 indicates some data on the split frequencies of one device using a current-carrying single wire loop with $q = 1.0$ and $\mu_0 M_r = 0.1200$ T with and without the remanence field. The value R obtained here is of the order of 0.8.

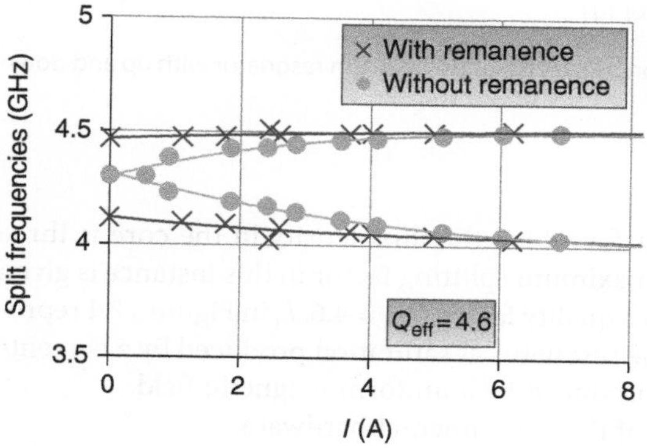

Figure 17.6 Experimental split frequencies of wire-operated half-wave long prism resonator with up and down magnetization ($\mu_0 M_s = 0.1200$ T).

17.7 The Complex Gyrator Circuit of the Three-port Junction Circulator

The gyrator circuit of any junction circulator for which the in-phase eigen-network may be idealized by a frequency-independent short-circuit boundary condition and for which the counterrotating ones are weakly split by the gyrotropy of the resonator is the one-port STUB-R circuit shown in Figure 17.7 (Helszajn 1994). A knowledge of its element values is a prerequisite for design. This circuit is usually described in terms of its susceptance slope parameter (b'), gyrator conductance (g), and the normalized split frequencies of the degenerate counterrotating eigenvalues $[(\omega_+ - \omega_-)/\omega_0]$. The gyrator conductance is set by the split frequencies of the resonator once the susceptance slope parameter is fixed by the choice of the resonator shape (Helszajn 1994).

Its normalized conductance is here

$$g_{\text{eff}} = \sqrt{3}b'k_{\text{f}}\left(\frac{\omega_+ - \omega_-}{\omega_0}\right) \tag{17.12}$$

b' is the normalized susceptance slope parameter of the STUB. The other quantities have the usual meaning.

A first-order approximation to the gyrator conductance may be obtained by assuming that the susceptance slope parameter of the junction is a property of the resonator and not of the gyrotropy. A typical value obtained on a similar prism resonator is

$$b' = 7$$

The gyrator conductance g_{eff} may also be written in terms of Q_{L} of the homogeneous resonator and the splitting factor k_{f} previously defined as

$$g_{\text{eff}} = \frac{b'}{Q_{\text{L}}}k_{\text{f}}$$

Figure 17.7 One-port complex gyrator circuit of junction circulator.

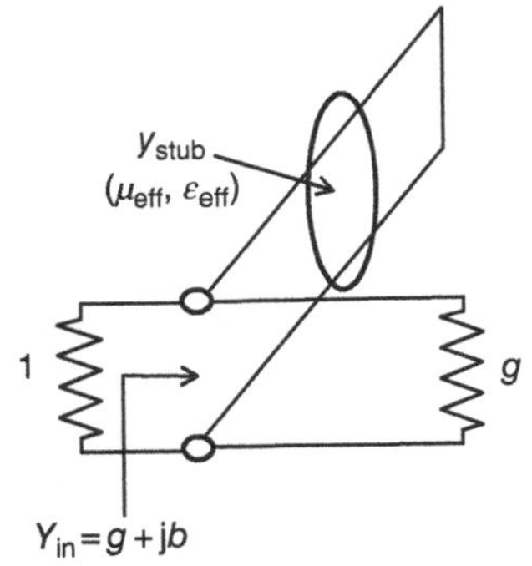

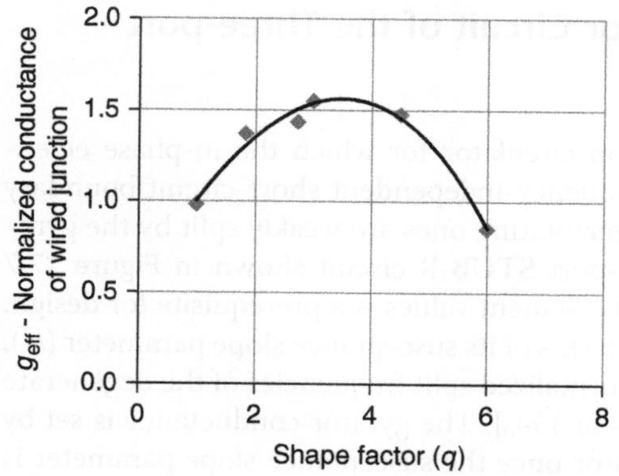

Figure 17.8 Normalized conductance of wired junction.

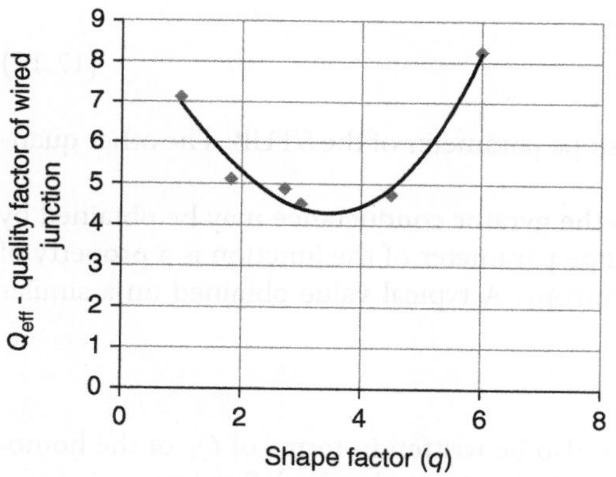

Figure 17.9 Quality factor of wired junction.

Introducing this value of b' in the real part condition with $q = 3$ gives

$$g_{\mathrm{eff}} = 1.52$$

Figure 17.8 illustrates the relationship between the gyrator conductance obtained here and the shape factor.

Figure 17.9 shows Q_{eff} versus q.

17.8 The Alternate Line Transformer

A suitable matching network is here an alternate line transformer. The topology of the matching arrangement under consideration is indicated in Figure 17.10. It consists of one short unit element (UE) adjacent to the load with the admittance

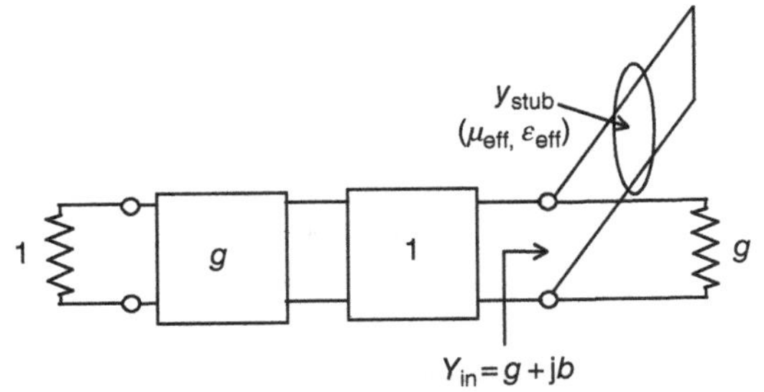

Figure 17.10 Topology of alternate line-matched complex gyrator of junction circulator.

of the generator in cascade with a second UE with the admittance of the load. The implementation of this arrangement is outside this work.

17.9 Effective Complex Gyrator Circuit

A circulator switch may be characterized in terms of the flux density produced by an electro magnet, the current through a magnetizing loop giving up and down magnetization in the vicinity of the loop, or in terms of the current through the wire loop at the remnant magnetization. The effect of each arrangement may be readily compared by plotting in each instance the gyrator conductance versus the split frequencies of the resonator. The slope of such a typical curve is the susceptance slope parameter of the junction, so that the three descriptions may be superimposed on a single graph for comparison purposes. A typical graph separately defines the real part condition of the complex gyrator circuit.

Bibliography

Akaiwa, Y. (1974). Operation modes of a waveguide Y-circulator. *IEEE Trans. Microw. Theory Tech.* **MTT-22**: 954–959.

Clavin, A. (1963). Reciprocal and nonreciprocal switches utilizing ferrite junction circulators. *IEEE Trans. Microw. Theory Tech.* **MTT-11**: 217–218.

Freiberg, L. (1961). Pulse operated circulator switch. *IRE Trans. Microw. Theory Tech.* **MTT-9**: 266.

Goodman, P.C. (1965). A latching ferrite junction circulator for phased array switching applications. *IEEE, GMTT Symposium*, Clearwater, FL (5–7 May 1965).

Green, J.J. and Sandy, F. (1974). Microwave characterization of partially magnetized ferrites. *IEEE Trans. Microw. Theory Tech.* **MTT-22**: 541–645.

Helszajn, J. (1994). Experimental evaluation of junction circulators: a review. *Proc. IEE Microw. Antennas Propag.* **141** (5): 351–358.

Helszajn, J. and Sharp, J. (2012). Cut-off space of a gyromagnetic planar disk resonator with a triplet of stubs with up and down magnetization. *IET Microw. Antennas Propag.* **6** (5): 569–576.

Katoh, I., Konishi, H., and Sakamoto, K. (1980). A 12 GHz broadband latching circulator. *Conference Proceedings, European Microwaves*, Poland, pp. 360–364.

Passaro, W.C. and McManus, J.W. (1966). A 35 GHz latching switch. *Presented at the IEEE International Microwave Symposium*, Palo Alto, CA (19 May 1966).

18

Numerical Adjustment of Waveguide Ferrite Switches Using Tri-toroidal Resonators

Joseph Helszajn[1] *and Mark McKay*[2]

[1] *Heriot Watt University, Edinburgh, UK*
[2] *Honeywell, Edinburgh, UK*

18.1 Introduction

The direction of circulation in a circulator is determined by the polarity of the direct magnetic field intensity utilized to magnetize its gyromagnetic resonator. It may therefore be employed to switch an input signal at one port to either one of the other two. One practical means of doing so is to internally or externally latch the resonator by using a current-carrying wire loop to switch between the two remanent states of its hysteresis loop. Figure 18.1 depicts two possible half-wave long latched gyromagnetic resonators met in the design of waveguide switches. The geometries under consideration here are strictly speaking examples of inverted reentrant turnstile structures using doublets of quarter-wave long resonators separated by virtual electric walls.

The waveguide circulator dealt with here is often but not exclusively referred to as the turnstile arrangement. Its adjustment involves a two-step procedure. The first fixes the degeneracy between a pair of degenerate counterrotating modes and a quasi in-phase mode. The second replaces the dielectric resonator by a gyromagnetic one in order to remove the degeneracy between the counter-rotating modes. The resonator, in the case of a cylindrical cavity, comprises a doublet of quarter-wave long $HE_{1,1,\delta}$ resonators separated by a virtual electric wall. The in-phase mode is a two-layer quasi-planar structure supporting a TM_{010} mode with no propagation along the axis of the resonator.

Microwave Polarizers, Power Dividers, Phase Shifters, Circulators, and Switches,
First Edition. Joseph Helszajn.

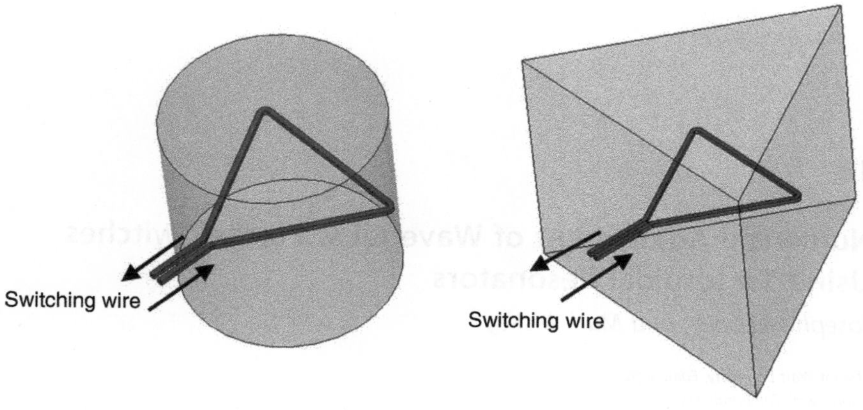

Figure 18.1 Topologies of two half-wave long gyromagnetic resonators.

18.2 The Tri-toroidal Resonator

The internal direct magnetization of the tri-toroidal gyromagnetic resonator is established by either winding a single switching wire around the core of the tri-toroid or by wrapping individual wires around its vertical back members. The chapter includes calculations on the flux on either side of the current-carrying wire loop. It also includes calculations on the splitting between the degenerate counterrotating modes of the tri-toroidal cavity. Figure 18.2 shows schematic diagrams of wire-activated gyromagnetic prism cavities in a waveguide junction in more detail.

An important property of the tri-toroidal geometry is that the direct magnetic flux in a typical back vertical branch of the arrangement is one third that of the central core. The aforementioned feature is readily understood by having recourse to the symmetries of the geometries. Figure 18.3 indicates two possible wire arrangements.

An approximate model divides the volumes of the geometry into one gyromagnetic region with a magnetic flux density in one direction, a middle isotropic dielectric region, and one gyromagnetic region with a uniform magnetic flux density in the opposite direction, but, in general, with a smaller or negligible value from that of the inner core. Here, the outer region only serves to complete the magnetic return path of the core of the tri-toroidal circuit. The way to explore the distribution of the direct magnetic flux density within the tri-toroidal circuit is to have recourse to a Finite Element (FE) magnetostatic package.

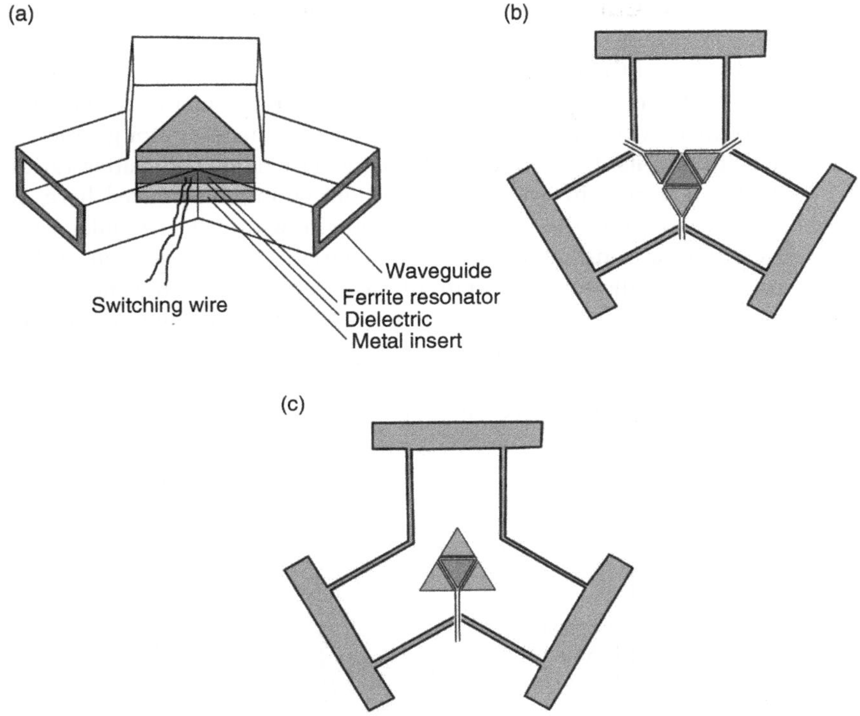

Figure 18.2 (a–c) Schematic diagrams of waveguide circulators showing different switching wire configurations.

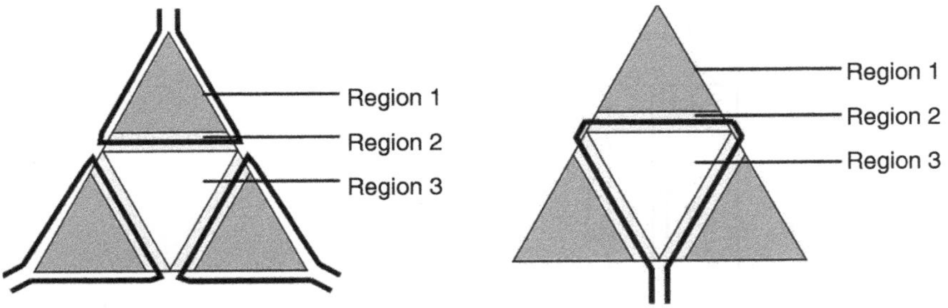

Figure 18.3 Practical configurations met in the design of latched prism cavities.

18.3 The Wire Carrying Slot Geometry

The size and shape of the wire carrying slots are, in practice, fixed by the corresponding direct magnetic flux density in the magnetic insulator. A difficulty with the introduction of a triplet of oversized slots into the resonator geometry, in the calculation of its degenerate cutoff frequencies, is that the latter varies with geometry. The way to overcome this shortcoming, introduced here, is to pot the slot containing the current-carrying wire with dielectric filler with the relative dielectric constant of the ferrite material. The introduction of this filler ensures that the dielectric constant of each region is the same and allows the shaping of the magnetic flux density to proceed without having to reset the frequency with each and every iteration of the geometry.

The geometry of the wire carrying slots in a prism cavity met in connection with triangular and irregular cores are depicted in Figure 18.4. A typical slot

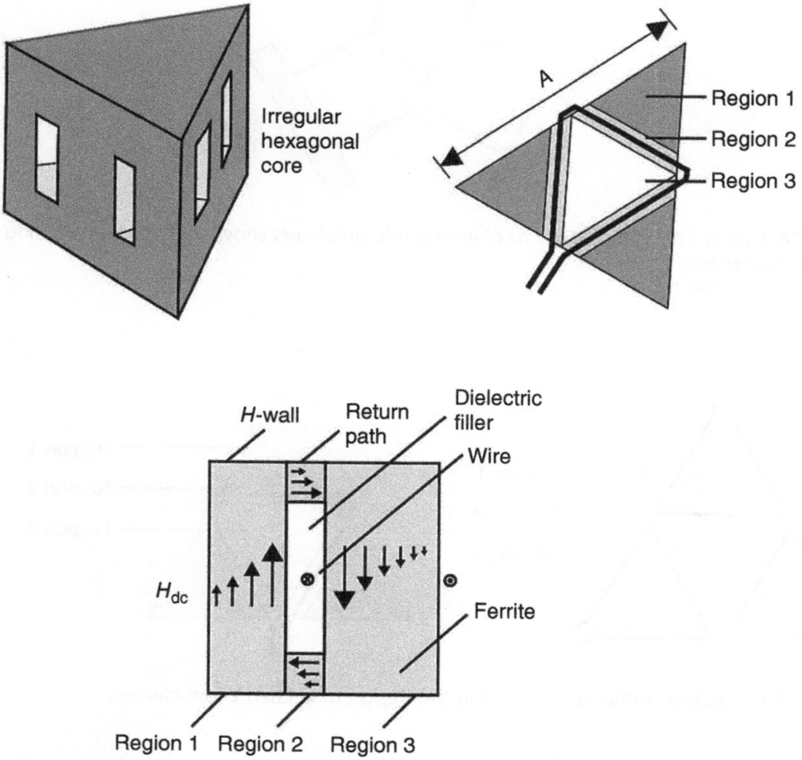

Figure 18.4 Tri-toroidal gyromagnetic cavity.

configuration is, in practice, set by the need to shape the direct magnetic flux densities in the inner and outer cores.

One approximate model adopted for the microwave circuit retains the vertical element in a typical tri-toroidal circuit but disregards the magnetization of the two parallel ones in the slot region. This approximation allows the regions containing the wire carrying slots to be essentially represented by dielectric ones with the dielectric constant of the ferrite material. The cross-section for the purpose of the calculation of the split cutoff space consists of an inner region with one value of magnetization, an intermediate dielectric region, and one outer region with a magnetization of one-third that of the inner one. One way to explore the internal direct magnetization of the magnetic insulator in the presence of one or more loops is to have recourse to a magnetostatic solver.

18.4 The Magnetostatic Problem

The main problem under consideration is that of an equilateral triangle divided into two regions by an irregular hexagonal wire. The surfaces of the inner and a typical outer region are A_{in} and A_{out}, respectively. The symmetry of the tri-toroidal problem indicates that the direct magnetic flux ϕ_{out} in each of the outer ribs of the tri-toroidal geometries is one-third that of the core ϕ_{in}. The flux density, however, in a typical outer region may be more or less than that of the inner one. The governing equation for the magnetic flux ϕ_{out} in a tri-toroidal typical return path due to a flux ϕ_{in} in the core of the toroid is

$$\phi_{out} = \frac{1}{3}\phi_{in} \tag{18.1}$$

The magnetic flux densities B_{in} and B_{out} in the two regions are related by

$$B_{out}A_{out} = \frac{1}{3}B_{in}A_{in} \tag{18.2}$$

The magnetic flux densities in a typical outer path is

$$B_{out} = \frac{1}{3}qB_{in} \tag{18.3}$$

where q is the shape factor:

$$q = \frac{A_{\text{in}}}{A_{\text{out}}} \tag{18.4}$$

The ratio of the gyrotropies in the two regions is here assumed to be equal to that of the flux densities. The relationship between the two can be described in terms of the shape factor, q, by

$$\left(\frac{\kappa}{\mu}\right)_{\text{outer}} = -\frac{q}{3}\left(\frac{\kappa}{\mu}\right)_{\text{inner}} \tag{18.5a}$$

where

$$\kappa = \frac{\gamma M_0}{\omega_0} \tag{18.5b}$$

$$\mu = 1 \tag{18.5c}$$

$\gamma = 2.21 \times 10^5$ (rad s^{-1} per A m^{-1}) is the gyromagnetic ratio, M_0 is the magnetization (A m^{-1}), and ω_0 is the radian frequency (rad s^{-1}).

18.5 Quality Factor of Junction Circulators with Up and Down Magnetization

The removal of the degeneracy between the counterrotating modes of the resonator, the main endeavor of this work, is dependent not only on the gyrotropies in the two regions but also on the distribution of the alternating magnetic field within each region. The alternating magnetic field is, of course, more intense on the axis of the resonator than on its periphery and correctly forms part of the FE calculation.

The development of a circulator or switch using an internally latched resonator differs only from that of a conventional one in that the degree of splitting that is realizable between the degenerate counterrotating modes of the resonator is in this situation degraded by the up and down magnetization on either side of the wire. This feature may be accounted for by replacing the gyrotropy constant C_{11} in the definition of the split phase constants of the gyromagnetic waveguide by an effective constant C'_{11}.

The perturbed gyrotropy constant may be deduced by introducing a filling factor (k_f) in the classic relationship between the quality factor Q_{eff} of the complex gyrator circuit in this type of device and $\omega_\pm$, the split frequencies of the homogeneous resonator:

$$\frac{1}{Q_{\text{eff}}} = \sqrt{3}k_f\left(\frac{\omega_+ - \omega_-}{\omega_0}\right) \tag{18.6}$$

where

$$k_f = \frac{\omega'_+ - \omega'_-}{\omega_+ - \omega_-} \tag{18.7}$$

ω'_+ and ω'_- are the split frequencies of the resonator with its inner region magnetized along the positive z direction and its outer one in the opposite direction. The filling factor k_f may therefore be deduced by evaluating the ratio of the split frequencies of the homogeneous and inhomogeneous circuits, respectively.

18.6 Split Frequencies of Planar and Cavity Gyromagnetic Resonators

The split cutoff frequencies of the planar gyromagnetic resonator are

$$\frac{\omega_\pm}{c} = \frac{4\pi}{3A\sqrt{\varepsilon_f}}\left(1 \mp \frac{C_{11}\kappa}{2}\right) \tag{18.8}$$

provided

$$\frac{1}{1-x} = 1 + x \tag{18.9a}$$

$$\sqrt{1+x} = 1 + \frac{x}{2} \tag{18.9b}$$

The split propagation constants of the half-wave long gyromagnetic cavity resonator are given by

$$\beta_\pm = \frac{\pi}{2L} \tag{18.10}$$

Application of this condition gives the split frequencies of the cavity resonator.

$$\frac{\omega'_\pm}{c} = \sqrt{\frac{(\pi/2L)^2 + (4\pi/3A)^2(1 \mp C_{11}\kappa/2)}{\varepsilon_f}} \tag{18.11}$$

An important quantity that enters in the development of both planar and cavity resonators to be dealt with in the next section is the difference between the split frequencies. An inspection of the above two results indicates that

$$\frac{\omega_+ - \omega_-}{\omega_0} = \frac{\omega'_+ - \omega'_-}{\omega'_0} = C_{11}\kappa \tag{18.12}$$

This result suggests that the quality factor of a cavity resonator may be deduced from a calculation of the split cutoff frequencies of the cavity.

18.7 The Split Frequencies of Prism Resonator with Up and Down Magnetization

A typical three-dimensional calculation on the split frequencies of a prism resonator in WR229 waveguide, with the inner core magnetized in one sense and the outer region in the other, is given in Figure 18.5. This is done for various values of the shape factor of the cavity.

The physical geometry under consideration is summarized below.

$$k_0L = 0.445$$

$$A/L = 4.448$$

$$S/L = 0.338$$

$$k_0 = 0.0792\ \text{rad mm}^{-1}$$

The corresponding result in the case of a homogeneously magnetized resonator is superimposed in Figure 18.5 for the purpose of comparison. The magnetization employed to obtain this result is $\mu_0 M_0 = 0.1200$ T. The relative dielectric constants of the ferrite material and that of the gap are 15.1 and 2.1 respectively. The illustration here assumes that the fluxes on either side of the current-carrying wire loop are uniform but that the flux in a typical outside region is one-third that of the inner region.

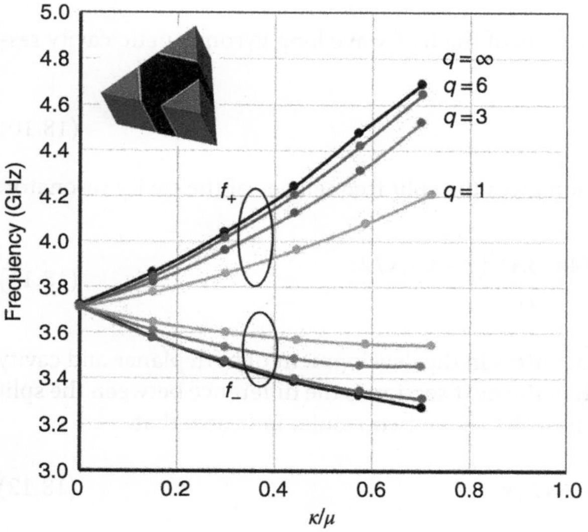

Figure 18.5 Split frequencies of up and down magnetized tri-toroidal prism resonator for parametric values of shape factor (q = 1, 3, 6 and ∞).

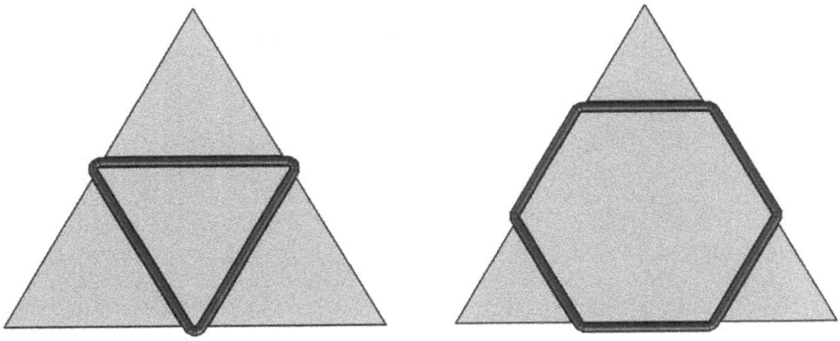

Figure 18.6 Uniform up and down magnetizations in a planar resonator with top and bottom electric walls, and a magnetic sidewall, for $q = 1$ and $q = 6$.

18.8 Exact Calculation of Split Frequencies in Tri-toroidal Cavity

The task of this section is to develop the exact relationship between the split frequencies and the shape factor of a typical tri-toroidal cavity. This may be done by having recourse to three-dimensional magnetostatic and high frequency FE solvers. Figure 18.6 depicts the wire configurations of two typical shape factors. A typical FE discretization is indicated in Figure 18.7.

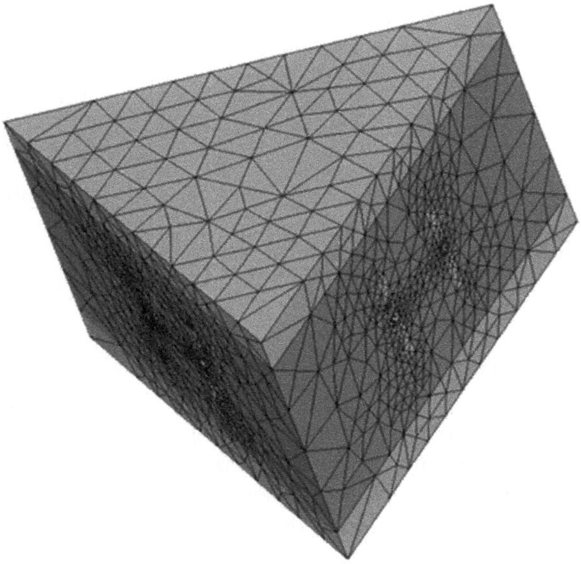

Figure 18.7 Discretization of prism resonator.

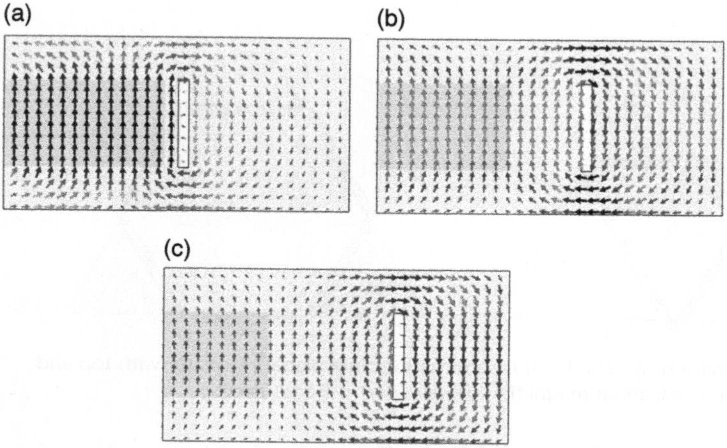

Figure 18.8 Magnetic flux density in prism resonator latched by wire loops: (a) $q = 1$, (b) $q = 3$, and (c) $q = 6$.

The magnetostatic problem in the case of the prism geometry with an irregular switching wire is indicated in Figure 18.8 for $q = 1$, 3, and 6. The squareness ratio (R) of the magnetic insulator is 0.80. The wire used is 24 SWG and the current in the wire is 7 A.

The magnetic flux density ($\bar{B}$) and the magnetic field intensity ($\bar{H}$) in a typical region are related by the details of the hysteresis loop of the material in question.

$$\bar{B} = \mu_0 \bar{H} + \bar{M} \tag{18.13a}$$

and

$$\kappa = \frac{\gamma M}{\mu_0 \omega} \tag{18.13b}$$

The split frequencies obtained in this way are indicated in Figure 18.9.

18.9 Calculation and Experiment

The three-dimensional FE calculation of the splitting factor (k_f) of the tri-toroidal cavity, on the assumption of up and down uniform magnetized regions on either side of the current-carrying wire loop, is summarized in Figure 18.10 for a number of different shape factors. The exact description of the arrangement considered here is also illustrated in the same diagram. The two results are separately compared with some experimental data in the same figure. The up and down magnetized model of the latched resonator adopted here is, in practice, inadequate.

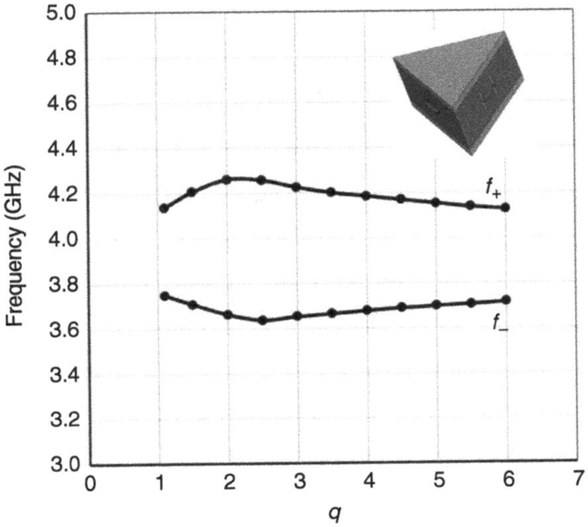

Figure 18.9 Exact split frequencies versus q in latched prism resonator.

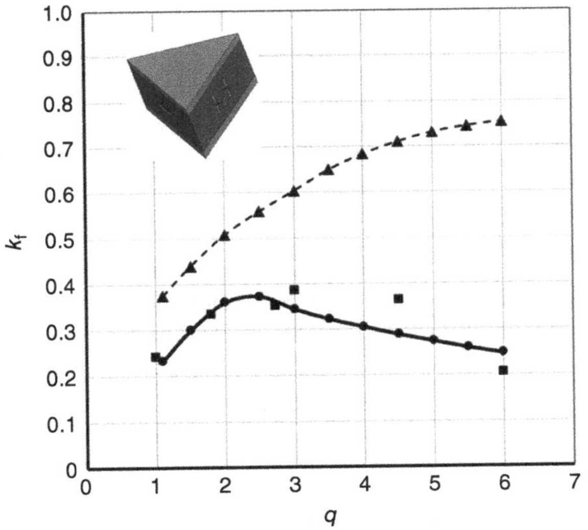

Figure 18.10 k_f versus q in latched prism resonator (simulation: ▲ up/down magnetization, ● exact; ■ experiment).

18.10 Tri-toroidal Composite Prism Resonator

The design of a composite resonator using two different ferrite materials is discussed in this section. The arrangement under consideration consists of an outer region with a lower magnetization than the inner core and the exact characterization requires again a three-dimensional FE package with both

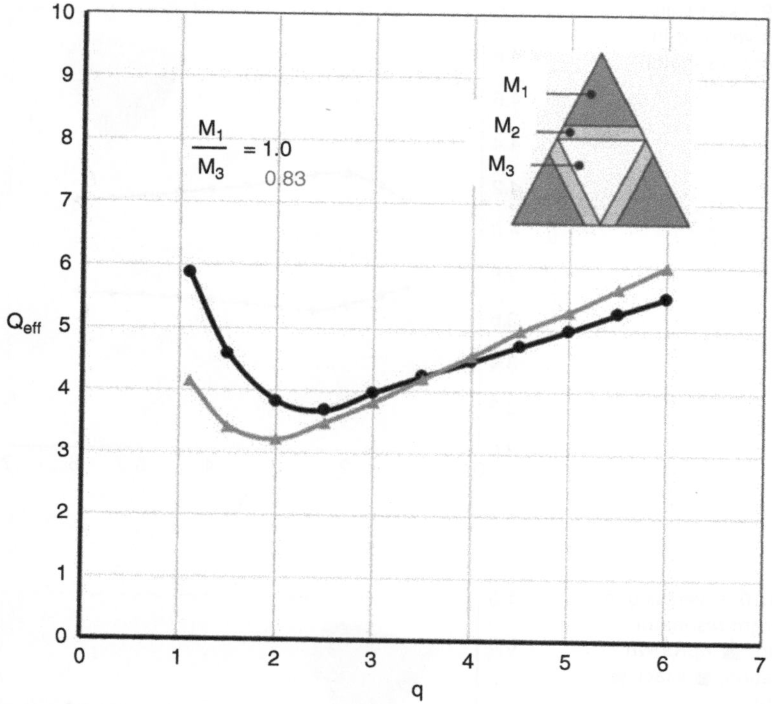

Figure 18.11 Q_{eff} versus q in latched composite prism resonator ($M_2 = 0$).

magnetostatic and high frequency solvers. A comparison of Q_{eff} for two different ratios of magnetization is given in Figure 18.11. A first-order approximation to the split frequencies of a gyromagnetic cavity with up and down magnetization may be deduced by combining a planar description of the flux on either side of a current wire loop with a high frequency FE discretization of the geometry.

18.11 Tri-toroidal Wye Resonator with Up and Down Magnetization

The cutoff space and the split frequencies of the tri-toroidal half-wave long wye cavity structure composed of a cylindrical region and a triplet of gyromagnetic ribs incorporating a switching wire is discussed in the section. This geometry depicted in Figure 18.12, unlike the prism arrangement, has no closed form solution. Its split cutoff space is described, in addition to the location of the wire,

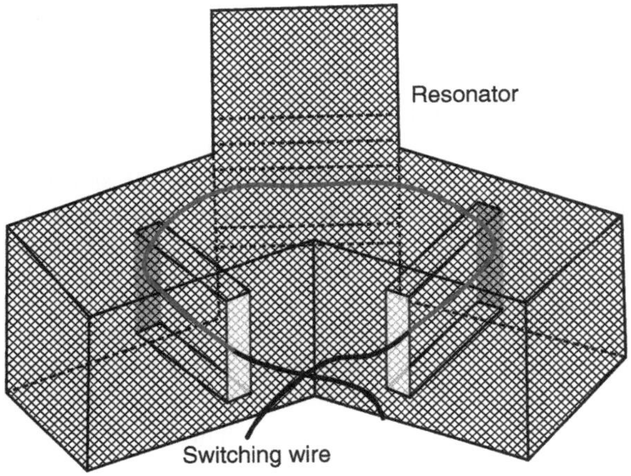

Figure 18.12 Tri-toroidal wye resonator.

by its coupling angle, ψ, and its interior radial wavenumber, k_0r. The radial wavenumber k_0R, or the electrical length, θ, of the ribs or the ratio of the internal and external radii, r/R, is the unknown of the problem. The wire geometry in a typical resonator is shown in Figure 18.13. The attraction of this arrangement is that it permits the switching wire to be located closer to the axis of the resonator compared to what is possible with either a cylindrical or prism configuration. One specific family of solutions is obtained by placing the wires at the ports

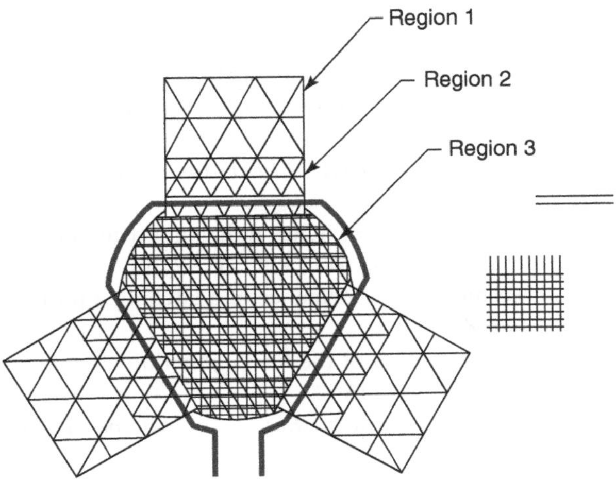

Figure 18.13 First wire configuration.

of the circular gyromagnetic region and adjusting the coupling angle at the same boundary for each and every value of $k_0 r$.

A separate calculation in Chapter 12 indicates that the opening between the split cutoff frequencies of this sort of circuit deteriorates rapidly for r/R below about 0.50. The up and down problem is therefore restricted to the interval $0.50 \le r/R \le 1.0$. The physical parameters of the degenerate cutoff space are obtained by having recourse to the data in Chapter 12. One solution at 4.0 GHz is

$$k_0 = 0.0838\,\text{rad}\,\text{mm}^{-1}$$

$$k_0\sqrt{\varepsilon_f}r = 1.2\,\text{rad}$$

$$\psi = 0.35\,\text{rad}$$

The unknown of the problem is $k_0 R$, given in Chapter 12. The result is

$$k_0\sqrt{\varepsilon_f}R = 2.0$$

The electrical length of the unit element (UE) is then given by

$$\theta = k_0\sqrt{\varepsilon_f}(R - r) \tag{18.14}$$

There is no closed-form expression for the general cutoff number; each geometry must be dealt with separately. The length $2L$ must also be obtained numerically. The split cutoff frequencies of the arrangements with $\psi = 0.20$ and 0.40 are available in Chapter 12.

Bibliography

Betts, J., Temme, D.H., and Weiss, J.A. (1966). A switching circulator S band; stripline; 15 kilowatts; 10 microseconds; temperature stable. *IEEE Trans. Microw. Theory Tech.* **MTT-14**: 665–669.

Bosma, H. (1964). On stripline Y-circulation at UHF. *IEEE Trans. Microw. Theory Tech.* **MTT-12** (1): 61–72.

Clavin, A. (1963). Reciprocal and nonreciprocal switches utilizing ferrite junction circulators. *IEEE Trans. Microw. Theory Tech.* **MTT-11**: 217–218.

Davis, L.E. (1966). Computed phaseshift and performance of a latching 3 port waveguide circulator. *NEREM Rec.* **8**: 96.

Fay, C.E. and Comstock, R.L. (1965). Operation of the ferrite junction circulator. *IEEE Trans. Microw. Theory Tech.* **MTT-13**: 15–27.

Freiberg, L. (1961). Pulse operated circulator switch. *IEEE Trans. Microw. Theory Tech.* **MTT-9**: 266–266.

Goodman, P.C. (1965). A latching ferrite junction circulator for phased array switching applications. *IEEE MTT Symp.* **65**: 123–126.

Helszajn, J. (1981). Standing wave solutions of planar irregular hexagonal and wye resonators. *IEEE Trans. Microw. Theory Tech.* **MTT-29**: 562–567.
Helszajn, J. (2003). Standing wave solutions and cut-off numbers of planar wye resonators. *Microwave Engineering Europe* (October 2003), pp. 31–34, 36, 38, 41.
Helszajn, J. and D'Orazio, W. (unpublished). Adjustment of prism turnstile resonators latched by wire loops.
Helszajn, J. and Hines, M.L. (1968). A high speed TEM junction ferrite modulator using a wire loop. *Radio Electron. Eng.* **35**: 81–82.
Helszajn, J. and Nisbet, W.T. (1981). Circulators using planar wye resonator. *IEEE Trans. Microw. Theory Tech.* **MTT-29**: 689–699.
Helszajn, J. and Sharp, J. (2012). Cut-off space of a gyromagnetic planar disk resonator with a triplet of stubs with up and down magnetization. *IET Microw. Antennas Propag.* **6**: 569–576.
Helszajn, J. and Tan, F.C.F. (1975). Mode charts for partial-height ferrite waveguide circulators. *IEE Proc. Microw. Antennas Propag.* **122**: 34–36.
Hines, M.E. (1973). Microwave measurement techniques for below resonance junction circulators. *IEEE Trans. Microw. Theory Tech.* **MTT-21**: 347–351.
Ito, Y. and Yochouchi, H. (1969). Microwave junction circulators. *Fujitsu Sci. Tech. J.* **5**: 55–90.
Katoh, I., Konishi, H., and Sakamoto, K. (1980). A 12 GHz broadband latching circulator. *Proceedings of European Microwaves Conference*, Warsaw, Poland (8–12 September 1980), pp. 360–364.
Kroening, A.M. (2016). Advances in ferrite redundancy switching for Ka-band receiver applications. *IEEE Trans. Microw. Theory Tech.* **MTT-64**: 1911–1917.
Levy, R. and Helszajn, J. (1982). Specified equations for one and two section quarter-wave matching networks for stub resistor loads. *IEEE Trans. Microw. Theory Tech.* **MTT-30**: 55–63.
Mlinar, M.J., Piotrowski, W.S., and Raue, J.E. (1981). A 19 GHz high power low loss latching switch for space applications. *Proceedings of European Microwaves Conference*, Amsterdam, the Netherlands (7–11 September 1981), pp. 399–404.
Passaro, W.C. and McManus, J.W. (1966). A 35 GHz latching switch. *IEEE MTT Symp.* **66**: 270–274.
Siekanowicz, W.W. and Schilling, W.A. (1968). A new type of latching switchable ferrite junction circulator. *IEEE Trans. Microw. Theory Tech.* **MTT-16**: 177–183.
Siekanowicz, W.W., Paglione, R.W., and Walsh, T.E. (1970). A latching ring and post ferrite waveguide circulator. *IEEE Trans. Microw. Theory Tech.* **MTT-18**: 212–216.

19

The Waveguide *H*-plane Tee Junction Circulator Using a Composite Gyromagnetic Resonator

Joseph Helszajn

Heriot Watt University, Edinburgh, UK

19.1 Introduction

The waveguide tee junction rather than the wye geometry is perhaps the optimum arrangement in the construction of compact waveguide circulators and switches. The geometry is here either a composite turnstile or a post geometry. The advantage of the post resonator is that its aspect ratio may be used together with a switching wire to adjust the profile of the direct magnetization. It may also be used to both internally and externally latch a switch. The three degrees of freedom in the design of a switch besides the option of the resonator are its aspect ratio, the ratio of the up and down magnetized regions, and the ratio of the magnetization of the inner and outer regions.

The waveguide tee junction differs from its wye counterpart in that its scattering matrix contains four instead of two parameters. Its symmetry may be recovered by either introducing a septum or a dielectric post along its symmetry plane. The chapter develops this problem as a preamble to constructing the two circulation conditions of the device. The resonator used here relies on a composite resonator with the magnetization of the inner regions different from the outer. The latter only provides a return path for the direct magnetization. A shortcoming of the conventional arrangement is that the sense of circulation is different on either side of the switching wire. The present arrangement avoids this problem. The split frequencies of a post resonator with up and down uniformly magnetized regions are available in the literature as well as that in a turnstile geometry. The split frequencies of a planar composite resonator with the outside gyromagnetic region replaced by a dielectric one is experimentally available in the literature.

Microwave Polarizers, Power Dividers, Phase Shifters, Circulators, and Switches,
First Edition. Joseph Helszajn.

Three different reference planes are met in the descriptions of tee junctions. These are the characteristic planes introduced by Williams and the Dicke and Altman ones. The Dicke terminals enter into the characterization of the unloaded tee junction. The Altman ones are met in the adjustment of the electrical symmetrical tee junctions. The latter coincide with its characteristic planes. The reference planes of the Dicke and Altman terminals differ by 90°.

19.2 Eigenvalue Problem of the *H*-plane Reciprocal Tee Junction

The eigenvalue problem of the *H*-plane tee junction in Figure 19.1 is complicated because two if its three eigenvectors are not unique. The matric relationships entering into its descriptions are summarized in this section. The square scattering matrix $\bar{S}$ has here four independent parameters.

$$\bar{S} = \begin{bmatrix} \alpha & \delta & \gamma \\ \delta & \alpha & \gamma \\ \gamma & \gamma & \beta \end{bmatrix} \tag{19.1}$$

One possible triplet, but not unique set of eigenvectors of the tee junction, in Figure 19.1 has been specified by Dicke.

$$\bar{U}_1 = \frac{1}{2}\begin{bmatrix} 1 \\ 1 \\ \sqrt{2x} \end{bmatrix} \tag{19.2a}$$

$$\bar{U}_2 = \frac{1}{2}\begin{bmatrix} 1 \\ 1 \\ -\sqrt{2}/x \end{bmatrix} \tag{19.2b}$$

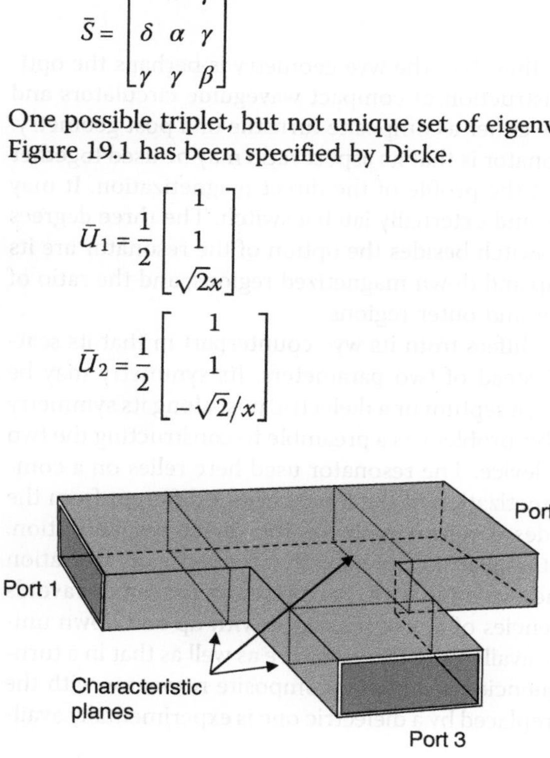

Figure 19.1 *H*-plane waveguide tee junction indicating the position of Dicke's planes.

$$\bar{U}_3 = \frac{1}{2}\begin{bmatrix} \sqrt{2} \\ -\sqrt{2} \\ 0 \end{bmatrix} \tag{19.2c}$$

The eigenvectors are here not orthogonal unless x is unity

$$\bar{U}_i \bar{U}_i^{\mathrm{T}} \neq 1 \quad i = 1,2,3 \tag{19.3}$$

U_1 is an in-phase eigenvector, $U_{2,3}$ are even and odd eigenvectors, respectively. The unknown parameter, x, in the definition of the eigenvector has been introduced by Dicke to deal with the lack of uniqueness in the choice of the side port in the tee junction. It takes on a unique value at the so-called Dicke planes at which the eigenvalues have unit amplitude.

The connections between the scattering parameters and the eigenvalues are given by expanding the similarity transformation in the usual way. The result is (Allanson et al. 1946; Chait and Curry 1959)

$$\alpha = \frac{1}{2[(1/x)+x]}\left[\frac{s_1}{x} + xs_2\right] + \frac{1}{2}s_3 \tag{19.4a}$$

$$\beta = \frac{1}{2[(1/x)+x]}\left[2xs_1 + \frac{2s_2}{x}\right] \tag{19.4b}$$

$$\gamma = \frac{1}{2[(1/x)+x]}\left[\sqrt{2}s_1 - \sqrt{2}s_2\right] \tag{19.4c}$$

$$\delta = \frac{1}{2[(1/x)+x]}\left[\frac{s_1}{x} + xs_2\right] - \frac{1}{2}s_3 \tag{19.4d}$$

A property of matrices is that the trace or spur of the diagonal matrix with the eigenvalues of the junction is equal to the sum of the reflection parameters of the scattering matrix. This condition is satisfied here

$$2\alpha + \beta = s_1 + s_2 + s_3 \tag{19.5}$$

The scattering matrix separately meets the unitary condition.

The reference terminals in this work are measured from the openings of the square box formed by the junction of the three waveguides illustrated in Figure 19.1.

The main task of this chapter is to construct the eigenvalue diagram of the *H*-plane tee junction at the Dicke plane, which has not been done so far and to introduce a new boundary condition together with a symmetric septum.

The value of x is that which ensures that the eigenvalues in Eq. (19.4) have unit amplitude. The eigenvalues appearing in Eq. (19.4) may be deduced by making use of the connection between the scattering matrix $\bar{S}$, its eigenvalues s_i, and the transformation matrix $[\bar{U}]$ containing the eigenvectors of the geometry as in

Allanson et al. (1946) and Chait and Curry (1959). The approach utilized here is to take linear combinations of Eq. (19.4):

$$s_1 = \beta + \frac{\sqrt{2}\gamma}{x} \tag{19.6a}$$

$$s_2 = \beta - \sqrt{2}x\gamma \tag{19.6b}$$

$$s_3 = \alpha - \delta \tag{19.6c}$$

The above equation clearly indicates a relationship between the phase angles of the scattering parameters (α, β, γ, and δ) and the amplitudes of the eigenvalues (s_1, s_2, and s_3). The planes at which the eigenvalues have unit amplitude have been established by Dicke and are reproduced, with some simplification, in the next section. The nature of the eigenvectors may also be verified at the Dicke planes once the eigenvalues are available. This may be done by constructing the matrix relationship below, one at a time.

$$\bar{S}\bar{U}_i = s_i\bar{U}_i \tag{19.7}$$

The standing wave patterns produced by these eigenvectors are illustrated in Figure 19.2 for $x = 1$.

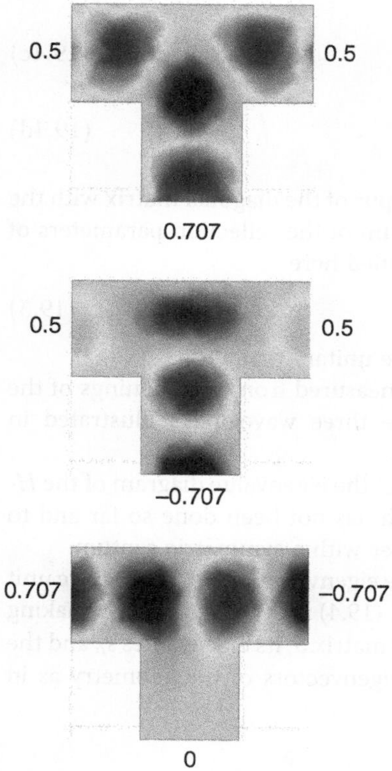

Figure 19.2 Standing wave solutions produced by eigenvectors of waveguide tee junction (arbitrary plane).

The complication of deducing the reference planes of the Dicke tee plane junction may be avoided by recognizing that these are 90° away from the Altman ones or, what is the same thing, 90° away from the characteristic planes of the junction defined by Allanson.

19.3 Electrically Symmetric *H*-plane Junction at the Altman Planes

One purpose of this section is to summarize the boundary conditions of the electrically symmetric *H*-plane tee junction proposed by Altman (Allanson et al. 1946). The scattering matrix is

$$\bar{S} = \begin{bmatrix} \alpha & \delta & \gamma \\ \delta & \alpha & \gamma \\ \gamma & \gamma & \beta \end{bmatrix} \tag{19.8}$$

The proposed boundary condition postulated by Altman is defined by

$$\alpha = \beta = -\frac{1}{3} \tag{19.9a}$$

$$\gamma = \delta = \frac{2}{3} \tag{19.9b}$$

This boundary condition cannot be met at the Dicke planes, in that β and γ are not orthogonal, but can at the characteristic planes of the junction with an inductive vane or a suitable dielectric resonator. The unknown parameter in the bilinear relationship between the scattering parameters and the eigenvalues is in this instance (Allanson et al. 1946)

$$x = \frac{1}{\sqrt{2}} \tag{19.10}$$

Introducing this condition in Eqs. (19.4a)–(19.4d) yields:

$$\alpha = \frac{1}{3}\left(s_1 + \frac{s_2}{2}\right) + \frac{s_3}{2} \tag{19.11a}$$

$$\beta = \frac{1}{3}(s_1 + 2s_2) \tag{19.11b}$$

$$\gamma = \frac{1}{3}(s_1 - s_2) \tag{19.11c}$$

$$\delta = \frac{1}{3}\left(s_1 + \frac{s_2}{2}\right) - \frac{s_3}{2} \tag{19.11d}$$

The above scattering parameters satisfy the unitary condition. The bilinear transformation between the elements of the scattering and diagonal matrix containing the eigenvalues of the problem may be simplified by making use of the fact that the eigenvalues s_2 and s_3 are degenerate. The result is

$$\gamma = \delta = \frac{1}{3}(s_1 - s_2) \tag{19.12}$$

The reflection eigenvalues are separately related to the scattering parameters by

$$s_1 = \beta + 2\gamma \tag{19.13a}$$

$$s_2 = \beta - \gamma \tag{19.13b}$$

$$s_3 = \alpha - \delta \tag{19.13c}$$

Introducing the assumed boundary conditions into the above relationships gives

$$s_1 = 1 \tag{19.14a}$$

$$s_2 = s_3 = -1 \tag{19.14b}$$

The corresponding eigenvalue diagram is shown in Figure 19.3.

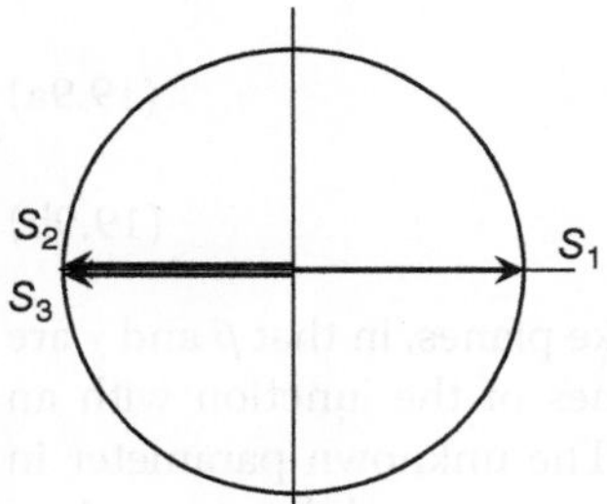

Figure 19.3 Eigenvalue diagram of electrically symmetrical *H*-plane waveguide tee junction at its characteristic planes.

19.4 Characteristic Planes

The adjustment of the Altman boundary conditions is a two-step process. The first step fixes the magnitudes of the scattering parameters of the junction. The second fixes the angles. It is satisfied, provided the reference planes of the junction coincide with the characteristic planes defined in Figure 19.4. A property of such planes is that when a wave incident on the junction at one port is totally reflected by the location of a piston at the second port, the electric field vanishes at all characteristic planes of a waveguide tee junction

The reflection coefficients at the characteristic plane are

$$\rho_1 = \rho_2 = \alpha - \delta \tag{19.15a}$$

$$\rho_3 = \beta - \frac{\gamma^2}{\delta} \tag{19.15b}$$

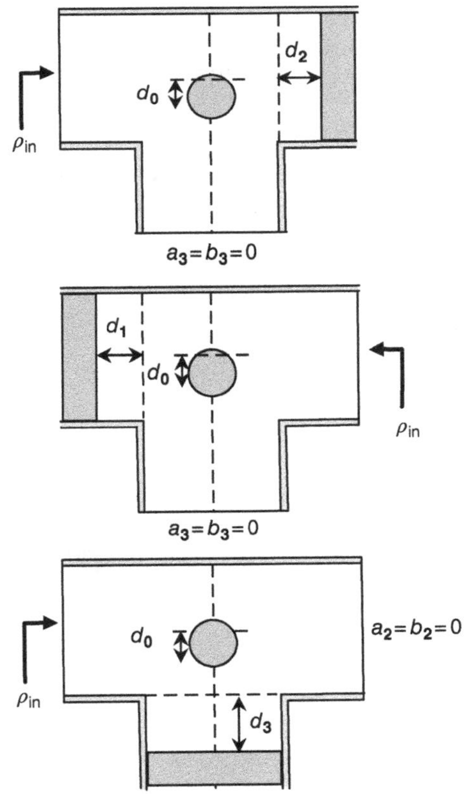

Figure 19.4 Definition of characteristic planes.

A property of the Altman boundary condition is that

$$\rho_1 = \rho_2 = \rho_3 \tag{19.16}$$

A comparison between the eigenvalue s_3 of the Altman solution and the reflection coefficients of the junction at its characteristic planes is the same.

The characteristic planes referred to the opening of the inner box of the junction are separately specified in the main and the side waveguide by

$$\phi_{1,2} = \frac{2\pi d_{1,2}}{\lambda_g} \tag{19.17a}$$

$$\phi_3 = \frac{2\pi d_3}{\lambda_g} \tag{19.17b}$$

The standing wave solutions produced by each generator setting are indicated in Figure 19.5 for the Altman turnstile tee junction using a cylindrical resonator.

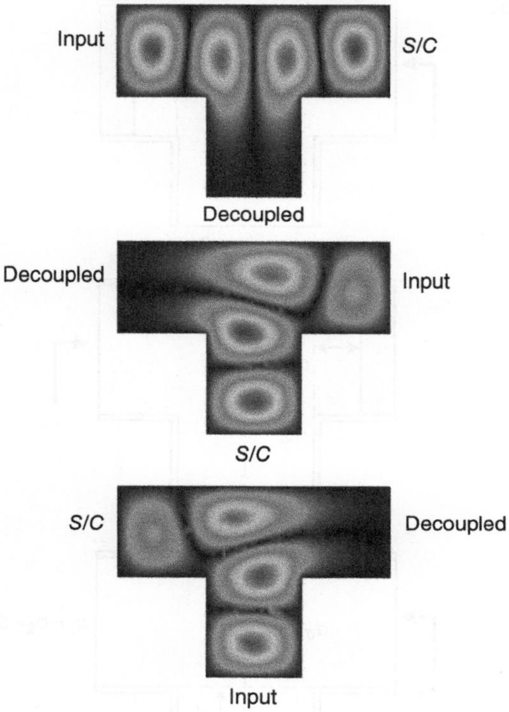

Figure 19.5 Standing wave patterns produced by pistons at the characteristic planes of the junction.

19.5 The Septum-loaded *H*-plane Waveguide

The geometry of the septum-loaded *H*-plane tee junction discussed here is depicted in Figure 19.6. The introduction of a septum in this waveguide produces a perturbation of the field patterns of the eigenvector $\bar{U}_1$ but leaves those produced by $\bar{U}_2$ and $\bar{U}_3$ unchanged. This feature is indicated in Figure 19.7 in

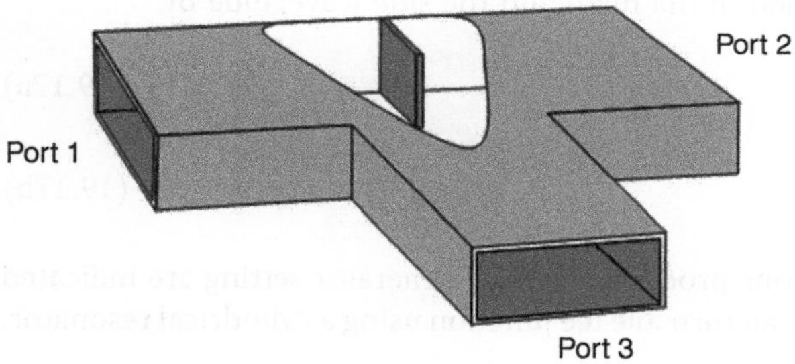

Figure 19.6 Inductive vane-loaded *H*-plane tee junction.

Figure 19.7 Standing wave solution of *H*-plane waveguide tee junction with symmetry vane ($s/a = 0.25$).

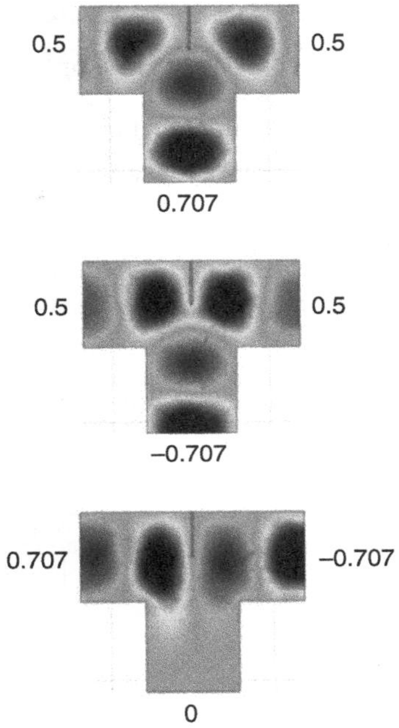

the case of a typical septum. The connection between the scattering variables and the penetration of the vane in WR75 waveguide at a frequency of 13.25 GHz is illustrated separately in Figure 19.8. It clearly satisfies the solution postulated by Altman.

Figure 19.8 Experimental results of fabricated *H*-plane tee junction.

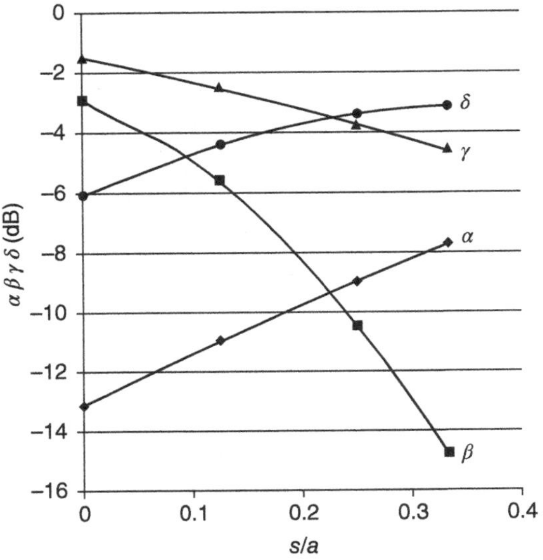

19.6 The Waveguide Tee Junction Using a Dielectric Post Resonator: First Circulation Condition

The waveguide tee junction has a symmetry plane across its side waveguide. It may be loaded with either a septum or a dielectric resonator in order to recover the electrical symmetry of a wye junction. Figure 19.9 illustrates the latter arrangement. The relationship between the reflection coefficients α and β and the transmissions γ and δ versus the position of a dielectric resonator are shown in Figures 19.10 and 19.11.

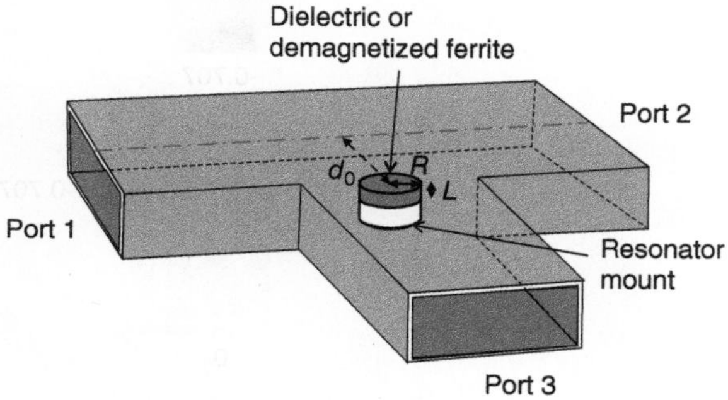

Figure 19.9 Topology of *H*-plane waveguide tee junction using a dielectric post resonator.

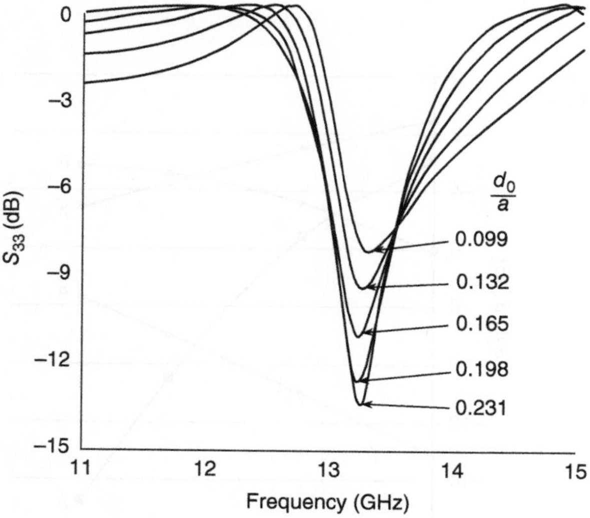

Figure 19.10 Return loss at port 3 versus frequency for parametric values of d_0/a.

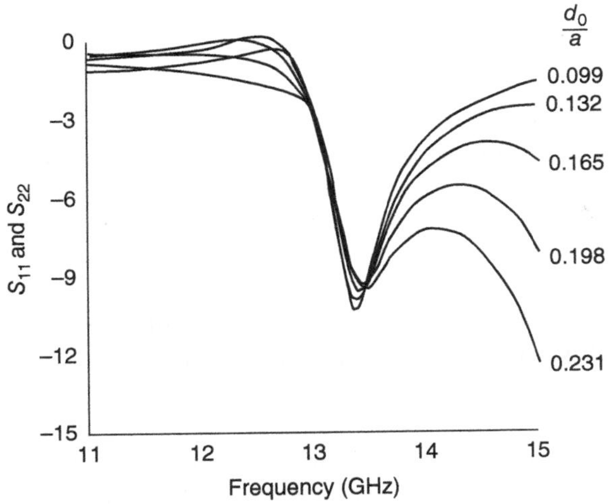

Figure 19.11 Return loss at ports 1 and 2 versus frequency for parametric values of d_0/a.

The resonator has been separately optimized in a symmetrical wye junction in WR75 waveguide at a frequency of 13.25 GHz. It is defined by the length (L) and radius (R) of the post resonator. Its length has been arbitrarily fixed to one-third the narrow dimension (b) of the waveguide.

$$L = \frac{1}{3} \tag{19.18}$$

The radial wavenumber of the resonator is

$$k_0\sqrt{\mu_{\text{demag}}\varepsilon_f}R \tag{19.19}$$

k_0 at 13.25 GHz is

$$k_0 = \frac{2\pi}{\lambda_0} = 277.5\,\text{rad}\,\text{m}^{-1} \tag{19.20}$$

The relative dielectric constant of the resonator is 15.0. The demagnetized permeability of the magnetic insulator is

$$\mu_{\text{demag}} = \frac{1}{3} + \frac{2}{3}\left[1 - \left(\frac{\gamma M_0}{\mu_0 \omega}\right)^2\right]^{\frac{1}{2}} \tag{19.21}$$

where

$$\gamma = 2.21 \times 10^5\,\text{rad}\,\text{s}^{-1}\,\text{per}\,\text{A}\,\text{m}^{-1}$$

and

$$\mu_0 = 4\pi \times 10^{-7}\,\mathrm{H\,m^{-1}}$$

The ferrite material employed in this work is a yttrium iron garnet with a magnetization of $\mu_0 M_0 = 0.1780$ T. The former quantity gives a value for the demagnetized permeability of 0.94.

19.7 The Waveguide Tee Junction Circulator Using a Gyromagnetic Post Resonator: Second Circulation Condition

The second circulation condition of any junction circulator may be established without ado by replacing the dielectric resonator with a gyromagnetic one. Figure 19.12 depicts the experimental scattering parameters of the geometry in the previous section. The standing wave circulation solution, using a commercial solver, with an input signal at ports 1, 2, and 3 one at a time is illustrated in Figure 19.13. The magnetization of the magnetic insulator $\mu_0 M_0$ is here 0.1780 T and its dielectric constant ε_f is 15.0.

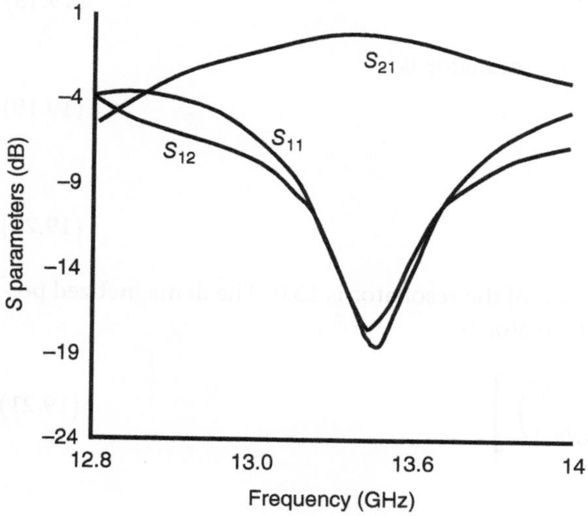

Figure 19.12 Experimental frequency response of tee junction circulator.

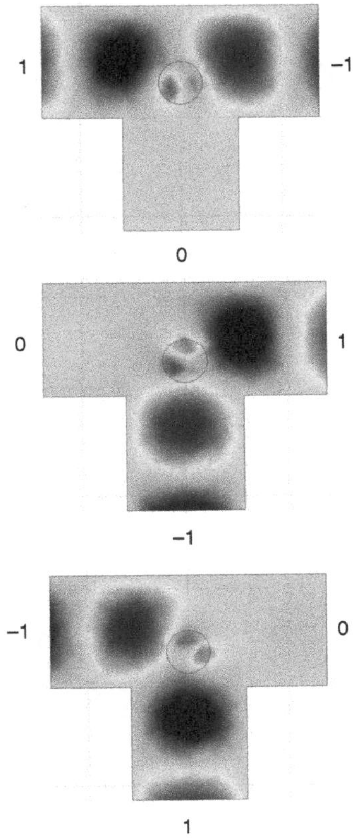

Figure 19.13 Standing wave solution of turnstile circulator with a gyromagnetic resonator.

19.8 Composite Dielectric Resonator

The way to avoid the shortcoming of the up and down magnetization on either side of the switching wire is to have the magnetization of the outer region small compared to the inner one. The outer region in this arrangement merely serves to close the magnetic circuit. One approximate way to investigate this geometry is to assume that the magnetization of the outside region can be neglected compared to that of the inner one as shown in Figure 19.14.

The split frequencies for ring stripline junction circulator are illustrated in Figure 19.15.

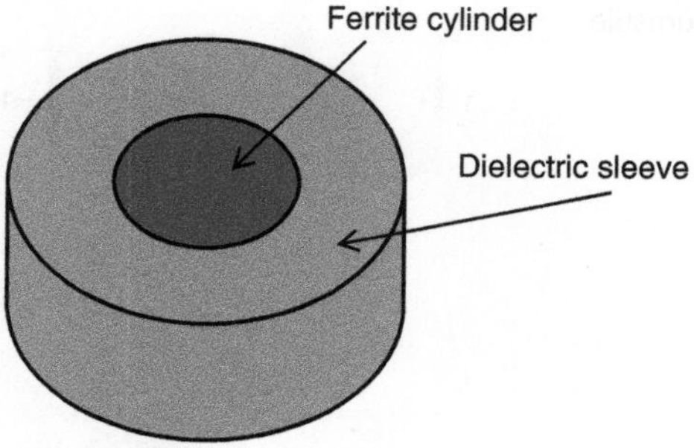

Figure 19.14 Radial composite resonator.

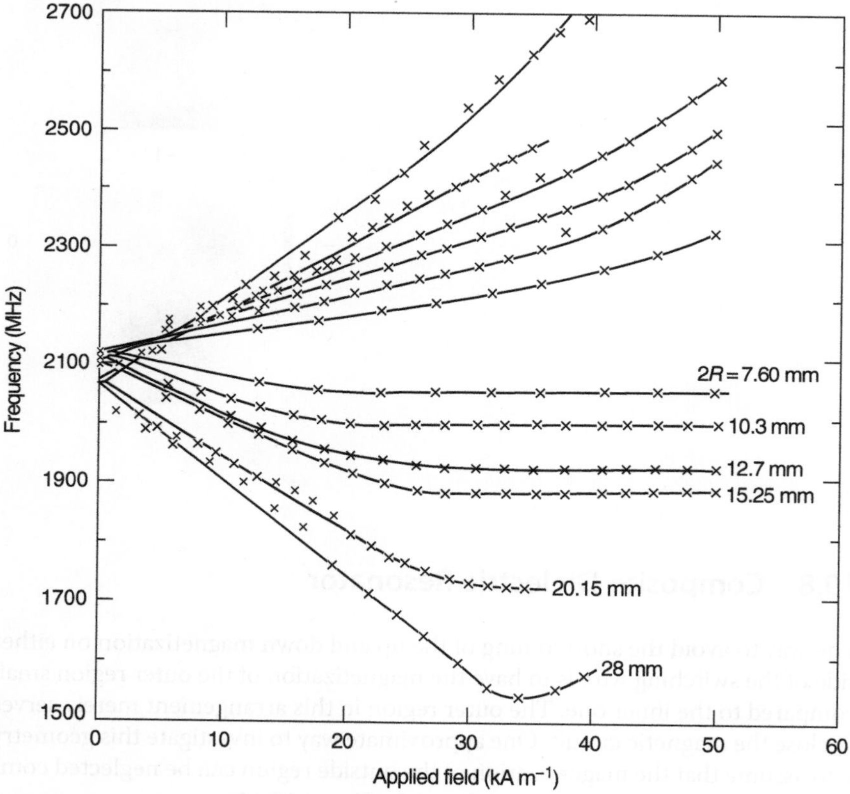

Figure 19.15 Split frequencies for ring stripline junction circulator.

Bibliography

Allanson, J.T., Cooper, R., and Cowling, T.G. (1946). The theory and experimental behaviour of ight-angled junctions in rectangular-section wave guides. *IEE Proc.* **93** (23): 177–187.

Buchta, G. (1966). Miniturized broadband E-tee circulator at X-band. *Proc. IEEE* **54**: 1607.

Casanueva, A., Leon, A., Mediavilla, A., and Helszajn, J. (2013). Characteristic planes of microstrip and unilateral finline tee-junctions. *Progress in Electromagnetics Research Symposium (PIERS), Proceedings of a Meeting*, Stockholm, Sweden (12–15 August 2013), pp. 173–179.

Chait, H.N. and Curry, R.T. (1959). Y-circulator. *J. Appl. Phys.* **30**: 152.

Helszajn, J. (2015). Electrically symmetric solution of the 3-port H-plane waveguide tee junction at the Dicke ports. *IET Microw. Antennas Propag.* **9** (6): 561–568.

Helszajn, J. and Tan, C.F.C. (1975). Design data for radial-waveguide circulators using partial height ferrite resonators. *IEEE Trans. Microw. Theory Tech.* **MTT-23** (3): 288–298.

Helszajn, J., Hocine, M., and D'Orazio, W. (2014a). Adjustment of H-plane waveguide tee-junction circulator using a circular post gyromagnetic resonator. *Int. J. RF Microw. CAE* **24** (1): 55–58.

Helszajn, J., Caplin, M., Frenna, J., and Tsounis, B. (2014b). Characteristic planes and scattering of E and H-plane waveguide tee junctions. *IEEE Microw. Wireless Compon. Lett.* **24** (4): 209–214.

Montgomery, C.G., Dicke, R.H., and Purcell, E.M. (1948). *Principles of Microwave Circuits*, MIT Radiation Laboratory Series, vol. **VIII**, 432. McGraw-Hill Book Co.

Yoshida, S. (1962). E-type T circulator. Unpublished memorandum, Raytheon Co., Walttham, MA.

20

0°, 90°, and 180° Passive Power Dividers

Joseph Helszajn[1] *and Mark McKay*[2]

[1] *Heriot Watt University, Edinburgh, UK*
[2] *Honeywell, Edinburgh, UK*

20.1 Introduction

Hybrid directional couplers may be classified according to whether their output waves are in-phase, in-quadrature, or out-of-phase. One property of such networks is that all the ports are matched. The in-phase arrangement is a Wilkinson geometry, strictly speaking, a power divider rather than a directional coupler, the quadrature one is either a coupled line or branch line structure, the 180° one is often referred to as a Magic-Tee. Hybrid networks are widely used as elements in balanced mixers, microwave discriminators, and switches, to name a few applications. The properties of directional couplers are almost exclusively expressed in terms of the scattering matrix, and this is the approach adopted in this chapter. The permissible symmetries are established by having recourse to the unitary condition. The operation and design of this class of device is often based on the odd and even mode description of the circuit. The odd and even mode circuits of some typical structures are included for completeness. A generalization of the quadrature hybrid is the $2n$-port Butler circuit which consists of 3 dB couplers and various fixed delays. It has the property that power at any one of its n-input ports is divided equally between its n-output ports with various delays. Four-port networks in which a signal at one port is equally divided into equal out-of-phase signals are also standard building blocks.

Microwave Polarizers, Power Dividers, Phase Shifters, Circulators, and Switches,
First Edition. Joseph Helszajn.

20.2 Wilkinson Power Divider

One useful power divider met in microwave engineering is the Wilkinson three-port circuit illustrated in Figure 20.1. It has the property that its three ports are matched, that the power at the input port is equally divided between the other two and that the two output ports are isolated. The scattering matrix that satisfies these conditions is described by

$$\bar{S} = \frac{1}{\sqrt{2}} \begin{bmatrix} 0 & 1 & 1 \\ 1 & 0 & 0 \\ 1 & 0 & 0 \end{bmatrix} \tag{20.1}$$

The presence of a resistive element in this circuit suggests that this matrix does not satisfy the unitary condition. This remark is readily verified by forming

$$\frac{1}{2} \begin{bmatrix} 0 & 1 & 1 \\ 1 & 0 & 0 \\ 1 & 0 & 0 \end{bmatrix} \begin{bmatrix} 0 & 1 & 1 \\ 1 & 0 & 0 \\ 1 & 0 & 0 \end{bmatrix} \neq \begin{bmatrix} 1 & 0 & 0 \\ 0 & 1 & 0 \\ 0 & 0 & 1 \end{bmatrix} \tag{20.2}$$

It may be separately demonstrated that the only matched three-port circuit is that of the junction circulator.

20.3 Even and Odd Mode Adjustment of the Wilkinson Power Divider

A general matrix which satisfies the symmetry of the Wilkinson network is

$$\begin{bmatrix} S_{11} & S_{21} & S_{21} \\ S_{21} & S_{22} & S_{23} \\ S_{21} & S_{23} & S_{22} \end{bmatrix} \tag{20.3}$$

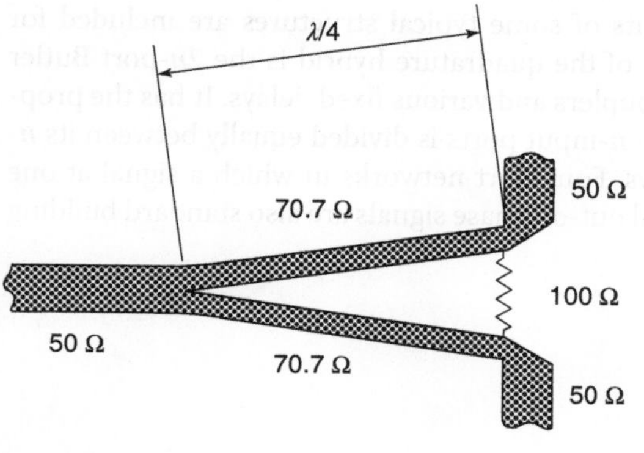

Figure 20.1 Ideal Wilkinson two-way power divider.

An understanding of the adjustment of this device is facilitated by identifying the odd and even mode circuit of Figure 20.2. The odd and even modes in Figure 20.3a and b are obtained by introducing magnetic and electric walls along the symmetry plane. The relationship between the input and output waves of the circuit for in-phase signals at ports 2 and 3 is

$$\begin{bmatrix} b_1 \\ b_2 \\ b_3 \end{bmatrix} = \begin{bmatrix} S_{11} & S_{21} & S_{21} \\ S_{21} & S_{22} & S_{23} \\ S_{21} & S_{23} & S_{22} \end{bmatrix} \begin{bmatrix} 0 \\ 1 \\ 1 \end{bmatrix} \tag{20.4}$$

This gives

$$b_2 = (S_{22} + S_{23}).1 \tag{20.5a}$$

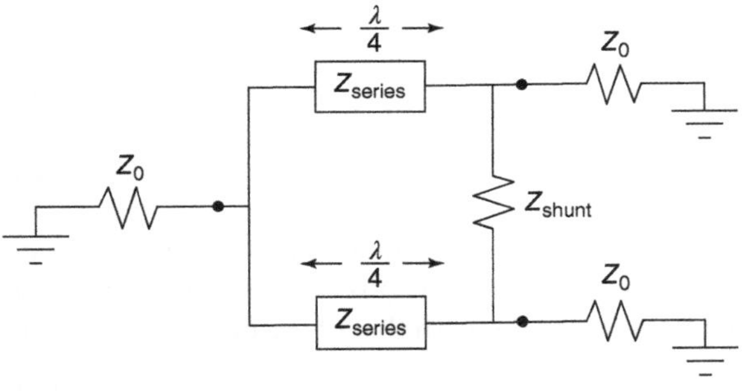

Figure 20.2 Circuit diagram of Wilkinson two-way power divider.

Figure 20.3 (a) In-phase and (b) out-of-phase circuits of Wilkinson two-way power divider.

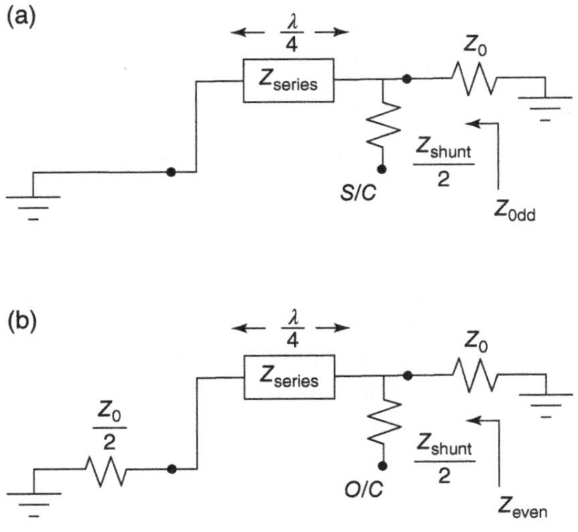

$$b_3 = (S_{23} + S_{22}).1 \tag{20.5b}$$

Scrutiny of the two relationships indicates the possibility of defining an even mode reflection coefficient at each port:

$$\rho_{\text{even}} = \frac{b_2}{a_2} = \frac{b_3}{a_3} = S_{22} + S_{23} \tag{20.6}$$

Application of out-of-phase signals at the same ports indicates the possibility of defining an odd mode reflection coefficient at each port:

$$\rho_{\text{odd}} = \frac{b_2}{a_2} = \frac{b_3}{a_3} = S_{22} - S_{23} \tag{20.7}$$

Combining these last two relationships suggests that S_{22} and S_{23} are simple linear combinations of the even and odd mode reflection coefficients of the network.

$$S_{22} = \frac{\rho_{\text{even}} + \rho_{\text{odd}}}{2} \tag{20.8a}$$

$$S_{23} = \frac{\rho_{\text{even}} - \rho_{\text{odd}}}{2} \tag{20.8b}$$

The nature of S_{11} may be identified by recognizing that if $I_2 = I_3$, then there is no current through the shunt resistor as shown in Figure 20.1 or 20.2. The input impedance of the circuit, therefore, coincides with that of the even mode circuit:

$$Z_{\text{in}} = Z_{\text{even}} \tag{20.9}$$

Since Z_{in} has the form of Z_{even} it follows that

$$S_{11} = \rho_{\text{even}} \tag{20.10}$$

It may be separately demonstrated that

$$|S_{21}| = 0.70\,|\tau_{\text{even}}| \tag{20.11}$$

The even mode transmission parameter is deduced by noting that the even mode circuit is a reactance two-port network:

$$\tau_{\text{even}}^2 = 1 - \rho_{\text{even}}^2 \tag{20.12}$$

The odd mode circuit is not a reactance network and so its transmission parameter cannot be deduced by having recourse to the unitary condition. Since it does not enter into the description of the circuit, it is not considered further.

The element values of the ideal Wilkinson power divider is obtained, provided both the even and odd mode coefficients are identically equal to zero.

$$\rho_{\text{even}} = 0 \tag{20.13a}$$

$$\rho_{\text{odd}} = 0 \tag{20.13b}$$

These two conditions are met, provided the even and odd circuits are separately matched to the uniform transmission lines.

$$Z_{\text{even}} = Z_0 \tag{20.14a}$$

$$Z_{\text{odd}} = Z_0 \tag{20.14b}$$

The detailed adjustment of the circuit in Figure 20.2 proceeds once the even mode and the odd mode circuits are established. The required two port circuits are illustrated in Figure 20.3a and b. The even mode impedance at midband is therefore defined by

$$Z_{\text{even}} = \frac{Z_{\text{series}}^2}{2Z_0} \tag{20.15}$$

The odd mode impedance at midband is separately obtained by recognizing that the short-circuit boundary condition at port 1 is transformed to an open-circuit at the odd mode port. The result is

$$Z_{\text{odd}} = \frac{Z_{\text{shunt}}}{2} \tag{20.16}$$

Combining the preceding two relationships with the boundary conditions of the ideal Wilkinson power divider gives the solution indicated in Figure 20.1.

$$Z_{\text{series}} = \sqrt{2}\,Z_0 \tag{20.17a}$$

$$Z_{\text{shunt}} = 2\,Z_0 \tag{20.17b}$$

20.4 Scattering Matrix of 90° Directional Coupler

The directional coupler is a four-port circuit having an input port, two mutually isolated output ports, and one port isolated from the input port. The device is also reciprocal and all its ports are matched. The relationship between the entries of its scattering matrix is deduced by having recourse to the unitary condition subject to those boundary conditions and its symmetry. The derivations start with the most general description of a four-port circuit in terms of its scattering properties:

$$\bar{S} = \begin{bmatrix} S_{11} & S_{12} & S_{13} & S_{14} \\ S_{21} & S_{22} & S_{23} & S_{24} \\ S_{31} & S_{32} & S_{33} & S_{34} \\ S_{41} & S_{42} & S_{43} & S_{44} \end{bmatrix} \tag{20.18}$$

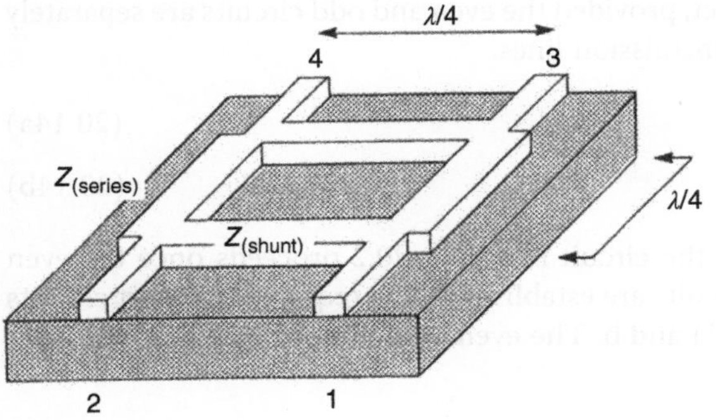

Figure 20.4 Schematic diagram of microstrip branch line coupler.

The port designation employed here is illustrated in Figure 20.4. Introducing the assumption that one adjacent port is isolated from the input port gives

$$S_{12} = S_{21} = S_{34} = S_{43} = 0 \tag{20.19}$$

and the scattering matrix becomes

$$\bar{S} = \begin{bmatrix} S_{11} & 0 & S_{13} & S_{14} \\ 0 & S_{22} & S_{23} & S_{24} \\ S_{31} & S_{32} & S_{33} & 0 \\ S_{41} & S_{42} & 0 & S_{44} \end{bmatrix} \tag{20.20}$$

If all the ports are assumed matched, then

$$S_{11} = S_{22} = S_{33} = S_{44} = 0 \tag{20.21}$$

and

$$\bar{S} = \begin{bmatrix} 0 & 0 & S_{13} & S_{14} \\ 0 & 0 & S_{23} & S_{24} \\ S_{31} & S_{32} & 0 & 0 \\ S_{41} & S_{42} & 0 & 0 \end{bmatrix} \tag{20.22}$$

Making use of the fact that the directional coupler is reciprocal requires that the scattering matrix be symmetric about the main diagonal:

$$S_{13} = S_{31} \tag{20.23a}$$

$$S_{14} = S_{41} \tag{20.23b}$$

$$S_{23} = S_{32} \tag{20.23c}$$

$$S_{24} = S_{42} \tag{20.23d}$$

Introducing these conditions into the scattering matrix for the directional coupler simplifies its description still further. The result is

$$\bar{S} = \begin{bmatrix} 0 & 0 & S_{31} & S_{41} \\ 0 & 0 & S_{32} & S_{42} \\ S_{31} & S_{32} & 0 & 0 \\ S_{41} & S_{42} & 0 & 0 \end{bmatrix} \tag{20.24}$$

It is here recognized that S_{13} connects ports 3 and 1 and that S_{31}connects ports 1 and 3. The symmetry of the junction may now be used to further reduce the number of entries in the scattering matrix. If it is completely symmetric, then the following additional relationships apply:

$$S_{31} = S_{42} \tag{20.25a}$$

$$S_{41} = S_{32} \tag{20.25b}$$

The required matrix, therefore, involves only two independent variables and is given by

$$\bar{S} = \begin{bmatrix} 0 & 0 & S_{31} & S_{41} \\ 0 & 0 & S_{41} & S_{31} \\ S_{31} & S_{41} & 0 & 0 \\ S_{41} & S_{31} & 0 & 0 \end{bmatrix} \tag{20.26}$$

To establish whether such a matrix is realizable as a lossless network, it is necessary to invoke the unitary condition discussed:

$$\bar{S}^{\mathrm{T}}\left(\bar{S}^*\right) = \bar{I} \tag{20.27}$$

$$\begin{bmatrix} 0 & 0 & S_{31} & S_{41} \\ 0 & 0 & S_{41} & S_{31} \\ S_{31} & S_{41} & 0 & 0 \\ S_{41} & S_{31} & 0 & 0 \end{bmatrix} \begin{bmatrix} 0 & 0 & S_{31}^* & S_{41}^* \\ 0 & 0 & S_{41}^* & S_{31}^* \\ S_{31}^* & S_{41}^* & 0 & 0 \\ S_{41}^* & S_{31}^* & 0 & 0 \end{bmatrix} = \begin{bmatrix} 1 & 0 & 0 & 0 \\ 0 & 1 & 0 & 0 \\ 0 & 0 & 1 & 0 \\ 0 & 0 & 0 & 1 \end{bmatrix} \tag{20.28}$$

This gives

$$|S_{31}|^2 + |S_{41}|^2 = 1 \tag{20.29a}$$

$$S_{31}S_{41}^* + S_{31}^*S_{41} = 0 \tag{20.29b}$$

The first equation satisfies energy conservation, whereas the second suggests that one possible solution at a suitable pair of terminals is

$$S_{31} = \alpha \tag{20.30a}$$

$$S_{41} = j\beta \tag{20.30b}$$

where α and β are real numbers. The matrix of the symmetrical directional coupler is therefore

$$\bar{S} = \begin{bmatrix} 0 & 0 & \alpha & j\beta \\ 0 & 0 & j\beta & \alpha \\ \alpha & j\beta & 0 & 0 \\ j\beta & \alpha & 0 & 0 \end{bmatrix} \tag{20.31}$$

The above development indicates that all directional couplers are perfectly matched. One important property of the symmetrical arrangement is that there is a 90° phase difference between the waves in the two output ports.

A special class of directional couplers is the 3 dB hybrid, for which there is equal power division between ports 3 and 4. The symmetric 3 dB coupler is in fact a hybrid junction.

$$\bar{S} = \frac{1}{\sqrt{2}} \begin{bmatrix} 0 & 0 & 1 & j \\ 0 & 0 & j & 1 \\ 1 & j & 0 & 0 \\ j & 1 & 0 & 0 \end{bmatrix} \tag{20.32}$$

Figure 20.5 is a schematic diagram of a broadband multi-branch coupler.

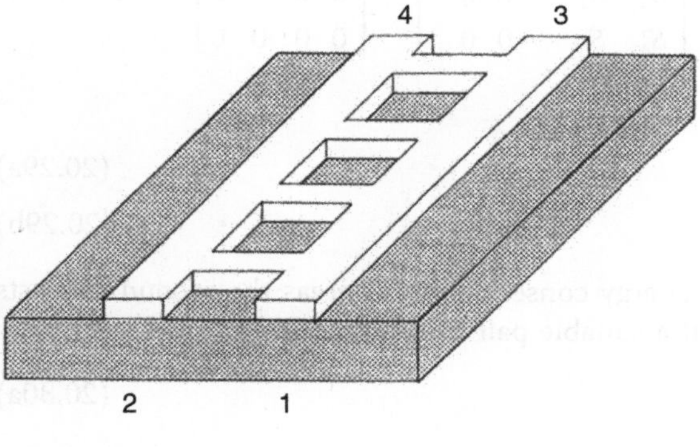

Figure 20.5 Schematic diagram of microstrip multi-branch coupler.

20.5 Even and Odd Mode Theory of Directional Couplers

While the unitary condition may be employed to fix the permissible relationships between the ports of a circuit, its adjustment is a different matter. One approach to this task is to express the entries of the scattering matrix in terms of its eigenvalues. The eigenvalues of this problem are the reflection coefficients associated with the four possible ways of exciting the network that will give the eigenvalues at any port. Another approach, which is equally applicable in the case of the symmetrical directional coupler, is to express these quantities in terms of linear combinations of even and odd mode transmission and reflection parameters. The even mode variables are obtained by applying equal in-phase waves at ports 1 and 2 of the network. The odd mode variables are obtained by applying out-of-phase waves there. These two situations are illustrated in Figure 20.6a and b. The even mode displays an open-circuit boundary at the plane of symmetry. The odd mode imposes a short-circuit boundary there.

The derivation of the scattering matrix of the network in terms of the odd and even mode parameters proceeds by taking each boundary condition one at a time as the input waves of the directional coupler and constructing the output

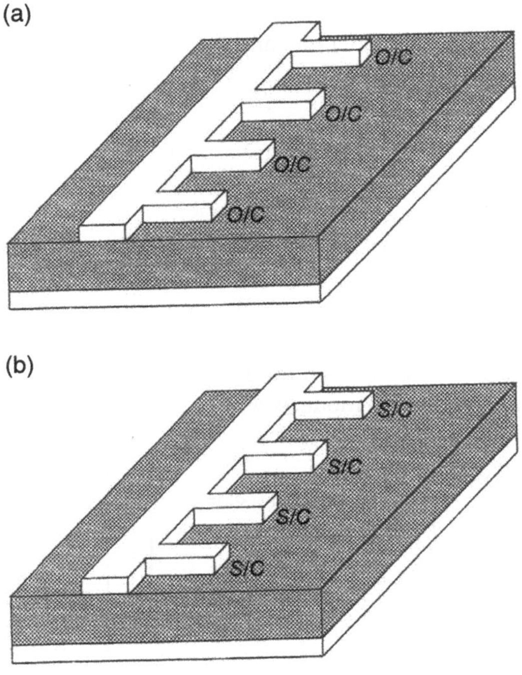

Figure 20.6 (a) Even and (b) odd mode circuits for branch line coupler.

ones. For the even mode excitation, the input/output relation of the directional coupler is

$$\begin{bmatrix} b_1 \\ b_2 \\ b_3 \\ b_4 \end{bmatrix} \begin{bmatrix} S_{11} & S_{21} & S_{31} & S_{41} \\ S_{21} & S_{11} & S_{41} & S_{31} \\ S_{31} & S_{41} & S_{11} & S_{21} \\ S_{41} & S_{31} & S_{21} & S_{11} \end{bmatrix} = \begin{bmatrix} \frac{1}{2} \\ \frac{1}{2} \\ 0 \\ 0 \end{bmatrix} \tag{20.33}$$

The above scattering matrix assumes that the device is reciprocal and symmetrical, but no assumption is made about the boundary conditions of the ideal directional coupler. Expanding the above matrix relation gives

$$b_1 = \frac{S_{11} + S_{21}}{2} \tag{20.34a}$$

$$b_2 = \frac{S_{11} + S_{21}}{2} \tag{20.34b}$$

$$b_3 = \frac{S_{31} + S_{41}}{2} \tag{20.34c}$$

$$b_4 = \frac{S_{31} + S_{41}}{2} \tag{20.34d}$$

Even mode reflection and transmission coefficients may now be defined for each waveguide as follows:

$$\rho_{\text{even}} = \frac{b_1}{a_1} = \frac{b_2}{a_2} = S_{11} + S_{21} \tag{20.35}$$

$$\tau_{\text{even}} = \frac{b_3}{a_1} = \frac{b_4}{a_2} = S_{31} + S_{41} \tag{20.36}$$

where ρ_{even} and τ_{even} are the reflection and transmission coefficients for each waveguide. Since there is no coupling between the two waveguides for this set of incident waves, the coupled waveguides may be replaced by a single waveguide with an even mode field pattern.

For the odd mode excitation, the input/output relation of the network becomes

$$\begin{bmatrix} b_1 \\ b_2 \\ b_3 \\ b_4 \end{bmatrix} \begin{bmatrix} S_{11} & S_{21} & S_{31} & S_{41} \\ S_{21} & S_{11} & S_{41} & S_{31} \\ S_{31} & S_{41} & S_{11} & S_{21} \\ S_{41} & S_{31} & S_{21} & S_{11} \end{bmatrix} = \begin{bmatrix} \frac{1}{2} \\ -\frac{1}{2} \\ 0 \\ 0 \end{bmatrix} \tag{20.37}$$

Odd mode reflection and transmission coefficients for each waveguide are here defined by

$$\rho_{\text{odd}} = \frac{b_1}{a_1} = \frac{b_2}{a_2} = S_{11} - S_{21} \tag{20.38}$$

$$\tau_{\text{odd}} = \frac{b_3}{a_1} = \frac{b_4}{a_2} = S_{31} - S_{41} \tag{20.39}$$

The reflection and transmission coefficients are again identical for each waveguide section so that the four-port network may be once more replaced by a two-port one for this excitation. The above two solutions may now be combined to give

$$S_{11} = \frac{\rho_{\text{even}} + \rho_{\text{odd}}}{2} \tag{20.40a}$$

$$S_{21} = \frac{\rho_{\text{even}} - \rho_{\text{odd}}}{2} \tag{20.40b}$$

$$S_{31} = \frac{\tau_{\text{even}} + \tau_{\text{odd}}}{2} \tag{20.40c}$$

$$S_{41} = \frac{\tau_{\text{even}} - \tau_{\text{odd}}}{2} \tag{20.40d}$$

Scrutiny of the relationships indicates that the boundary conditions of an ideal directional coupler may be established in one of two ways.

One solution requires that the odd and even mode reflection coefficients are zero while the transmission ones are different. These boundary conditions lead to the scattering matrix of the ideal directional coupler.

$$S_{11} = 0 \tag{20.41a}$$

$$S_{21} = 0 \tag{20.41b}$$

$$S_{31} = \frac{\tau_{\text{even}} + \tau_{\text{odd}}}{2} \tag{20.41c}$$

$$S_{41} = \frac{\tau_{\text{even}} - \tau_{\text{odd}}}{2} \tag{20.41d}$$

20.6 Power Divider Using 90° Hybrids

One of the simplest variable power dividers is obtained by placing a variable phase-shifter between two 3 dB hybrid couplers. If the phase setting of the phase-shifter is 0°, then the power is transmitted to the port diagonally opposite the input port; if it is set to 180°, it is emergent at the other port. Any other power division can be accommodated with this device by suitably adjusting the phase-shifter.

The operation of this type of device may be understood by forming the transmission matrices of each of its three sections in terms of its scattering ones one at a time. The scattering matrix of the input and output hybrid has been given in Eq. (20.32).

The scattering matrix of a four-port circuit formed by two decoupled lines with phase constants θ_{31} and θ_{42}, respectively, is

$$\bar{S} = \begin{bmatrix} 0 & 0 & S_{31} & 0 \\ 0 & 0 & 0 & S_{42} \\ S_{31} & 0 & 0 & 0 \\ 0 & S_{42} & 0 & 0 \end{bmatrix} \tag{20.42}$$

where

$$S_{31} = 1\exp(-j\theta_{31}) \tag{20.43a}$$

$$S_{42} = 1\exp(-j\theta_{42}) \tag{20.43b}$$

The operation of the overall device is readily understood by forming the input–output relationships of each section one at a time or by forming the overall transmission matrix.

The solution is

$$b_1 = 0 \tag{20.44a}$$

$$b_2 = 0 \tag{20.44b}$$

$$b_3 = \sin\left(\frac{\theta_{31} - \theta_{42}}{2}\right) \exp j\left(\frac{\theta_{31} + \theta_{42}}{2}\right) \tag{20.44c}$$

$$b_4 = \cos\left(\frac{\theta_{31} - \theta_{42}}{2}\right) \exp j\left(\frac{\theta_{31} + \theta_{42}}{2}\right) \tag{20.44d}$$

Setting $\theta_{31} = \theta_{42}$ gives

$$b_3 = 0 \tag{20.45a}$$

$$b_4 = 1 \tag{20.45b}$$

If $\theta_{31} = \theta_{42} + \pi$, the result is

$$b_3 = 1 \tag{20.46a}$$

$$b_4 = 0 \tag{20.46b}$$

The arrangement in Figure 20.5 may therefore be used to divide an incident wave at port 1 between ports 3 and 4 by varying the angle of the phase-shifter.

20.7 Variable Power Dividers

A generalization of the simple Butler power divider using 3 dB hybrids may be used by employing variable phase-shift networks between two $2n$-port Butler ones. A typical Butler network in the case of $n = 4$ is illustrated in Figure 20.7. It consists of 3 dB hybrids, fixed phase-shifters, and crossed links. The use of n-way dividers in phased array equipment is illustrated in Figure 20.8.

Figure 20.7 Butler network.

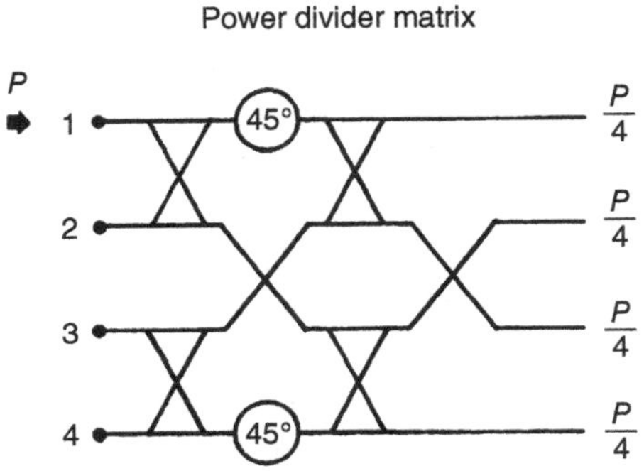

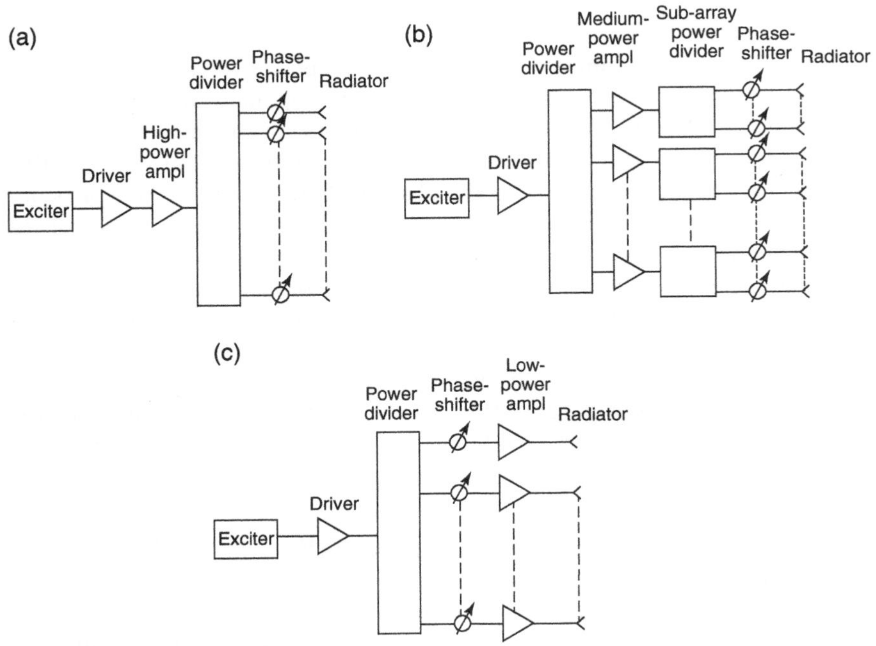

Figure 20.8 Use of Butler networks in phased array equipment: (a) nondistributed (passive), (b) semidistributed and (c) fully distributed (active).

(a)

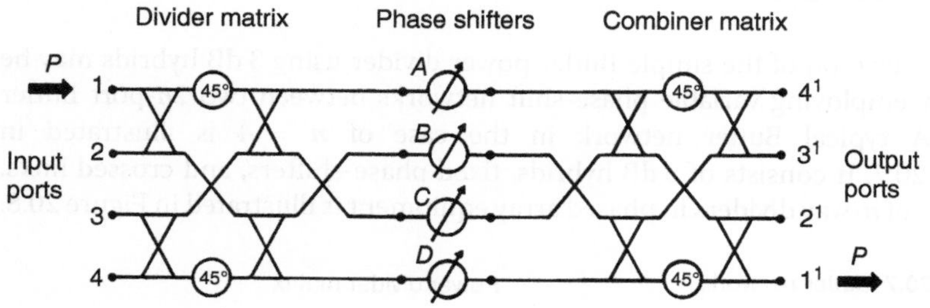

(b)

Input to output	Phase shift			
	A	B	C	D
1^1	0	0	0	0
2^1	−180°	0	−180°	0
3^1	−180°	−270°	0	−90°
4^1	−270°	−180°	−90°	0

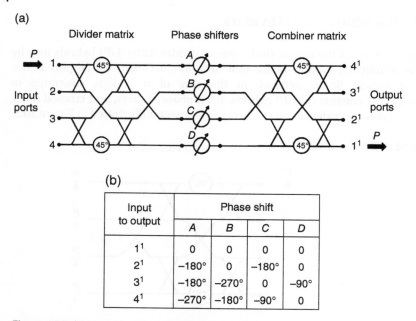

Figure 20.9 (a) Four-channel Butler switch. (b) Phase setting for $n = 4$ Butler switch.

If two such arrays are connected back to back the power will be recombined in the second Butler network and routed to the port diagonally opposite the input port. If, on the other hand, *n*-variable phase-shifters are connected between the two dividing networks, then the power can be recombined at any of the other output ports by properly adjusting the phase-shifters. Figure 20.9a depicts one arrangement for $n = 4$; Figure 20.9b indicates the appropriate values of the phase-shift sections in order to switch the power to any output port.

The original Butler network relies on fixed phase delays and some crossed lines for its operation. However, in some applications, the fixed phase-shifters can be omitted. One possible configuration in the case of an $n = 4$ network is depicted in Figure 20.10a; the required phase-shift settings are given in Figure 20.10b. The advantage of this circuit is that it allows the fixed phase-shifters in the original Butler circuit to be replaced by only 0° and 180° sections.

20.8 180° Waveguide Hybrid Network

The possibility of realizing 3 dB hybrids with in-phase or out-of phase instead of in-phase quadrature output signals by imposing a single symmetry plane on a four-port network will now be established. One possible solution has the

(a)

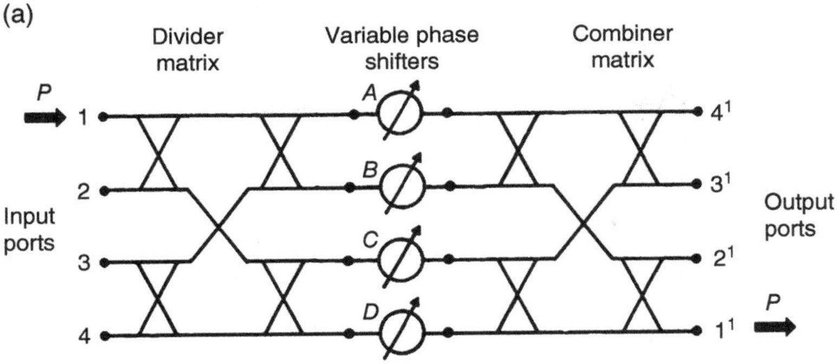

(b)

Switch paths established	Phase shift			
	A	B	C	D
1↔1¹, 2↔2¹, 3↔3¹, 4↔4¹	0	0	0	0
1↔2¹, 2↔1¹, 3↔4¹, 4↔3¹	0	0	−180°	−180°
1↔3¹, 2↔4¹, 3↔1¹, 4↔2¹	0	−180°	0	−180°
1↔4¹, 2↔3¹, 3↔2¹, 4↔1¹	−180°	0	0	−180°

Figure 20.10 (a) Four-channel hybrid matrix switch. (b) Phase setting for $n = 4$ hybrid network switch.

property that all its ports are matched, that one port is decoupled from an input port, and that a signal at one input port produces equal in-phase signals at two output ports and that a signal at another input port produces out-of-phase signals at two output ports. One classic waveguide version is the Magic-Tee one illustrated in Figure 20.11. If the *H*-port is taken as port 1, then it is appropriate to take the *E*-port as port 2. This nomenclature ensures that the entries connected with ports 1 and 2 of the scattering matrix of the junction have the symmetry of the 90° hybrid. If this port nomenclature is adopted, then

$$S_{12} = S_{21} = 0 \tag{20.47}$$

It is also assumed in the first instance that ports 3 and 4 are decoupled

$$S_{34} = S_{43} = 0 \tag{20.48}$$

and that the network is matched

$$S_{11} = S_{22} = S_{33} = S_{44} = 0 \tag{20.49}$$

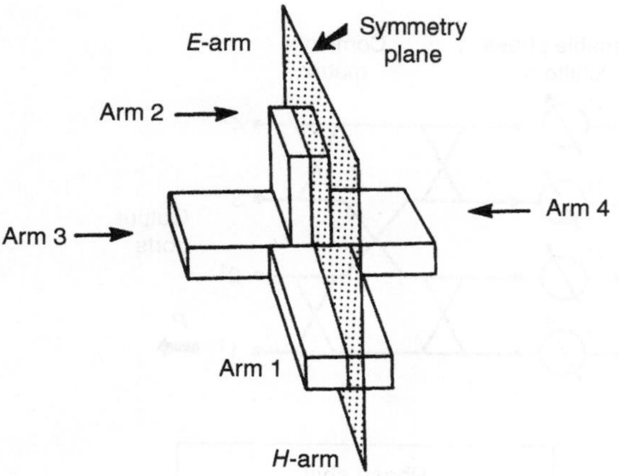

Figure 20.11 Symmetry plane of Magic-Tee.

The scattering matrix of the device under these assumptions is described by

$$\bar{S} = \begin{bmatrix} 0 & 0 & S_{31} & S_{41} \\ 0 & 0 & S_{32} & S_{42} \\ S_{31} & S_{32} & 0 & 0 \\ S_{41} & S_{42} & 0 & 0 \end{bmatrix} \tag{20.50}$$

In order to verify whether these assumptions are valid, recourse is made to the unitary condition. This gives

$$|S_{31}|^2 + |S_{41}|^2 = 0 \tag{20.51a}$$

$$|S_{32}|^2 + |S_{42}|^2 = 0 \tag{20.51b}$$

$$|S_{31}|^2 + |S_{32}|^2 = 0 \tag{20.51c}$$

$$|S_{41}|^2 + |S_{42}|^2 = 0 \tag{20.51d}$$

and

$$S_{31}^* S_{32} + S_{41}^* S_{42} = 0 \tag{20.52a}$$

$$S_{31}^* S_{41} + S_{32}^* S_{42} = 0 \tag{20.52b}$$

Scrutiny of Eqs. (20.51a) and (20.51c) indicates that one relation between S_{41} and S_{32} is

$$|S_{41}|^2 = |S_{32}|^2 \tag{20.53a}$$

A separate inspection of Eqs. (20.51b) and (20.51c) also gives one between S_{31} and S_{42}:

$$|S_{31}|^2 = |S_{42}|^2 \tag{20.53b}$$

One solution that meets both relationships is that encountered in connection with the 90° hybrid solution.

$$S_{41} = S_{32} \tag{20.54a}$$

$$S_{31} = S_{42} \tag{20.54b}$$

Introducing these two conditions into either Eq. (20.52a) or Eq. (20.52b) also gives

$$S_{31}^* S_{41} + S_{31} S_{41}^* = 0 \tag{20.55}$$

One possible solution is therefore

$$S_{11} = 0 \tag{20.56a}$$

$$S_{21} = 0 \tag{20.56b}$$

$$S_{31} = \alpha \tag{20.56c}$$

$$S_{41} = j\beta \tag{20.56d}$$

This solution corresponds to the description of the quadrature coupler and has already been dealt with.

Another possibility which also satisfies both Eqs. (20.53a) and (20.53b) but which has not been considered so far is achieved, provided

$$S_{32} = S_{41} \tag{20.57a}$$

$$S_{42} = -S_{31} \tag{20.57b}$$

Introducing these two conditions into either Eq. (20.52a) or Eq. (20.52b) now indicates that

$$S_{31}^* S_{41} - S_{31} S_{41}^* = 0 \tag{20.58}$$

One solution that meets both the preceding equation as well as Eq. (20.51a) for an input at port 1 is

$$S_{31} = \frac{1}{\sqrt{2}} \tag{20.59a}$$

$$S_{41} = \frac{1}{\sqrt{2}} \tag{20.59b}$$

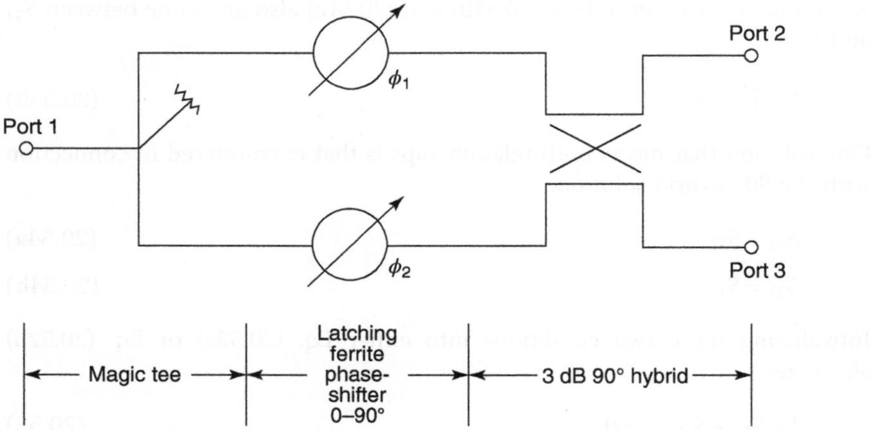

Figure 20.12 Power divider using 3 dB hybrids and variable phase-shifter.

The relationship between S_{42} and S_{32} is now met by satisfying Eq. (20.52a) or Eqs. (20.52b) and (20.51b). One possibility is

$$S_{32} = \frac{1}{\sqrt{2}} \tag{20.60a}$$

$$S_{42} = \frac{-1}{\sqrt{2}} \tag{20.60b}$$

The matrix description of the 180° class of network is, therefore, specified by

$$\bar{S} = \frac{1}{\sqrt{2}} \begin{bmatrix} 0 & 0 & 1 & 1 \\ 0 & 0 & 1 & -1 \\ 1 & 1 & 0 & 0 \\ 1 & -1 & 0 & 0 \end{bmatrix} \tag{20.61}$$

Figure 20.12 depicts one application of the Magic-tee.

Bibliography

Cohn, S.B. (1968). A class of broadband three-port TEM mode hybrids. *IEEE Trans. Microw. Theory Tech.* **MTT-16**: 110–116.

Davis, R.S. and Schrank, H.E. (1965). Application of the Butler matrix to high-power multichannel switching. *G-MTT Symposium Digest*, Clearwater, FL (5–7 May 1965).

Levy, R. (1984). A high-power X-band Butler matrix. *Microw. J.* **27**: 135–141.
Luzzato, G. (1967). An N-way hybrid combiner. *Proc. IEEE* **55**: 470–471.
MacNamara, T. (1987). Simplified design procedures for Butler matrices incorporating 90° or 180° hybrids. *IEE Proc., Pt. H* **134** (1): 50–54.
Malherbe, J.A.G. (1988). *Microwave Transmission Line Couplers*. London: Artech House.
Moody, J.H. (1964). The systematic design of the Butler matrix. *IEEE Trans.* **AP-11**: 786–788.
Wilkinson, E. (1960). A N-way hybrid power divider. *IEEE Trans. Microw. Theory Tech.* **MTT-8**: 116–118.
Wilkinson, E.J. and Sommers, D.J. (1978). Variable multiport power combiners. *Microw. J.* **21**: 59–65.

Index

Microwave Polarizers, Power Dividers, Phase Shifters, Circulators, and Switches,
First Edition. Joseph Helszajn.

g